Big Data in Bioeconomy

Caj Södergård · Tomas Mildorf ·
Ephrem Habyarimana · Arne J. Berre ·
Jose A. Fernandes · Christian Zinke-Wehlmann
Editors

Big Data in Bioeconomy

Results from the European DataBio Project

 Springer

Editors
Caj Södergård
VTT Technical Research Centre of Finland
Espoo, Finland

Ephrem Habyarimana
Consiglio per la Ricerca in Agricoltura e
l'Analisi dell'Economia Agraria (CREA)
Rome, Italy

Jose A. Fernandes
AZTI, Marine Research
Basque Research and Technology
Alliance (BRTA)
Pasaia, Gipuzkoa, Spain

Tomas Mildorf
University of West Bohemia
Univerzitni 8, 301 00 Plzen, Czech
Republic

Arne J. Berre
SINTEF
Oslo, Norway

Christian Zinke-Wehlmann
Institut für Angewandte Informatik
Leipzig University
Leipzig, Sachsen, Germany

ISBN 978-3-030-71071-2   ISBN 978-3-030-71069-9   (eBook)
https://doi.org/10.1007/978-3-030-71069-9

This Springer imprint is published by the registered company Springer Nature Switzerland AG
The registered company address is: Gewerbestrasse 11, 6330 Cham, Switzerland

Foreword

The European Union is today, in October 2020, a very different place from where it was in October 2015, when a call for proposals was launched that resulted in the funding of the DataBio project (among many others). We are now a Union that is planning its recovery from the most severe health emergency in our history, an emergency that has affected our economy very negatively.

And yet, the outcomes of the DataBio project, as described in this book, are all excellent examples of the potential of the 2020 European strategy for data,[1] which is central to the digital plans that form the second pillar in the Union's growth strategy[2] underpinning the recovery.

In its European strategy for data, the Commission has committed to promoting the development of common European data spaces in strategic economic sectors and domains of public interest. It has specifically committed to a common European agriculture data space, to enhance the sustainability performance and competitiveness of the agricultural sector through the processing and analysis of production and other data, allowing for precise and tailored application of production approaches at farm level. The EU will also contribute data and infrastructure from the Copernicus Earth Observation programme to underpinning the European data spaces where relevant.

In this book, we see the seeds of many of the technologies that are likely to play a prominent role in the European agriculture data space and the interplay of these technologies with some of the issues that will need to be addressed to put in place a trusted and efficient data space governance.

In Chaps. 4 and 21, we see first-hand the importance of the reuse of Earth Observation data from the Copernicus programme for the purpose of both improving efficiency and verifying compliance with EU regulations.

The ability to share data in a way that preserves not only personal privacy but also commercial confidentiality (both necessary prerequisites for the proper functioning and governance of a data space) is addressed in Chap. 12.

---

[1] https://eur-lex.europa.eu/legal-content/EN/TXT/?qid=1593073685620&uri=CELEX%3A52020DC0066.

[2] 16 September 2020 State of the Union speech, https://ec.europa.eu/commission/presscorner/detail/en/SPEECH_20_1655.

The growing importance of agricultural data from sensors, which the European strategy for data addresses in the context of a growing Internet of things connected by advanced telecommunication networks, is discussed in Chaps. 3, 15 and 19.

Data standards, a central concern of the European strategy, are addressed in Chaps. 2, 7, 8 and 9 as a crucial requirement for independently developed data resources and tools to come together in pipelines where different parties could bring different analytic skills to extract insights and valuable predictions from data assets.

And, of course, DataBio being a research and innovation project from the Horizon 2020 programme, the book contains a wealth of insights on the research frontier of it all, showing cutting-edge concrete results but also pointing at how more research there still remains to do in the upcoming Horizon Europe funding programme.

It is a privilege to be able to write the introduction to a volume such as the present one, which shows in great detail how important policy directions of the European Union are often preceded by years of work of our best scientists and technology developers. These identify both opportunities and technical challenges for the benefit of the technology adopters and policy-makers who can then form better informed opinions on what is possible and what is necessary to bring the greatest collective benefits to the citizens they serve.

Brussels, Belgium   Mrs. Gail Kent
                                                       Director Data at European Commission
                                                                             DG CONNECT

The original version of the book was revised: The Editor Tomas Mildorf's affiliation has been corrected. The correction to the book is available at https://doi.org/10.1007/978-3-030-71069-9_33

Introduction

DataBio was one of the first two lighthouse projects on big data awarded in the Horizon 2020 framework programme (two more were awarded a year later), running from January 2017 to December 2019. Its focus was on utilizing big data technologies to improve productivity and sustainability in bioeconomy.

The benefits stemming from big data applications have long been recognized, and concerted efforts like the Big Data Value Public–Private Partnership (BDV PPP) were put in action over 5 years ago. Bioeconomy is important in Europe, not just because it is worth €2, 3 trillion per annum and employs over 18 million people, but also because it is critical for the environment, food production and the development of rural areas. DataBio was the first initiative working on meshing these two domains of big data and bioeconomy on a large enough scale to produce significant impact.

The project was driven by the development, use and evaluation of 27 diverse pilots: 13 in agriculture, eight in forestry and six in fishery. Overall, 95 big data and Earth Observation technology components and 38 data sets were handled in DataBio. Most of them were applied in the pilot trials, and several were enhanced in DataBio. Sixteen major big data pipelines were formed with great potential to be exploited.

Besides its large scale, DataBio, as a lighthouse project, spent great effort in contributing to and engaging the research and bioeconomy communities. DataBio was the lead project in defining the BDVA[3] Reference Model, while also contributing to public OGC[4] Engineering Reports on the standardization of Earth Observation services. It organized or participated in over 180 events, including high-profile conferences, stakeholder events, training sessions and hackathons.

This book summarizes some of the main results from the breadth of the DataBio activities. It is divided into eight parts: the first four parts represent the relevant big data technologies that are the foundation for building bioeconomy solutions. The next three parts describe the applications in each of the three domains addressed: agriculture, forestry and fishery. The final part provides a summary and outlook for big data exploitation in bioeconomy.

---

[3] Big Data Value Association, www.bdva.eu.
[4] Open Geospatial Consortium, www.ogc.org.

I would like to thank the authors and, in particular, the editors of this book. They volunteered and spent substantial effort with great motivation to compile and bring to you the results of this project, working after the project contract had concluded.

But most of the thanks go to the tens of people who worked continuously for 3 years in this project to actually produce these results. Their efforts have already been recognized, as their work has been followed up by new research and commercialization activities.

I can only hope that you also find our results interesting and useful for your work.

Athanasios (Thanasis) Poulakidas
DataBio Project Coordinator
INTRASOFT International

Glossary

Big data	extensive data sets—primarily in the data characteristics of volume, variety, velocity, and/or variability— that require a scalable technology for efficient storage, manipulation, management, and analysis *Note 1 to entry: Big data is commonly used in many different ways, for example, as the name of the scalable technology used to handle big data extensive data sets. Source* ISO/IEC 20546:2019
Biomarine modelling	modelling oceanography and biology and their interactions in the same model. *Source*DataBio
Data Analytics	composite concept consisting of data acquisition, data collection, data validation, data processing, including data quantification, data visualization, and data interpretation *Note 1 to entry: Data analytics is used to understand objects represented by data, to make predictions for a given situation, and to recommend on steps to achieve objectives. The insights obtained from analytics are used for various purposes such as decision-making, research, sustainable development, design and planning. Source* ISO/IEC 20546:2019
Data Type	defined set of data objects of a specified data structure and a set of permissible operations, such that these data objects act as operands in the execution of any one of these operations *Note 1 to entry: Example: An integer type has a very simple structure, each occurrence of which, usually called value, is a representation of a member of a specified range of whole numbers and the permissible operations include the usual arithmetic operations on these integers. Source* ISO/IEC 2382:2015
Data Variability	changes in transmission rate, format or structure, semantics, or quality of data sets. *Source* ISO/IEC 2382:2015

Data Variety	range of formats, logical models, timescales, and semantics of a data set Note 1 to entry: Data variety refers to irregular or heterogeneous data structures, their navigation, query, and data typing. *Source* ISO/IEC 2382:2015
Data Velocity	rate of flow at which data is created, transmitted, stored, analysed, or visualized. *Source* ISO/IEC 2382:2015
Data Veracity	completeness and/or accuracy of data *Note 1 to entry: Data veracity refers to descriptive data and self-inquiry about objects to support real-time decision-making.* *Source* ISO/IEC 2382:2015
Earth observation	the gathering of information about planet Earth's physical, chemical, and biological systems. It involves monitoring and assessing the status of, and changes in, the natural and man-made environment Note: In recent years, Earth observation has become more and more sophisticated with the development of remote-sensing satellites and increasingly high-tech "in situ" instruments. Today's Earth observation instruments include floating buoys for monitoring ocean currents, temperature, and salinity; land stations that record air quality and rainwater trends; sonar and radar for estimating fish and bird populations; seismic and Global Positioning System (GPS) stations; and over 60 high-tech environmental satellites that scan the Earth from space. *Source* Group on Earth Observation (https://www.earthobservations.org/g_faq.html)
Enterprise Architecture	The fundamental organization of a system, embodied in its components, their relationships to each other and the environment, and the principles governing its design and evolution. *Source* ISO/IEC 42010
Internet of things	integrated environment, interconnecting anything, anywhere at anytime. *Source* ISO/IEC JTC 1 SWG 5 Report:2013
Partially Structured Data	data that has some organization *Note 1 to entry: Partially structured data is often referred to as semi-structured data by industry.Note 2 to entry: Examples of partially structured data are records with free text fields in addition to more structured fields. Such data is frequently represented in computer interpretable/parsible formats such as XML or JSON.* *Source* ISO/IEC 2382:2015
Pelagic fisheries	fisheries targeting fish in the pelagic zone of the oceans, as opposed to demersal fish living close to the bottom. *Source* DataBio
Pipeline	a reusable schema of interoperable software components coupled in order to create new services and/or

data, including description of mutual interfaces between the components *Note 1 to entry: In DataBio, pipelines fulfil pilot functionalities that cannot be supported by a single software component Note 2 to entry: A pipeline can be seen as a white box showing internal wiring and data flow between single components of the pipeline, thus providing technical guidance for configuration and deployment Note 3 to entry: Pipelines enable new software components to be easily and effectively combined with open source, standards-based big data, and proprietary components and infrastructures based on the use of generic and domain specific components Note 4 to entry: deployed pipelines become platform services. Source* DataBio

Platform
type of computer or hardware device and/or associated operating system, or a virtual environment, on which software can be installed or run *Note 1 to entry: A platform is distinct from the unique instances of that platform, which are typically referred to as devices or instances. Source* ISO/IEC 19770-5:2015

Platform Services
providers of functionalities to users that typically need to know the usability of the service, but do not need to understand the inner wiring, inner components, nor where the service is deployed *Note 1 to entry: these services are typically accessed via standardized interfaces like application programming interfaces, e.g. web services or library interfaces, interactive user interfaces, standard data transfer, or remote call protocols Note 2 to entry: services often refer to end points that are "black boxes" activated remotely and executed in the cloud. Source* DataBio

Sensor
sensor device that observes and measures a physical property of a natural phenomenon or man-made process and converts that measurement into a signal. *Source* ISO/IEC 29182-2:2013

Small pelagic fisheries
pelagic fisheries targeting the small pelagic species, such as mackerel and herring. *Source* DataBio

Structured Data
data which is organized based on a pre-defined (applicable) set of rules *Note 1 to entry: The predefined set of rules governing the basis on which the data is structured needs to be clearly stated and made known. Note 2 to entry: A pre-defined data model is often used to govern the structuring of data. Source* ISO/IEC 2382:2015

Unstructured Data
data which is characterized by not having any structure apart from that record or file level *Note 1 to entry: On*

the whole unstructured data is not composed of data elements.EXAMPLE: An example of unstructured data is free text. Source ISO/IEC 2382:2015

Contents

Part I
Technological Foundation: Big Data Technologies for BioIndustries

Chapter 1
Big Data Technologies in DataBio

Caj Södergård, Tomas Mildorf, Arne J. Berre, Aphrodite Tsalgatidou, and Karel Charvát

Abstract In this introductory chapter, we present the technological background needed for understanding the work in DataBio. We start with basic concepts of Big Data including the main characteristics volume, velocity and variety. Thereafter, we discuss data pipelines and the Big Data Value (BDV) Reference Model that is referred to repeatedly in the book. The layered reference model ranges from data acquisition from sensors up to visualization and user interaction. We then discuss the differences between open and closed data. These differences are important for farmers, foresters and fishermen to understand, when they are considering sharing their professional data. Data sharing is significantly easier, if the data management conforms to the FAIR principles. We end the chapter by describing our DataBio platform that is a software development platform. It is an environment in which a piece of software is developed and improved in an iterative process providing a toolset for services in agriculture, forestry and fishery. The DataBio assets are gathered on the DataBio Hub that links to content both on the DataBio website and to Docker software repositories on clouds.

1.1 Basic Concepts of Big Data

When we want to utilize data and computers to make raw material gathering more efficient and sustainable in bioeconomy, we will have to deal with vast amounts of heterogeneous data at high speeds, i.e. *Big Data*. This is because of the enormous

C. Södergård (✉)
VTT Technical Research Centre of Finland, Espoo, Finland
e-mail: Caj.Sodergard@vtt.fi

T. Mildorf
University of West Bohemia, Pilsen, Czech Republic

A. J. Berre · A. Tsalgatidou
SINTEF Digital, Forskningvejen 1, Oslo, Norway

K. Charvát
Lesprojekt-služby Ltd., Zaryby, Czech Republic

© The Author(s) 2021
C. Södergård et al. (eds.), *Big Data in Bioeconomy*,
https://doi.org/10.1007/978-3-030-71069-9_1

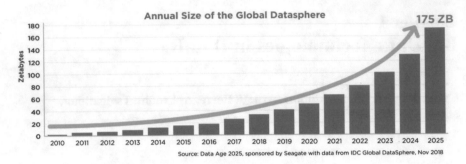

Fig. 1.1 Global data sphere grows exponentially. *Source* [3]

and all the time increasing flow of data from a variety of sensors and measurement devices, like cameras on satellites, aeroplanes and drones as well as measurement data from sensors in the air, in the soil and in the oceans. Moreover, the resolution and frequency of data acquisition from those sensors are exponentially increasing. Many industrial sectors benefit from Big Data, which were coined "the new oil" [1]. The term Big Data has been in use since 2001, when Doug Laney introduced the 3V characteristics: *volume, velocity* and *variety* [2]. The 3V's have the following meanings:

Volume is the amount of generated data. The global data sphere grows exponentially (Fig. 1.1). IDC has predicted that it will grow from 45 ZettaBytes (=10^{21} bytes) in 2020 till 175 ZettaBytes in 2025 [3]. This is mainly due to the growth in unstructured data, like multimedia (audio, images and video) as well as social media content. This puts a lot of pressure on Big Data technologies.

Velocity is the speed of generating and processing data. The development has gone from batch, periodic, near real time, to fully real time/streaming data, which requires a massive throughput.

Variety is the type of generated data (text, tables, images, video, etc.). Unstructured data is more and more dominating over semi-structured and unstructured data. The issue is to manage the heterogeneity of data.

Later, the Big Data concept has expanded with more V dimensions. Data has both social and economic *values*. Value is typically extracted from data with analytical methods, including predictive analytics, visualization and artificial intelligence tools. *Variability* refers to changes in data rate, format/structure, semantics, and/or quality that impact the supported application, analytic or problem [4]. Impacts can include the need to refactor architectures, interfaces, processing/algorithms, integration/fusion, storage, applicability or use of the data. Finally, *veracity* refers to the noise, biases and abnormality in Big Data. There is always the need of checking if the available data is relevant to the problem being studied.

Data *quality* is central in all processing. With low quality in the input data, we will get uncertain results out. *Metadata (=data about data)* allows identification of information resources. Metadata is needed for describing among other things data types, geographic extent and temporal reference, quality and validity, interoperability of spatial data sets and services, constraints related to access and use, and the organization responsible for the resource.

In the DataBio project, the data handling specifically aimed at the following sectors:

- *Agriculture*: The main goal was to develop smart agriculture solutions that boost the production of raw materials for the agri-food chain in Europe while making farming sustainable. This includes optimized irrigation and use of fertilizers and pesticides, prediction of yield and diseases, identification of crops and assessment of damages. Such smart agriculture solutions are based on the use data from satellites, drones, IoT sensors, weather stations as well as genomic data.
- *Forestry*: Big Data methods are expected to bring the possibility to both increase the value of the forests as well as to decrease the costs within sustainability limits set by natural growth and ecological aspects. The key technology is to gather more and more accurate information about the trees from a host of sensors including new generations of satellites, drones, laser scanning from aeroplanes, crowdsourced data collected from mobile devices and data gathered from machines operating in forests.
- *Fisheries*: The ambition is to herald and promote the use of Big Data analytical tools to improve the ecological and economic sustainability, such as improved analysis of operational data for engine fault detection and fuel reduction, tools for planning and operational choices for fuel reduction when searching and choosing fishing grounds, as well as crowdsourcing methods for fish stock estimation.

1.2   Pipelines and the BDV Reference Model

When processing streaming time-dependent data from sensors, data is put to travel through *pipelines*. The term pipeline was used in the DataBio project to describe the data processing steps. Each step has its input and output data. A pipeline is created by chaining individual steps in a consecutive way, where the output from the preceding processing step is fed into the succeeding step. Typically in Big Data applications, the pipeline steps include data gathering, processing, analysis and visualization of the results. The US National Institute of Standards NIST describes this process in their Big Data Interoperability Framework [5]. In DataBio, we call these steps for a *generic pipeline* (Fig. 1.2). This generic pipeline is adapted to the agricultural,

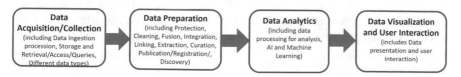

Fig. 1.2 Top-level generic pipeline

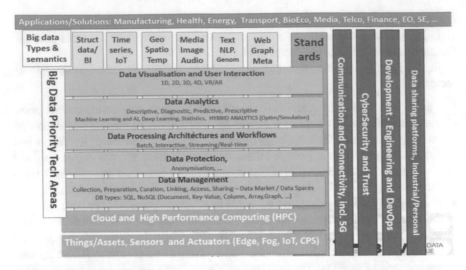

Fig. 1.3 DataBio project structured technologies as vertical pipelines crossing the horizontal layers in the BDV Reference Model

forestry and fisheries domains.

In order to describe the Big Data Value chains in more detail, the Big Data Value (BDV) Reference Model was adopted in DataBio (Fig. 1.3). The BDV Reference Model has been developed by the industry-led Big Data Value Association (BDVA). This model takes into account input from technical experts and stakeholders along the whole Big Data Value chain, as well as interactions with other industrial associations and with the EU. The BDV Reference Model serves as a common reference framework to locate Big Data technologies on the overall IT stack. It addresses the main concerns and aspects to be considered for Big Data Value systems in different industries. The BDV Reference Model is compatible with standardized reference architectures, most notably the emerging standards ISO JTC1 SC42 AI and Big Data Reference Architecture.

The steps in the generic pipeline and the associated layers in the reference model are:

Data acquisition from things, sensors and actuators: This layer handles the interface with the data providers and includes the transportation of data from various sources to a storage medium where it can be accessed, used and analysed. A main source of Big Data is sensor data from an IoT context and actuator interaction in cyberphysical systems. Tasks in this layer, depending on the type of collected data and on application implementation, include accepting or performing specific collections of data, pulling data or receiving pushes of data from data providers and storing or buffering data. Initial metadata can also be created to facilitate subsequent aggregation or look-up methods. Security and privacy considerations can also be included in this step, since authentication and authorization activities as well as recording and maintaining data provenance activities are usually performed during data collection.

Cloud, high performance computing (HPC) and data management: Effective Big Data processing and data management might imply the effective usage of cloud and HPC platforms. Traditional relational databases (RDB) do not typically scale well, when new machines are added to handle vast amounts of data. They are also not especially good at handling unstructured data like images and video. Therefore, they are complemented with non-relational databases like key-store, column-oriented, document and graph databases [6]. Of these, column-oriented architectures are used, e.g. in the Apache Cassandra and Hbase software for storing big amounts of data. Document databases have seen an enormous growth in recent years. The most used document database recently is MongoDB, that also was used in the DataBio project, e.g. in the DataBio Hub for managing the project assets and in the GeoRocket database component.

Data preparation: Tasks performed in this step include data validation, like checking formats, data cleansing, such as removing outliers or bad fields, extraction of useful information and organization and integration of data collected from various sources. In addition, the tasks consist of leveraging metadata keys to create an expanded and enhanced dataset, annotation, publication and presentation of the data to make it available for discovery, reuse and preservation, standardization and reformatting, as well as encapsulating. Source data is frequently persisted to archive storage and provenance data is verified or associated. Optimization of data through manipulations, like data deduplication and indexing, can also be included here.

Data processing and protection: The key to processing Big Data volumes with high throughput, and sometimes, complex algorithms is arranging the computing to take place *in parallel*. Hardware for parallel computing comprises 10, 100 or several thousands processors, often collected into graphical processing unit (GPU) cards. GPUs are used especially in machine learning and visualization. Parallelizing is straightforward in image and video processing, where the same operations typically are applied to various parts of the image. Parallel computing on GPU´s is used in DataBio, e.g. for visualizing data. *Data protection* includes privacy and anonymization mechanisms to facilitate protection of data. This is positioned between data management and processing, but it can also be associated with the area of cybersecurity.

Data analytics: In this layer, new patterns and relationships are discovered to provide new insights. The extraction of knowledge from the data is based on the requirements of the vertical application, which specify the data processing algorithms. Data analytics is a crucial step as it gives suggestions and makes decisions. Hashing, indexing and parallel computing are some of the methods used for Big Data analysis. Machine learning techniques and other artificial intelligence methods are also used in many cases.

Analytics utilize data both from the past and from the present.

- Data from the **past** is used for descriptive and diagnostic analytics, and classical querying and reporting. This includes performance data, transactional data, attitudinal data, behavioural data, location-related data and interactional data.

- Data from the **present** is harnessed in monitoring and real-time analytics. This requires fast processing many times handling data in real-time, for triggering alarms, actuators, etc.
- Harnessing data for the **future** includes prediction and recommendation. This typically requires processing of large data volumes, extensive modelling as well as combining knowledge from the past and present, to provide insight for the future.

Data visualization and user interaction: Visualization assists in the interpretation of data by creating graphical representations of the information conveyed. It thus adds more value to data as the human brain digests information better, when it is presented in charts or graphs rather than on spreadsheets or reports. In this way, users can comprehend large amounts of complex data, interact with the data, and make decisions. Effective data visualization needs to keep a balance between the visuals it provides and the way it provides them so that it attracts users' attention and conveys the right messages.

In the book chapters that follow, the above steps have been specialized based on the different data types used in the various project pilots. Solutions are set up according to different processing architectures, such as batch, real-time/streaming or interactive. See e.g. the pipelines for

- the real-time IoT data processing and decision-making in Chaps. 3 and 11,
- linked data integration and publication in Chap. 8,
- data flow in genomic selection and prediction in Chap. 16,
- farm weather insurance assessment in Chap. 19,
- data processing of Finnish forest data in Chap. 23.
- forest inventory in Chap. 24.

Vertical topics, that are relevant for all the layers in the reference model in Fig. 1.3, are:

- *Big Data Types and Semantics*: 6 Big Data types are identified, based on the fact that they often lead to the use of different techniques and mechanisms in the horizontal layers: (1) structured data; (2) time series data; (3) geospatial data; (4) media, image, video and audio data; (5) text data, including natural language processing data and genomics representations; and (6) graph data, network/web data and metadata. In addition, it is important to support both the syntactic and semantic aspects of data for all Big Data types.
- *Standards*: Standardization of Big Data technology areas to facilitate data integration, sharing and interoperability. Standards are advanced at many fora including communities like BDVA, and W3C as well as standardization bodies like ISO and NIST.
- *Communication and Connectivity*: Effective communication and connectivity mechanisms are necessary in providing support for Big Data. Especially important is wireless communication of sensor data. This area is advanced in various

communication communities, such as the 5G community as well as in telecom standardization bodies.

- *Cybersecurity*: Big Data often needs support to maintain security and trust beyond privacy and anonymization. The aspect of trust frequently has links to trust mechanisms such as blockchain technologies, smart contracts and various forms of encryption.
- *Engineering and DevOps* for building Big Data Value systems: In practise, the solutions have to be engineered and interfaced to existing legacy IT systems and feedback gathered about their usage. This topic is advanced especially in the Networked European Software and Service Initiative NESSI.
- *Marketplaces, Industrial Data Platforms (IDP) and Personal Data Platforms (PDPs), Ecosystems for Data Sharing and Innovation Support*: Data platforms include in addition to IDPs and PDPs, also Research Data Platforms (RDPs) and Urban/City Data Platforms (UDPs). These platforms facilitate the efficient usage of a number of the horizontal and vertical Big Data areas, most notably data management, data processing, data protection and cybersecurity.

1.3 Open, Closed and FAIR Data

Open and closed data

Open data means that data is freely available to everyone to use and republish, without restrictions from copyright, patents or other limiting mechanisms [7]. The access to closed datasets is restricted. Data is closed because of policies of data publishers and data providers. Closed data can be private data and/or personal data, valuable exploitable data, business or security sensitive data. Such data is usually not made accessible to the rest of the world. Data sharing is the act of certain entities (e.g. people) passing data from one to another, typically in electronic form [8].

Data sharing is central for bioeconomy solutions, especially in agriculture. At the same time, farmers need to be able to trust that their data is protected from unauthorized use. Therefore, it is necessary to understand that sharing data is different from the open data concept. Shared data can be closed data based on a certain agreement between specific parties, e.g. in a corporate setting, whereas open data is available to anyone in the public domain. Open data may require attribution to the contributing source, but still be completely available to the end user.

Data is constantly being shared between employees, customers and partners, necessitating a strategy that continuously secures *data stores* and users. Data moves among a variety of public and private storage locations, applications and operating environments and is accessed from different devices and platforms. That can happen at any stage of the data security lifecycle, which is why it is important to apply the right security controls at the right time. Trust of data owners is a key aspect for data sharing.

Generally, open data differs from closed data in three ways (see, e.g. www.opendatasoft.com).

1. Open data is accessible, usually via a data warehouse on the internet.
2. It is available in a readable format.
3. It is licenced as open source, which allows anyone to use the data or share it for non-commercial or commercial gain.

Closed data restricts access to the information in several potential ways:

1. It is only available to certain individuals within an organization.
2. The data is patented or proprietary.
3. The data is semi-restricted to certain groups.
4. Data that is open to the public through a licensure fee or other prerequisite.
5. Data that is difficult to access, such as paper records, that have not been digitized.

Examples of closed data are information that requires a security clearance; health-related information collected by a hospital or insurance carrier; or, on a smaller scale, your own personal tax returns.

FAIR data and data sharing

The FAIR data principles (Findable, Accessible, Interoperable, Reusable) ensure that data can be discovered through catalogues or search engines, is accessible through open interfaces, is compliant to standards for interoperable processing of that data and therefore can be easily reused also for other purposes than it was intitally created for [9]. This reuse improves the cost-balance of the initial data production and allows cross-fertilization across communities. The FAIR principles were adopted in DataBio through its data management plan [10].

1.4   The DataBio Platform

An application running on a Big Data platform can be seen as a pipeline consisting of multiple components, which are wired together in order to solve a specific Big Data problem (see https://www.big-data-europe.eu/). The components are typically packaged in Docker containers or code libraries, for easy deployment on multiple servers. There are plenty of commercial systems from known vendors like Microsoft, Amazon, SAP, Google and IBM that market themselves as Big Data platforms. There are also open-source platforms like Apache Hadoop for processing and analysing Big Data.

The DataBio platform was not designed as a monolithic platform; instead, it combines several existing platforms. The reasons for this were several:

*   The project sectors of agriculture, forestry and fishery are very diverse and a single monolithic platform cannot serve all users sufficiently well.
*   It is unclear who would take the ownership of such a new platform and maintain and develop it after the project ends.

- Several consortium partners had already at the outset of the project their own platforms. Therefore, DataBio should not compete with these partners by creating a new separate platform or by building upon a certain partner platform.
- Platform interoperability (public/private), data and application sharing were seen as more essential than creating yet another platform.

The DataBio platform should be understood in a strictly technical sense as a *software development platform* [11]. It is an environment in which a piece of software is developed and improved in an iterative process where after learning from the tests and trials, the designs are modified, and a new circle starts (Fig. 1.4). The solution is finally deployed in hardware, virtualized infrastructure, operating system, middleware or a cloud. More specifically, DataBio produced a Big Data *toolset* for *services* in agriculture, forestry and fishery [12]. The toolset enables new software *components* to be easily and effectively combined with open-source, standard-based, and proprietary components and infrastructures. These combinations typically form reusable and deployable *pipelines* of interoperable components.

The DataBio sandbox uses as resources mainly the *DataBio Hub*, but also the project *web site* and *deployed software* on public and private clouds [13]. The Hub links to content both on the DataBio website (deliverables, models) and to the Docker repositories on various clouds. This environment has the potential to make it easier and faster to design, build and test digital solutions for the bioeconomy sectors in future.

The *DataBio Hub* (https://databiohub.eu/) helps to manage the DataBio project assets, which are pilot descriptions and results, software components, interfaces, component pipelines, datasets, and links to deliverables and Docker modules (Fig. 1.5). The Hub has helped the partners *during* the project and has the potential to guide third party developers *after* the project in integrating DataBio assets into

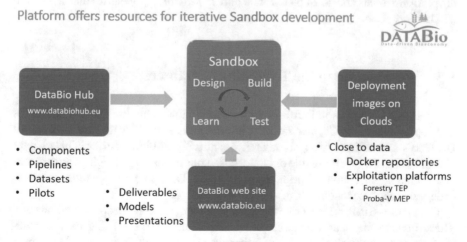

Fig. 1.4 Platform developed in DataBio consists of a network of resources for the interactive development of bioeconomy applications

Fig. 1.5 DataBio Hub provides searchable information on the assets developed in DataBio and helps the external developer to develop their own applications

new digital services for the bioeconomy sectors. The service framework at the core of the Hub is available as open source on GitHub (https://github.com/digitalservices hub/serviceregistry).

We identified 95 components, mostly from partner organizations, that could be used in the pilots. They covered all layers of the previously mentioned BDVA Reference Model (Fig. 1.6).

In total, 62 of the components were used in one or more of the pilots. In addition, the platform assets consist of 65 *datasets* and 25 *pipelines* (7 generic) that served the 27 DataBio pilots (Fig. 1.7).

1.5   Introduction to the Technology Chapters

The following chapters in Part I–Part IV describe the technological foundation for developing the pilots. Chapter 2 covers international *standards* that are relevant for DataBio's aim of improving raw material gathering in bioindustries. This chapter also discusses the emerging role of cloud-based platforms for managing Earth observation data in bioeconomy. The aim is to make Big Data processing a more seamless experience for bioeconomy data.

Chapters 3–6 in Part II describe the main *data types* that have been used in DataBio. These include the main categories sensor data and remote sensing data. Crowd-sourced and genomics data are also becoming increasingly important. The *sensor* chapter gives examples of in-situ IoT sensors for measuring atmospheric and soil

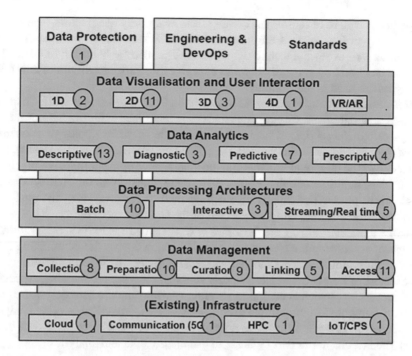

Fig. 1.6 Software components for use in DataBio pilots are in all parts of the BDV Reference Model, which here is presented in a simplified form. The number of components are given within the circles

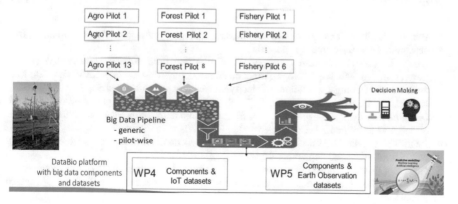

Fig. 1.7 DataBio platform served the pilots with components, IoT and Earth observation datasets and pipelines to demonstrate improved decision-making

properties as well as of sensor data coming from machinery like tractors. The *remote sensing* chapter lists relevant Earth observation (EO) formats, sources, datasets and services as well as several technologies used in DataBio for handling EO data. The chapters on *crowdsourced* and *genomics* data give illustrative examples of how these data types are used in bioeconomy.

  Data integration and modelling is dealt with in Chapters 7–9 in Part III. Chapter 7 explains how data from varying data sources is integrated with the help of a technology called *linked data*. Chapter 8 contains plenty of examples of *integrated linked data pipelines* in the various DataBio applications. Chapter 9 depicts how we *modelled* the pilot requirements and the architecture of the component pipelines. The models facilitate communication and comprehension among partners in the development phase. The chapter also defines metrics for *evaluating* the quality of the models and gives a quality assessment of the DataBio models.

  Analytics and visualizing are the topics of Chaps. 10–13 in Part IV. *Data analytics and machine learning* are treated in Chap. 10, which covers the data mining technologies, the mining process as well as the experiences from data analysis in the three sectors of DataBio. Chapter 11 deals with *real-time data processing*, especially event processing, which is central in several DataBio pilots, where dashboards and alerts are computed from multiple events in real-time. *Privacy preserving* analytics is described in Chapter 12. This is crucial, as parts of the bioeconomy data is not open. The last chapter in Part IV is about *visualizing* data and analytics results.

Literature

1. Palmer, M. (2006). *Data is the new oil.* https://ana.blogs.com/maestros/2006/11/data_is_the_ new.html. Accessed September 21, 2020.
2. Laney, D. (2001). *3D data management: Controlling data volume velocity and variety.* Gartner file No. 949. https://blogs.gartner.com/doug-laney/files/2012/01/ad949-3D-Data-Man agement-Controlling-Data-Volume-Velocity-and-Variety.pdf. Accessed September 21, 2020.
3. Reinsel, D., Gantz, J., & Rydning, J. (2018). Data age 2025: The digitization of the world. *International Data Corporation.* https://itupdate.com.au/page/data-age-2025-the-evolution-of-data-to-life-critical-. Accessed September 21, 2020.
4. ISO/IEC CD 205. (2019). Big Data overview and vocabulary. https://www.iso.org/standard/ 68305.html. Accessed September 28, 2020.
5. NIST Big Data Interoperability Framework: Volume 1, Definitions. (2019). https://www.nist. gov/publications/nist-big-data-interoperability-framework-volume-1-definitions.   Accessed September 21, 2020.
6. Drake, M. (2019). *A comparison of NoSQL database management systems and models.* https://www.digitalocean.com/community/tutorials/a-comparison-of-nosql-database-management-systems-and-models. Accessed September 21, 2020.
7. Auer, S. R., Bizer, C., Kobilarov, G., Lehmann, J., Cyganiak, R., & Ives, Z. (2007). DBpedia: A nucleus for a web of open data. The semantic web. *Lecture Notes in Computer Science* (4825, p. 722). https://doi.org/10.1007/978-3-540-76298-0_52. ISBN 978-3-540-76297-3. Accessed September 21, 2020.
8. Cambridge English Dictionary. (2020). Cambridge University Press. https://dictionary.cambri dge.org. Accessed September 22, 2020.

9. Ayris, P., Berthou, J-Y., Bruce, R., Lindstaedt, S., Monreale, A., Mons, B., Murayama, Y., Södergård, C., Tochtermann, K., & Wilkinson, R. (2016). *Realising the European open science cloud*. European Union. https://doi.org/10.2777/940154. Accessed December 9, 2020.
10. Habyarimana, E. et al. (2017). DataBio deliverable D6.2—data management plan. https://www.databio.eu/wp-content/uploads/2017/05/DataBio_D6.2-Data-Management-Plan_v1.0_2017-06-30_CREA.pdf . Accessed September 22, 2020.
11. Södergård, C. (2019). *DataBio platform*. Webinar: Big Data Breakthroughs for Global Bio-economy Business. https://es.slideshare.net/BDVA/bdv-webinar-series-caj-big-data-breakthroughs-for-global-bioeconomy-business. Accessed December 9, 2020.
12. Chaabouni, K., Bagnato, A., Walderhaug, S., Berre, A. J., Södergård, C., & Sadovykh, A. (2019) Enterprise architecture modelling with ArchiMate. *CEUR Workshop Proceedings, 2405*, 79–84. https://ceur-ws.org/Vol-2405/14_paper.pdf. Accessed December 7, 2020.
13. Plakia, M., Rousopoulos, K., Hara, S., Simarro, J. H., Palomares, M. A. E., Södergård, C., Siltanen, P., Kalaoja, J., Hayarimana, E., Kubo, B., Senner, I., Fournier, F., Berre, A. J., Tsalgatidou, A., Coene, Y., Auran, P. G., Kepka, M., Charvat, K., Charvat Jr., K., & Krommydas, S. (2019) *DataBio deliverable D4.4—service documentation*. https://www.databio.eu/wp-content/uploads/2017/05/DataBio_D4_4ServiceDocumentation_v1_2_2020_03_13_EXUS.pdf. Accessed December 7, 2020.

Chapter 2
Standards and EO Data Platforms

Ingo Simonis and Karel Charvát

Abstract In the digital bio-economy like in many other sectors, standards play an important role. With "Standards", we refer here to the protocols that describe how data and the data-exchange are defined to enable digital exchange of data between devices. This chapter evaluates how Big Data, cloud processing, and app stores together form a new market that allows exploiting the full potential of geospatial data. This chapter focuses on the essential cornerstones that help make Big Data processing a more seamless experience for bioeconomy data. The described approach is domain-independent, thus can be applied to agriculture, fisheries, and forestry as well as earth observation sciences, climate change research, or disaster management. This flexibility is essential when it comes to addressing real world complexities for any domain, as no single domain has sufficient data available within its own limits to tackle the major research challenges our world is facing.

2.1 Introduction

In the digital bio-economy like in many other sectors, standards play an important role. That is especially the case in exchanging digital data. With "*Standards*", we refer here to the protocols that describe how data and the data-exchange are defined to enable digital exchange of data between devices. Such standards enable interoperability between all participating players and ensure compatibility. Standards reduce transaction costs of sharing data and often promote competition, as users can easily change suppliers. Users are not 'locked in' to a closed system. Standards often support innovation, or provide a foundational layer that new innovation is built on.

This chapter evaluates how Big Data, cloud processing, and app stores together form a new market that allows exploiting the full potential of geospatial data. There

I. Simonis
OGC, Wayland, USA

K. Charvát (✉)
Lesprojekt-sluzby, Ltd., Zaryby, Czech Republic
e-mail: charvat@lesprojekt.cz

© The Author(s) 2021
C. Södergård et al. (eds.), *Big Data in Bioeconomy*,
https://doi.org/10.1007/978-3-030-71069-9_2

is a growing standards landscape for Big Data and cloud processing. There are new standards and industry agreements to handle orthogonal aspects such as security or billing. Still, an interoperable, secure, and publicly available Big Data exploitation in the cloud remains a challenge. It requires a set of standards to work together, both on the interface as well as the product exchange side. Related technologies for workflow and process orchestration or data discovery and access come with their own set of best practices, as well as emerging or existing standards.

Within the knowledge-based or data-driven bioeconomy, data and information sharing is an important issue. The complexity is high, as long supply chains with a variety of influencing factors need to be integrated. Often, bioeconomy information systems lack standardization and show a poorly organized exchange of information over the whole value and supply chain. Although arable and livestock farming, forestry and fishery have their own specific needs, there are many similarities in the need for an integrated approach.

DataBio identified a set of relevant technologies and requirements for the domains of agriculture, fisheries, and forestry. There is an extensive list of interfaces, interaction patterns, data models and modelling best practices, constraint languages, or visualization approaches. Together with the Open Geospatial Consortium, the worldwide leading organization for geospatial data handling, DataBio contributed to the development of emerging standards that help forming new data markets as described above. These markets are important for everyone from the individual farmer up to the Big Data provider. They will allow the exploitation of available data in an efficient way, with new applications allowing targeted analysis of data from the farm, fishery, or forest level, all the way up to satellite data from Earth Observation missions.

The underlying technology shifts have been implemented mostly independent of the (bioeconomy) domain. They have been driven by mass-market requirements and now provide essential cornerstones for a new era of geospatial data handling. The emerging standards define how the generic cornerstones need to be applied to Earth observation data discovery, access, processing, and representation.

This chapter focuses on the essential cornerstones that help make Big Data processing a more seamless experience for bioeconomy data. The described approach is domain-independent, thus can be applied to agriculture, fisheries, and forestry as well as earth observation sciences, climate change research, or disaster management. This flexibility is essential when it comes to addressing real world complexities for any domain, as no single domain has sufficient data available within its own limits to tackle the major research challenges our world is facing.

2.2 Standardization Organizations and Initiatives

ISO

ISO is the International Organization for Standardization, which develops and publishes international standards. ISO standards ensure that products and services are safe, reliable and of good quality. For businesses, they are strategic tools that reduce costs by minimising waste and errors and increasing productivity. They help companies to access new markets, level the playing field for developing countries and facilitate free and fair global trade. According to https://www.iso.orgl, "ISO standards for agriculture cover all aspects of farming, from irrigation and global positioning systems (GPS) to agricultural machinery, animal welfare and sustainable farm management. They help to promote effective farming methods while ensuring that everything in the supply chain—from farm to fork—meets adequate levels of safety and quality. By setting internationally agreed solutions to global challenges, ISO standards for agriculture also foster the sustainability and sound environmental management that contribute to a better future."

W3C

The World Wide Web Consortium (W3C, https://www.w3.org/) is an international community where member organisations,a full-time staff, and the public work together to develop Web standards. The W3C mission is to lead the World Wide Web to its full potential by developing protocols and guidelines that ensure the long term growth of the Web. According to W3C, the initial mission of the Agriculture Community Group (https://www.w3.org/community/agri/) is to gather and categorise existing user scenarios, which use Web APIs and services, in the agriculture industry from around the world, and to serve as a portal which helps both web developers and agricultural stakeholders create smarter devices, Web applications & services, and to provide bird's eye view map of this domain which enables.W3C and other SDOs to find overlaps and gaps of user scenarios and the Open Web Platform.

OASIS

OASIS (Organization for the Advancement of Structured Information Standards, https://www.oasis-open.org) is a not for-profit consortium that drives the development, convergence and adoption of open standards for the global information society. OASIS promotes industry consensus and produces worldwide standards for security, Cloud computing, SOA, Web services, the Smart Grid, electronic publishing, emergency management, and other areas. OASIS open standards offer the potential to lower costs, stimulate innovation, grow global markets, and protect the right of free choice of technology.

OGC

The Open Geospatial Consortium (OGC, https://www.ogc.org) is an international consortium of more than 500 businesses, government agencies, research organizations, and universities driven to make geospatial (location) information and services FAIR—Findable, Accessible, Interoperable, and Reusable. OGC's member-driven consensus process creates royalty free, publicly available geospatial standards. Existing at the cutting edge, OGC actively analyzes and anticipates emerging tech trends, and runs an agile, collaborative Research and Development (R&D) lab that builds and tests innovative prototype solutions to members' use cases. OGC members together form a global forum of experts and communities that use location to connect people with technology and improve decision-making at all levels. OGC is committed to creating a sustainable future for us, our children, and future generations.

The Agriculture DWG will concern itself with technology and technology policy issues, focusing on geodata information and technology interests as related to agriculture as well as the means by which those issues can be appropriately factored into the OGC standards development process. The **mission** of the Agriculture Working Group is to identify geospatial interoperability issues and challenges within the agriculture domain, then examine ways in which those challenges can be met through application of existing OGC standards, or through development of new geospatial interoperability standards under the auspices of OGC. The **role** of the Agriculture Working Group is to serve as a forum within OGC for agricultural geo-informatics; to present, refine and focus interoperability-related agricultural issues to the Technical Committee; and to serve where appropriate as a liaison to other industry, government, independent, research, and standards organizations active within the agricultural domain.

IEEE

IEEE, https://www.ieee.org/, is the world's largest professional association dedicated to advancing technological innovation and excellence for the benefit of humanity. IEEE and its members inspire a global community through IEEE's highly cited publications, conferences, technology standards, and professional and educational activities. IEEE, pronounced "Eye-triple-E," stands for the Institute of Electrical and Electronics Engineers. The association is chartered under this name and it is the full legal name.

VDMA—ISOBUS

ISOBUS (https://www.isobus.net/isobus/) was managed by the ISOBUS group in VDMA. The VDMA (Verband Deutscher Maschinen und Anlagenbau—German Engineering Federation) is a network of around 3,000 engineering industry companies in Europe and 400 industry experts. The ISOBUS standard specifies a serial data network for control and communications on forestry or agricultural tractors. It consists of several parts: General standard for mobile data communication, Physical layer, Data link layer, Network layer, Network management, Virtual terminal, Implement messages applications layer, Power train messages, Tractor ECU, Task controller and management information system data interchange, Mobile data element dictionary, Diagnostic, File Server. The work for further parts is ongoing. It is currently ISO standard ISO 11783.

agroXML

agroXML (https://195.37.233.20/about/) is a markup language for agricultural issues providing elements and XML data types for representing data on work processes on the farm including accompanying operating supplies like fertilizers, pesticides, crops and the like. It is defined using W3C's XML Schema. agroRDF is an accompanying semantic model that is at the moment still under heavy development. It is built using the Resource Descrition Framework (RDF).

While there are other standards covering certain areas of agriculture like e.g., the ISOBUS data dictionary for data exchange between tractor and implement or ISOagriNet for communication between livestock farming equipment, the purposes of agroXML and agroRDF are:

- exchange between on-farm systems and external stakeholders
- high level documentation of farming processes
- data integration between different agricultural production branches
- semantic integration between different standards and vocabularies
- a means for standardized provision of data on operating supplies

INSPIRE

In Europe a major recent development has been the entering in force of the INSPIRE Directive in May 2007, establishing an infrastructure for spatial information in Europe to support Community environmental policies, and policies or activities which may have an impact on the environment. INSPIRE is based on the infrastructures for spatial information established and operated by the all Member States of the European Union. The Directive addresses 34 spatial data themes needed for environmental applications, with key components specified through technical implementing rules. This makes INSPIRE a unique example of a legislative "regional" approach. For more details, see https://inspire.ec.europa.eu/about-inspire/563.

2.2.1   The Role of Location in Bioeconomy

Few activities are more tied to location, geography, and the geospatial landscape than farming. The farm business, farm supply chain, and public agricultural policies are increasingly tied as well to quantitative data about crops, soils, water, weather, markets, energy, and biotechnology. These activities involve sensing, analyzing, and communicating larger and larger scale geospatial data streams. How does farming become more, not less, sustainable as a business and as a necessity for life in the face of climate change, growing populations, scarcity of water and energy. Matching precision agricultural machinery with precision agricultural knowledge and promoting crop resiliency at large and small scales are increasing global challenges. As food markets grow to a global scale, worldwide sharing of information about food traceability and provenance, as well as agricultural production, is becoming a necessity. The situation is not much different from fishery or forestry. Both are geospatial disciplines to a good extent and require integration of location data.

2.2.2 The Role of Semantics in Bioeconomy

"Semantic Interoperability is usually defined as the ability of services and systems to exchange data in a meaningful/useful way." In practice, achieving semantic interoperability is a hard task, in part because the description of data (their meanings, methodologies of creation, relations with other data etc.) is difficult to separate from the contexts in which the data are produced. This problem is evident even when trying to use or compare data sets about seemingly unambiguous observations, such as the height of a given crop (depending on how height was measured, at which growth phase, under what cultural conditions, etc.). Another difficulty with achieving semantic interoperability is the lack of the appropriate set of tools and methodologies that allow people to produce and reuse semantically-rich data, while staying within the paradigm of open, distributed and linked data.

The use and reuse of accurate semantics for the description of data, datasets and services, and to provide interoperable content (e.g., column headings, and data values) should be supported as community resources at an infrastructural level. Such an infrastructure should enable data producers to find, access and reuse the appropriate semantic resources for their data, and produce new ones when no reusable resource is available.

2.3 Architecture Building Blocks for Cloud Based Services

To fully understand the architecture outlined below, this chapter introduces high level concepts for future data exploitation platforms and corresponding applications markets first. There is a growing number of easily accessible Big Data repositories hosted on cloud infrastructures. Most commonly known are probably earth observation satellite data repositories, with petabyte-sized data volumes, that are accessible to the public. These repositories currently transform from pure data access platforms towards platforms that offer additional sets of cloud-based products/services such as compute, storage, or analytic services. Experiences have shown that the combination of data and corresponding services is a key enabler for efficient Big Data processing. When the transport of large amounts of data is not feasible or cost-efficient anymore, processes (or applications) need to be deployed and executed as closely as possible to the actual data. These processes can either be pre-deployed, or deployed ad-hoc at runtime in the form of containers that can be loaded and executed safely. Key is to develop standards that allow packing any type of application or multi-application-based workflow into a container that can be dynamically deployed on any type of cloud environment. Consumers can discover these containers, provide the necessary parameterization and execute them online even easier than on their local machines, because no software installation, data download, or complex configuration is necessary.

Figure 2.1 illustrates the main elements of such an architecture. Data providers on

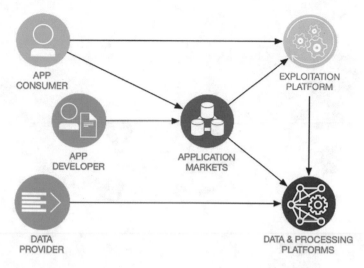

Fig. 2.1 High level architecture (*Source* [1])

the lower left make their data available at publicly accessible Data and Processing Platforms in the cloud. Ideally, these platforms provide access to larger sets of raw data and data products from multiple data providers. Application consumers (upper left), i.e. customers with specific needs that can be served by processing the data, identify the appropriate application(s) that produces the required results by processing (Big) data. The applications are produced by application developers and offered on application markets that work pretty similar to smart phone markets, with the difference that applications are deployed on demand on cloud platforms rather than downloaded and installed on smartphones. Exploitation platforms support the application consumers with single sign on, facilitate application chaining even across multiple Data and Processing Platforms, and ensure the most seamless user experience possible.

2.4   Principles of an Earth Observation Cloud Architecture for Bioeconomy

"Earth Observation Cloud Architecture" standardization efforts are underway that fulfill the aforementioned requirements to establish marketplaces for domain-specific and cross-domain Big data processing in the cloud. The architecture supports the "application to the data" paradigm for Big data that is stored and distributed on independent Data and Processing Platforms. The basic idea is that each platform provides a standardized interface that allows the deployment and parameterized execution of applications that are packaged as software containers. A logically second type of

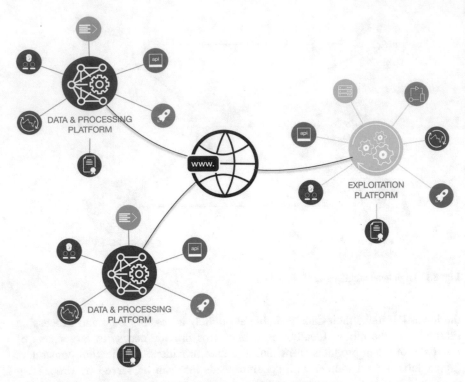

Fig. 2.2 Earth observation cloud architecture platforms (*Source* [1])

platform is called Exploitation Platform and allows chaining containers/applications into workflows with full support for quoting and billing.

Exploitation and Data & Processing platforms are built using a number of components to provide all required functionality. As illustrated in Fig. 2.2, any number of these platforms can co-exist. Both types of platform can be implemented within a single cloud environment. Given that they all support the same interface standards, applications can be deployed and chained into complex workflows as necessary.

Standards define key components, interaction patterns, and communication messages that allow the ad-hoc deployment and execution of arbitrary applications close to the physical storage location of data. The application developer can be fully independent of the data provider or data host. The applications become part of an application market similar to what is currently common practice for mobile phone applications. The major difference is that applications are not downloaded to cell phones, but deployed and executed on cloud platforms. This is fully transparent to the user, who selects and pays an application and only needs to wait for the results to appear.

The above-mentioned standardization efforts are mainly driven by the Open Geospatial Consortium (OGC). These standards are made through a consensus process and are freely available for anyone to use to improve sharing of the world's geospatial data. OGC standards are used in a wide variety of domains including Environment, Defense, Health, Agriculture, Meteorology, Sustainable Development and many more. OGC members come from government, commercial organizations, NGOs, academic and research organizations.

The OGC has worked for the last three years on a set of standards and software design principles that allow a vendor and platform neutral secure Big Data processing architecture. Supported by the space agencies ESA and NASA, the European Commission through H2020 co-funded projects (DataBio being one of them), and Natural Resources Canada, OGC has developed a software architecture that decouples the data and cloud operators from Earth Observation data application developers and end-consumers and provides all the essential elements for standards-based Big Data processing across domains and disciplines.

The Earth Observation Cloud Architecture defines a set of interface specifications and data models working on top of the HTTP layer. The architecture allows application developers and consumers to interact with Web services that abstract from the underlying complexity of data handling, scheduling, resource allocation, or infrastructure management.

2.4.1   Paradigm Shift: From SOA to Web API

Standards are the key pillar of any exchange or processing of information on the World Wide Web. Offering geospatial data and processing on the Web is often referred to as Spatial Data Infrastructure (SDI). These SDIs have been built following the Service Oriented Architecture (SOA) software paradigm. Nowadays, the focus is shifting towards Web Application Programming Interfaces (Web APIs). The differences for the end users are almost negligible, as client applications handle all protocol specific interactions. To the end user, the client may look the same, even though the underlying technology has changed.

At the moment, both approaches work next to each other to acknowledge the large number of existing operational SOA-based services. However, in the long run, Web APIs offer significant benefits, which is also reflected in OGCs Open API development activities. The architecture described in the following two sections, defines two 'logical' types of platforms. Both can be implemented using SOA-style Web services or Web-API-style REST interfaces. To the end user, it is most likely irrelevant.

2.4.2   Data and Processing Platform

The Data and Processing Platform illustrated in Fig. 2.3 has six major components: In addition to the actual data repository, the platform offers the application deployment and execution API. The API allows deployment, discovery, and execution of

Fig. 2.3 Data and processing platform (*Source* [1])

applications or to perform quoting requests. All applications are packaged as Docker containers to allow easy and secure deployment and execution within foreign environments (though alternative solutions based on other container technology are currently explored). The Docker daemon provides a Docker environment to instantiate and run Docker containers. The Billing and Quoting component allows obtaining quotes and final bills. This is important because the price of an application run is not necessarily easily calculated. Some applications feature a simple price model that only depends on parameters such as area of interest or time period. Other applications, or even more complex entire workflows with many applications, may require heuristics to calculate the full price of execution. The workflow runner can start the Docker container applications. It manages dynamic data loading and result persistency in a volatile container environment. Identity and Access Management provide user management functionalities.

2.4.3 *Exploitation Platform*

The Exploitation Platform is responsible for registration and management of applications and the deployment and execution of applications on Data and Processing Platforms. It further supports workflow creation based on registered applications, and aggregates quoting and billing elements that are part of these workflows. Ideally, the Exploitation Platform selects the best suited Data and Processing Platform based on consumer's needs. As illustrated in Fig. 2.4, the Exploitation Platform itself consists of seven major components.

The *Execution Management Service API* provides a Web interface to application developers to register their applications and to build workflows from registered applications. The *application registry* implementation (i.e. application catalog) allows managing registered applications (with create, read, update, and delete options), whereas the optional *workflow builder* supports the application developer to build workflows form registered applications. The *workflow runner* executes workflows and handles the necessary data transfers from one application to the other.

The *Application Deployment and Execution Client* interacts with the data and processing environments that expose the corresponding Application Deployment and Execution Service API. The *Billing & Quoting* component aggregates

Fig. 2.4 Exploitation platform (*Source* [1])

billing and quoting elements from the data and processing environments that are part of a workflow. *Identity and Access Management* provides user management functionalities.

2.5 Standards for an Earth Observation Cloud Architecture

The architecture described above builds primarily on three key elements: The Application Deployment and Execution Service (ADES), the Execution Management Service (EMS), and the Application Package (AP). The specifications for all three have been initially developed in OGC Innovation Program initiatives and are handed over gradually after maturation to the OGC Standards Program for further consideration. Applications are shared as Docker containers. All application details required to deploy and run an application are provided as part of the metadata container called Application Package. The following diagram illustrates the high-level view on the two separated loops application development (left) and application consumption (right) (Fig. 2.5).

The left loop shows the application developer, who puts the application into a container and provides all necessary information in the Application Package. The application will be made available at the cloud platform using the Application Deployment and Execution Service (ADES). Using the Execution Management Service (EMS), application developers can chain existing applications into processing chains. The right loop shows the application consumer, who uses the EMS to request an application to be deployed and executed. Results are made available through additional standards-based service interfaces such as OGC API-Features, -Maps, -Coverages, or web service such as Web Map Service, Web Feature Service,

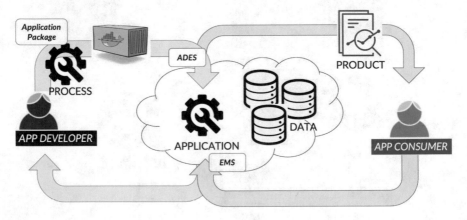

Fig. 2.5 Architecture elements in context

or Web Coverage Service. Alternatively, results can be provided at direct download links.

2.5.1   Applications and Application Packages

Any application can be executed as a Docker Container in a Docker environment that needs to be provided by the platform. The application developer needs to build the container with all libraries and other resources required to execute the application. This includes all data that will not be provided in the form of runtime parameters or be dynamic mounted from the platform's Big data repository. The Docker container image itself can be built from a Docker Build Context stored in a repository following the standard manual or Dockerfile-based scripting processes. To allow standards-based application deployment and execution, the application should be wrapped with a start-up script.

As described in Ref. [2], the Application Package (AP) serves as the application metadata container that describes all essential elements of an application, such as its functionality, required processing data, auxiliary data, input runtime parameters, or result types and formats. It stores a reference to the actual container that is hosted on a Docker hub independently of the Application Package. The Application Package describes the input/output data and defines mount points to allow the execution environment to serve data to an application that is actually executed in a secure memory space; and to allow for persistent storage of results before a container is terminated (Fig. 2.6).

The OGC has defined the OGC Web Services Context Document (OWS Context Document) as a container for metadata for service instances [3]. The context document allows to exchange any type of metadata for geospatial services and data offerings. Thus, the context document is perfectly qualified to serve as a basis for the

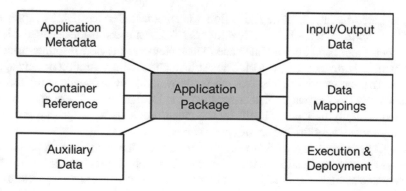

Fig. 2.6 Application package elements

Application Package. It can be used to define all application specific details required to deploy and execute an application in remote cloud environments.

2.5.2 Application Deployment and Execution Service (ADES)

Once application consumers request the execution of an app, the Exploitation forwards the execution request to the processing clouds and makes final results available at standardized interfaces again, e.g. at Web Feature Service (WFS) or Web Coverage Service (WCS) instances. In the case of workflows that execute a number of applications sequentially, the Exploitation realizes the transport of data from one process to the other. Upon completion, the application consumer is provided a data access service endpoint to retrieve the final results. All communication is established in a web-friendly way implementing the emerging next generation of OGC services known as WPS, WFS, and WCS 3.0.

2.5.3 Execution Management Service (EMS)

The execution platform, which offers EMS functionality to application developers and consumers, acts itself as a client to the Application Deployment and Execution Services (ADES) offered by the data storing cloud platforms. The cloud platforms support the ad-hoc deployment and execution of Docker images that are pulled from the Docker hubs using the references made available in the deployment request.

2.5.4 AP, ADES, and EMS Interaction

As illustrated in Fig. 2.7, the Execution Management Service (EMS) represents the front-end to both application developers and consumers. It makes available an OGC Web Processing Service interface that implements the new resource-oriented paradigm, i.e. provides a Web API. The API supports the registration of new applications. The applications themselves are made available by reference in the form of containerized Docker images that are uploaded to Docker Hubs. These hubs may be operated centrally by Docker itself, by the cloud providers, or as private instances that only serve a very limited set of applications.

The EMS represents a workflow environment that allows application developers to re-use existing applications and orchestrate them into sequential work-flows that can be made available as new applications again. This process is transparent to the application consumer.

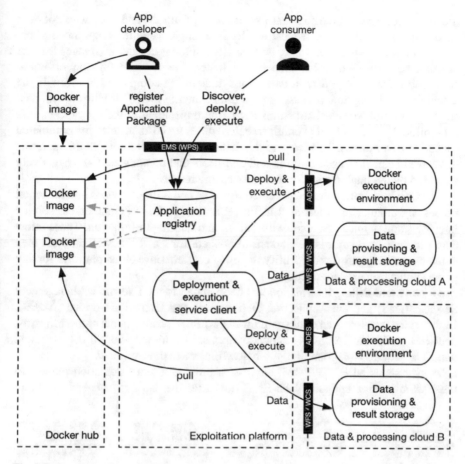

Fig. 2.7 Detailed software architecture (*Source* [4])

2.6   Standards for Billing and Quoting

Currently, lots of Big data and in particular satellite image processing still happens to a large extent on the physical machine of the end-user. This approach allows the end-user to understand all processing costs upfront. The hardware is purchased, prices per data product are known in advance, and actual processing costs are defined by the user's time required to supervise the process. The approach is even reflected in procurement rules and policies at most organizations that often require a number of quotes before an actual procurement is authorized.

The new approach outlined here requires a complete change of thinking. No hardware other than any machine with a browser (which could even be a cell phone) needs to be purchased. Satellite imagery is not purchased or downloaded anymore, but rented just for the time of processing using the architecture described above, and the final processing costs are set by the computational resource requirements of the process. Thus, most of the cost factors are hidden from the end-user, who does not necessarily know if his/her request results in a single satellite image process that can run on a tiny virtual machine, or a massive amount of satellite images that are processed in parallel on a 100+ machines cluster. The currently ongoing efforts to store Earth Observation data in data cubes adds to the complexity to estimate the actual data consumption, because the old unit "satellite image" is blurred with data stored in multidimensional structures not made transparent to the user. Often, it is even difficult for the cloud operator to calculate exact costs prior to the completed execution of a process. This leads to the difficult situation for both cloud operators that have to calculate costs upfront, and end-users that do not want to be negatively surprised by the final invoice for their processing request.

The OGC has started the integration of quoting and billing services into the cloud processing architecture illustrated in Fig. 2.8. The goal is to complement service interfaces and defined resources with billing and quoting information. These allow a user to understand upfront what costs may occur for a given service call, and they allow execution platforms to identify the most cost-effective cloud platform for any given application execution request.

Quoting and Billing information has been added to the Execution Management Service (EMS) and the Application Deployment and Execution Service (ADES). Both service types (or their corresponding APIs) allow posting quota requests against dedicated endpoints. A JSON-encoded response is returned with all quote related data. The sequence diagram in figure below illustrates the workflow.

A user sends an HTTP POST request to provide a quasi-execution request to the EMS/quotation endpoint. The EMS now uses the same mechanism to obtain

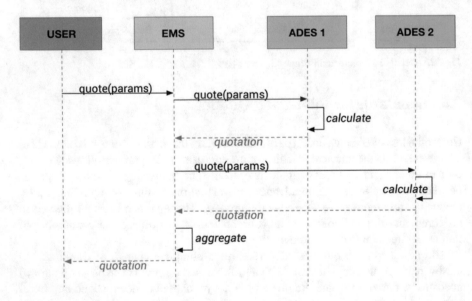

Fig. 2.8 Quoting process

quotes from all cloud platforms that offer deployment and execution for the requested application. In case of a single application that is deployed and executed on a single cloud only, the EMS uses the approach to identify the most cost-efficient platform. In case of a workflow that includes multiple applications being executed in sequence, the EMS aggregates involved cloud platforms to generate a quote for the full request. Identification of the most cost-efficient execution is not straightforward in this case, as cost efficiency can be considered a function of processing time and monetary costs involved. In all cases, a quote is returned to the user. The quote model is intentionally simple. In addition to some identification and description details, it only contains information about its creation and expiration date, currency and price-tag, and an optional processing time element. It further repeats all user-defined parameters for reference and optionally includes quotations for alternatives, e.g. at higher costs but reduced processing time or vice versa. These can for example include longer estimated processing times at reduced costs.

2.7   Standards for Security

Reliable communication within business environments requires some level of security. This includes all public interfaces as well as data being secured during transport. As shown in 4, the system uses identity providers to retrieve access tokens that can be used in all future communication between the application consumer, EMS, and ADES. The authentication loop is required to handle multiple protocols to support existing, e.g. eduGAIN, as well as emerging identity federations. Once an authentication token has been received, all future communication is handled over HTTPS and handles authorization based on the provided access token. Full details on the security solution are provided in OGC document OGC Testbed-14: Authorisation, Authentication, and Billing Engineering Report; OGC document OGC 18-057).

2.8   Standards for Discovery, Cataloging, and Metadata

DataBio's contribution to OGC standardization further includes metadata and service interfaces for service discovery. This includes Earth Observation (EO) products, services providing on-demand processing capabilities, and applications that are not deployed yet but waiting in an application store for their ad-hoc deployment and execution. The aforementioned OGC Innovation Program has developed an architecture that allows the containerization of any type of application. These applications can be deployed on demand and executed in cloud environments close to the physical location of the data.

From a catalog/discovery perspective, the following questions arise: How to discover EO applications? How to understand what data an application can be applied to? How to chain applications? How to combine applications with already deployed services that provide data and data processing capabilities? The following provides paragraphs provide a short overview of standardization efforts currently underway.

Catalog Service Specification

The discovery solution proposed by OGC comprises building blocks through which applications and related services can be exposed through a Catalogue service. It consists of the following interfaces:

- Service Interface: providing the call interface through which a catalogue client or another application can discover applications and services through faceted search and textual search, and then retrieve application/service metadata providing more detail.
- Service Management Interface: providing the call interface through which a catalog client or any other application can create, update and delete information about applications/services.

Each of the above interfaces is discussed in full detail in the OGC Testbed-15: Catalogue and Discovery Engineering Report [5]. This discussion includes the metadata model that provides the data structure through which the application and/or service is described and presented as a resource in the catalog.

The current standardization work builds on a series of existing standards as illustrated below (Figs. 2.9, 2.10 and 2.11).

These standards provide robust models and encodings for EO products and collections.

Now extended by OpenSearch specifications as illustrated below.

And integrated into a set of specifications as shown in figure below.

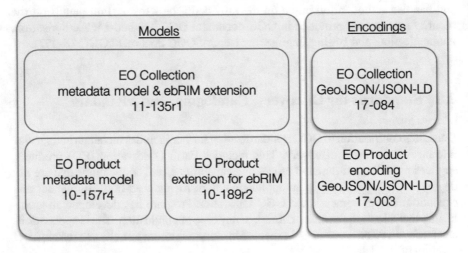

Fig. 2.9 Existing OGC Standards supporting discovery for EO data

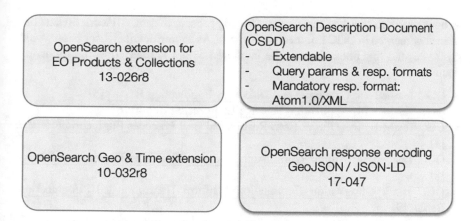

Fig. 2.10 OpenSearch extensions for existing OGC Standards

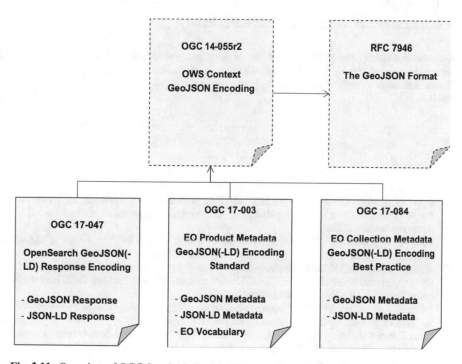

Fig. 2.11 Overview of OGC Standards for standards-based application discovery (*Source* [5])

2.9 Summary

This chapter provided an overview of currently ongoing standardization efforts executed by the Open Geospatial Consortium with support by DataBio to define an application-to-the-data environment for Big geospatial data. All work till date has been documented in OGC Engineering Reports. As a more detailed discussion would go far beyond this book chapter, the interested reader is referred to the following documents:

- OGC Testbed-15: Catalogue and Discovery Engineering Report [5]
- OGC Testbed-14: Application Package Engineering Report [6]
- OGC Testbed-14: ADES & EMS Results and Best Practices Engineering Report [7]
- OGC Testbed-14: Authorisation, Authentication, & Billing Engineering Report [8]
- OGC Earth Observation Exploitation Platform Hackathon 2018Engineering Report [9]
- OGC Testbed-13: EP Application Package Engineering Report [10]
- OGC Testbed-13: Application Deployment and Execution Service Engineering Report [11]
- OGC Testbed-13: Cloud Engineering Report [12]

Acknowledgements The authors would like to thank everybody who contributed to the work presented in this chapter. Particularly, we would like to thank the European Space Agency, ESA, and Natural Resources Canada for additional support in the development of the architecture documented herein.

References

1. Open Geospatial Consortium. (OGC). *OGC earth observation applications pilot: call for participation (CFP)*. Online publication: https://portal.ogc.org/files/?artifact_id=90733. Accessed January 5, 2020.
2. Simonis, I. Standardized Big Data processing in hybrid clouds. In *Proceedings of the 4th International Conference on Geographical Information Systems Theory, Applications and Management (GISTAM 2018)* (pp. 205–210).
3. Open Geospatial Consortium. (OGC). OGC OWS context conceptual model. OGC Implementation Standard. OGC Document 12-080r2.
4. Simonis, I. Quoting and billing: commercialization of Big Data analytics. In *Proceedings of the 2019 conference on Big Data from Space (BiDS'19)* (pp. 53–56).
5. Open Geospatial Consortium (OGC). *OGC Testbed-15: Catalogue and discovery engineering report, OGC Document 19-020r1*. Online publication. Accessed January 5, 2020.
6. Open Geospatial Consortium (OGC). *OGC testbed-14: Application package engineering report, OGC Document 18-049r1*. Online publication: https://docs.opengeospatial.org/per/18-049r1.html. Accessed January 5, 2020
7. Open Geospatial Consortium (OGC). *OGC testbed-14: ADES & EMS results and best practices engineering report, OGC Document 18-050r1*. Online publication: https://docs.opengeospatial.org/per/18-050r1.html. Accessed January 5, 2020.
8. Open Geospatial Consortium (OGC). *OGC earth observation exploitation platform hackathon 2018 engineering report, OGC Document 18-057*. Online publication: https://docs.opengeospatial.org/per/18-057.html. Accessed January 5, 2020

9. Open Geospatial Consortium (OGC). *OGC earth observation exploitation platform Hackathon 2018 engineering report, OGC Document 18-046*. Online publication: https://docs.opengeospatial.org/per/18-046.html. Accessed January 5, 2020
10. Open Geospatial Consortium (OGC). *OGC testbed-13: EP Application package engineering report, OGC Document 17-023*. Online publication: https://docs.opengeospatial.org/per/17-023.html. Accessed January 5, 2020
11. Open Geospatial Consortium (OGC). *OGC testbed-13: Application deployment and execution service engineering report, OGC Document 17-024*. Online publication: https://docs.opengeospatial.org/per/17-024.html. Accessed January 5, 2020
12. Open Geospatial Consortium (OGC). *OGC Testbed-13: Cloud engineering report, OGC Document 17-035*. Online publication: https://docs.opengeospatial.org/per/17-035.html. Accessed January 5, 2020

Part II
Data Types

Chapter 3
Sensor Data

Savvas Rogotis, Fabiana Fournier, Karel Charvát, and Michal Kepka

Abstract The chapter describes the key role that sensor data play in the DataBio project. It introduces the concept of sensing devices and their contribution in the evolution of the Internet of Things (IoT). The chapter outlines how IoT technologies have affected bioeconomy sectors over the years. The last part outlines key examples of sensing devices and IoT data that are exploited in the context of the DataBio project.

3.1 Introduction

Sensing devices have been introduced in order to bridge the gap between the physical and the digital world. In fact sensors are responsible for gathering and responding to physical stimulus originating from the environment. Different sensors respond to different environmental input signals such as light, heat, motion, humidity, pressure, sound, etc. Sensors translate the input signal to a digital one, so that it can be easily displayed, stored or transmitted over networks and processed in a more sophisticated way.

Gartner [1] defines The Internet of Things (IoT) as "a core building block for digital business and digital platforms. IoT is the network of dedicated physical objects that contain embedded technology to communicate and sense or interact with their internal states and/or the external environment. IoT comprises an ecosystem that includes assets and products, communication protocols, applications, and data and

S. Rogotis (✉)
NEUROPUBLIC SA, Methonis 6 & Spiliotopoulou, 18545 Piraeus, Greece
e-mail: s_rogotis@neuropublic.gr

F. Fournier
IBM Research—Haifa University Campus, Mount Carmel, 3498825 Haifa, Israel

K. Charvát
LESPROJEKT-SLUZBY Ltd., Martinov 197, 27713 Zaryby, Czech Republic

M. Kepka
University of West Bohemia, Univerzitni 8, Pilsen 30100, Czech Republic

© The Author(s) 2021
C. Södergård et al. (eds.), *Big Data in Bioeconomy*,
https://doi.org/10.1007/978-3-030-71069-9_3

41

analytics". Sensors are at the core of IoT systems and along with sensor connectivity and network they collect the information to be analyzed by an IoT application.

According to Ref. [2], at the start of 2019, the Internet of Things (IoT) remains a critical enabler helping organizations achieve their digital transformation goals. Alongside cloud, analytics, and mobile investments, IoT remains a top priority for organizations as they make technology decisions. Worldwide spending on the Internet of Things (IoT) is forecast to reach $745 billion in 2019, an increase of 15.4% over the $646 billion spent in 2018, according to a new update to the International Data Corporation (IDC) Worldwide Semiannual Internet of Things Spending Guide. IDC expects worldwide IoT spending will maintain a double-digit annual growth rate throughout the 2017–2022 forecast period and surpass the $1 trillion mark in 2022.

3.2   Internet of Things in Bioeconomy Sectors

With emergence of the Internet of Things (IoT), a hype in the proliferation of use and application of sensors in almost every vertical domain is being witnessed. One of the domains that has been taking advantage of sensor data is agriculture [3]. As smart machines and sensors crop up on farms and farm data grow in quantity and scope, farming processes will become increasingly data driven and data enabled [4]. The development of highly accurate embedded sensors measuring the environmental context inside farms has led to the enablement of precision agriculture [5]. Precision agriculture enables smart farming, which includes real-time data gathering, processing and analysis, as well as automation of the farming procedures, allowing improvement of the overall farming operations and management, and more data-driven decision making by the farmers. In smart farming, IoT extends conventional tools (e.g., rain gauge, tractor, notebook) by adding autonomous context-awareness by all kind of sensors, built-in intelligence, capable to execute autonomous actions or enabling their execution remotely. These smart devices provide the required data to drive real-time operational decisions. Real-time assistance is required to carry out agile actions, especially in cases of suddenly changed operational conditions or other circumstances (e.g., weather or disease alert) [4]. Farming is highly unpredictable, due to its large dependency on weather and environmental conditions (eg. rain, temperature, humidity, hail), unpredictable events (e.g. animal diseases, pests), as well as price volatility in agricultural markets. Combining and analyzing data streams provided by sensors in real-time, can help in more informed decision-making and enable fast reaction to changes and unpredictable events. For example, by combining sensor data about soil fertility, with web services for weather forecasting, better decisions could be made about more precise irrigation and fertilization of the crops. Sensor data can also be used to enable real time monitoring of agrifood parameters, such as pH, temperature, earth's moisture or oxygen flow.

With the aid of sensors connected to internet, it is possible to continuously monitor different crops and parcels, even if they are remotely located, as well as to predict and control yields and food quality. Georeferencing is another important aspect that allows agricultural machinery to accurately fill the daily needs of different crop types, without or with minimal human intervention [5]. Livestock management is following the IoT trends as well, with farmers in Australia being obliged to affix passive RFID ear tags to their cattle and to report movements between farms to an online national database [6]. RFID devices are very common and are used to track the geographical position of individual animals or items such as packages, pallets, shipping containers, or trucks, which are stationary or in movement during distribution.

3.3   Examples from DataBio

Within the Databio pilots, several key parameters have been monitored through various sensors. Sensor data have been collected along the way and made available in order to support the project activities. Especially, IoT sensor data is exchanged most commonly via wireless telecommunications technologies (i.e. ZigBee, Cellular), using various protocols (i.e. MQTT, Websocket) and data formats (i.e. JSON, binary). In DataBio the following sensor data categories are in use:

- IoT agro-climate/field sensors measuring crop status (ambient temperature, humidity, solar radiation, leaf wetness, rainfall volume, wind speed and direction, barometric pressure, soil temperature and humidity).
- IoT control data in the parcels/fields measuring sprinklers, drippers, metering devices, valves, alarm settings, heating, pumping state, pressure switches, etc..
- Machinery data associated with the operation of tractors, UVs and other actuators (fuel consumption, position, temperature, operation, etc.) conducting use case specific tasks in the field.
- Contact sensing data speeding up techniques which help to solve problems.
- Vessel and buoy-based sensor data for numerical measurements, typically of hydro acoustic, sonar and machinery data (Fig. 3.1).

3.3.1   Gaiatrons

Gaiatrons, designed and built by NEUROPUBLIC, are agro-climate IoT sensor stations involved in a number of agriculture pilots providing critical in-situ information for DataBio. Gaiatrons are telemetric autonomous stations which collect data from sensors installed in the field that monitor several atmospheric and soil parameters (air temperature, relative moisture, wind direction and velocity, rain, leaf wetness, soil temperature and moisture, etc.). They have reached an adequate maturity for outdoor operation (industrial grade) and have data exchange and control capabilities. Gaiatrons are considered "power starving" systems. They are energy-autonomous and remain in sleep mode for extended periods of time in order to minimize their energy consumption needs. They can connect to other stations and to

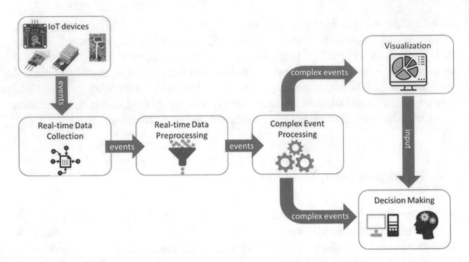

Fig. 3.1 Data flow for real-time IoT data processing and decision-making generic pipeline

cloud systems using different wireless connection technologies (GPRS, UHF). Gaiatrons are specially designed for providing exact fit to the operational requirements asked from modern smart farming infrastructures. Dense installation network under the canopy, large scale deployment, low operation cost and mobility are some of these operational requirements in order for Gaiatrons to be viable and commercially successful (Fig. 3.2).

Fig. 3.2 Gaiatron station for in-situ agro-climatic monitoring

Fig. 3.3 AgroNode unit

3.3.2   AgroNode

AgroNode is a radio based data logger device intended to be used in any scenario where sensor data is collected. AgroNode is used for online measuring of physical phenomena directly in terrain. The device interoperates with a wide spectrum of digital sensors and due to the modularity of architecture, it can be modified for a variety of data transmission technologies—GPRS, LoRa, Sigfox, NB-IoT. It is able to permanently save sensor measurement data and/or make them accessible online. Due to solar power, life span is from device point of view unlimited it has also a battery backup. AgroNode is designed and built by the Lesprojekt company and utilized in many projects and measuring campaigns in agriculture, forestry, water management, meteorology etc. (Fig. 3.3).

3.3.3   SensLog and Data Connectors

SensLog is a cloud-based sensor data management component that is receiving, storing, processing and publishing sensor data [7]. SensLog is storing data in its own relational data model based on the ISO Observations and Measurements standard [8] and extended to functionalities for sensor network metadata and system of alerts. A NoSQL version of the SensLog data model was tested during the DataBio project [9]. The SensLog interface provides receiving and publishing of sensor data in various formats. The main interface is a proprietary REST API with JSON data encoding

[10]. The core services of OGC Sensor Observation Service 1.0.0 [11] are providing a standardized interface for data publication.

SensLog is defined as a fully cloud-based environment. All components were developed as microservices, which in turn means independent components. The main objective is to separate all systems from each other. The software is designed to be deployed as a Docker container and managed by Kubernetes orchestrator which allows scaling each component easily. The individual microservices were written by modern constructs of the Java language using the Spring framework. SensLog environment contains 3 types of microservices (Fig. 3.4).

The first service group is connectors and feeders representing the data layer. The Data connector was created as a self-configured modular application. The main task is to integrate different data sources into the SensLog system. These data sources can be another system API, static files, databases, etc. Each module represents a data source that fetches raw data that is pushed to public APIs of SensLog-processing via HTTP. The Feeder component gets data directly from individual sensors. Each wireless telecommunications technology has its implementation.

Next micro service is the Processing component, which collects data from the first group. Data is authenticated, validated and saved to the data store. The Processing group provides a proprietary API with JSON data encoding optimized for pushing data to the Data storage.

The last service group consists of Provider and Analytics components. The Provider publishes access to stored data via public API which can be used for end-user applications such as client and visualization applications, smart device apps, etc. The Analytics component is prepared for real-time and batch processing of data stored in the Data warehouse.

The architecture of the SensLog environment can be seen in the following Fig. 3.4.

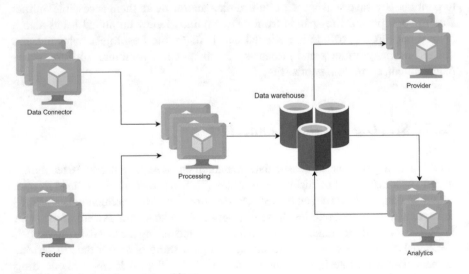

Fig. 3.4 SensLog environment architecture

3.3.4   Mobile/Machinery Sensors

The need of telemetry in mobile agricultural machinery including tractors and combined harvesters can be seen from different points of view. The authors in this chapter represent the producer view, particularly the Zetor tractor producer.

- <u>Design and reliability aspects during the development</u>. Tractors and other agriculture machines are difficult mechanical products which need to fulfill many mandatory safety, ecological, reliability and technical standards. The design of a new product, e.g. a tractor, takes many years. In order to speed up the design process, innovative technologies are used to make the process process cheaper and more efficient. An example of such technology is telemetry that can be used for:

  - Remote and real time observation of reliability tests.
  - Remote and real time observation of tractor CAN Bus communication, tractor control units analysis.

- <u>Commercial product for other markets</u>. Telemetry developed as a support for design and development phases can be very easily adopted for other commercial exploitation. The main two areas include:

  - Telemetric products for farmers. This type of telemetry is supporting work on the farm and farming functionality. It has a huge potential when implemented on a farm with a bigger number of tractors and is used for organization of work on the farm.
  - Telemetric products for tractor owners (e.g., banks). This is a relatively new functionality of telemetry. The reason for this functionality is that the owner wants to know what happened to the property in real time, in real position and whether it is well managed.

Machinery sensors and telemetry units can be useful not only for observing machinery in itself, but it can be used as a way to observe the status and conditions of fields. An example is yield prediction based on observations from yield sensors on combine harvesters [12].

References

1. Hype Cycle for Analytics and Business Intelligence. (2019). Hare J and Schlegel K Gartner report # G00369713 published on July 18, 2019.
2. MacGillivray, C. et al. (2019). Worldwide internet of things forecast update 2018–2022. Doc # US44755019 published in January 2019.
3. Ferreira, D. et al. (2017). Towards smart agriculture using FIWARE enablers. In *International Conference on Engineering, Technology and Innovation (ICE/ITMC)* (pp. 1544–1551). https://doi.org/10.1109/ICE.2017.8280066

4. Wolfert, S., Ge, L., Verdouw, C., & Bogaardt, M.-J. (2017). Big data in smart farming—A review. *Agricultural System, 153*, 69–80.
5. Kamilaris, A. et al. (2016). Agri-IoT: A semantic framework for internet of things-enabled smart farming applications. In: *Proceeding of the IEEE World Forum on Internet of Things (WF-IoT)*. https://doi.org/10.1109/WF-IoT.2016.7845467
6. Taylor, K., et al. (2013). Farming the web of things. *Intelligent Systems, 28*(6), 12–19.
7. Kepka, M. et al. (2017). The SensLog platform—A solution for sensors and citizen observatories. In J. Hřebíček, R. Denzer, G. Schimak, & T. Pitner (Eds.), *Environmental software systems. Computer science for environmental protection*. ISESS 2017. IFIP Advances in Information and Communication Technology, Vol. 507. Springer, Cham. https://doi.org/10.1007/978-3-319-89935-0_31
8. International Organization for Standardization. (2011). *ISO 19156:2011 Geographic information—Observations and measurements* (p. 46).
9. Dadhich, P., Sadovykh, A., Bagnato, A., Kepka, M., Kaas, O., & Charvát, K. (2018). Sensor-based database with SensLog: A case study of SQL to NoSQL migration. In: *Proceedings of the 7th International Conference on Data Science, Technology and Applications (DATA 2018)*. SCITEPRESS—Science and Technology Publications, Lda, Setubal, PRT (pp. 239–244). https://doi.org/10.5220/0006909202390244
10. SensLog API. https://www.senslog.org/api/v1/
11. 06-009r6 OpenGIS sensor observation service. Open Geospatial Consortium (2006). https://www.opengeospatial.org/standards/sos
12. Řezník, T., Jedlička, K., Lukas, V., Čerba, O., Kepka, M., Leitgeb, S., & Charvát, K. (2017). Yield potential cloud computing and applications based on sentinel, landsat and MODIS data. Remote Sens Basel (MDPI AG, Switzerland). ISSN 2072-4292.

Chapter 4
Remote Sensing

Miguel Ángel Esbrí

Abstract In this chapter we present the concepts of remote sensing and Earth Observation and, explain why several of their characteristics (volume, variety and velocity) make us consider Earth Observation as Big Data. Thereafter, we discuss the most commonly open data formats used to store and share the data. The main sources of Earth Observation data are also described, with particular focus on the constellation of Sentinel satellites, Copernicus Hub and its six thematic services, as well as other private initiatives like the five Copernicus-related Data and Information Access Services and Sentinel Hub. Next, we present an overview of representative software technologies for efficiently describing, storing, querying and accessing Earth Observation datasets. The chapter concludes with a summary of the Earth Observation datasets used in each DataBio pilot.

4.1 Introduction

Remote sensing is one of the most common ways to extract relevant information about the Earth and our environment. It can be defined as "the acquisition of information about an object or phenomenon without making physical contact with the object and thus in contrast to on-site observation, especially the Earth, including on the surface and in the atmosphere and oceans, based on propagated signals (e.g. electromagnetic radiation)" [1]. The term "remote sensing" was first utilized in the early 1960s to describe any means of observing the Earth from afar, particularly as applied to aerial photography, the main sensor used at that time. Today, as a result of rapid technological advances, we routinely survey our planet's surface from different platforms: low-altitude unmanned aerial vehicles (UAVs), airplanes and satellites. The surveillance of Earth's terrestrial landscapes, oceans and ice sheets constitutes the main goal of remote sensing techniques [2]. Remote sensing acquisitions, done through both active (synthetic aperture radar, LiDAR) and passive (optical and thermal range, multispectral and hyperspectral) sensors, provide a variety of information about the

M. Á. Esbrí (✉)
Atos Spain, Albarracin 25, 28004 Madrid, Spain
e-mail: miguel.esbri@atos.net

© The Author(s) 2021
C. Södergård et al. (eds.), *Big Data in Bioeconomy*,
https://doi.org/10.1007/978-3-030-71069-9_4

land and ocean processes. In a broader context, remote sensing activities include a wide range of aspects, from the physical basis to obtain information from a distance, to the operation of platforms carrying out the sensor system, and further to the data acquisition, storage and interpretation. Then, the remotely collected data are converted to relevant information, which is provided to a vast variety of potential end users: farmers, foresters, fishers, hydrologists, geologists, ecologists, geographers, etc.

The use of Earth observation data imposes a series of technological challenges to:

- Combine satellite data with in situ or enterprise data.
- Understand, select, download, conserve and process data.
- Harness a range of scientific and technical skills and manpower.
- Load and store petabytes of data.
- Deploy high-performance processing capabilities.

4.2   Earth Observation Relation to Big Data

Different types of Earth observation data have been produced over the last forty years, bringing significant changes in the context of the big data concept. Moreover, the precise and up-to-date worldwide Earth observation data are changing the way that Earth is interpreted. It is leading to the implementation of applications powered with humongous amounts of remote sensing information. In that regard, several of the remote sensing data characteristics allow us to consider remote sensing data as big data:

- *Volume*

Among the various areas where big data sets have become common, the ones related to remote sensing and information and communication technology are foremost, since the datasets involved have reached huge dimensions. This makes exceptionally complex their visualization, analysis and interpretation [2]. Besides, just in 2010, the satellite observation networks around the world had more than 200 on-orbit satellite sensors [3], capturing several gigabytes of information per second [3]. Nowadays, with the advent of the Copernicus programme with its Sentinel and contributing missions' satellites and with the entering into the commercial market of the US satellite operator Planet, the observation capacities dramatically increased, adding several petabytes of annual observations. According to Open Geospatial Consortium (OGC), the worldwide observation information currently most likely surpasses one exabyte.

- *Variety*

Variety refers to the number of types of data, and concerning remote sensing data, it is specifically linked to structured information such as images obtained by satellite

sensors. More specifically, in this context, variety depends on the different resolution (spectral, temporal, spatial and radiometric) of the captured data. Remote
sensing data variety is enormous. There are approximately 200 satellite sensors with
a huge variety of spatial, temporal, radiometric and spectral resolutions [3]. Thus,
for instance, satellites have a wide range of orbital altitudes, optics, and acquisition techniques. Consequently, the imagery acquired can be at very fine resolutions
(fine level of detail) of 1 m or less with very narrow coverage swaths, or the images
may have much larger swaths and cover entire continents at very coarse resolutions
(>1 km). In addition, the satellites are equipped with sensors capable of acquiring
data from portions of the electromagnetic spectrum that cannot be sensed by the
human eye or conventional photography. The ultraviolet, near-infrared, shortwave
infrared, thermal infrared and microwave portions of the spectrum provide valuable
information of critical environmental variables [1].

- *Velocity*

Velocity refers to the frequency of incoming data and the speed at which is generated,
processed and transmitted. In the case of remote sensing data, the orbital characteristics of most satellite sensors enable repetitive coverage of the same area of Earth's
surface on a regular basis with a uniform method of observation. The repeat cycle
of the various satellite sensor systems varies from 15 min to nearly a month. This
characteristic makes remote sensing ideal for multi-temporal studies, from seasonal
observations over an annual growing season to inter-annual observations depicting
land surface changes [2].

4.3 Data Formats, Storage and Access

4.3.1 Formats and Standards

Nowadays, remote sensing images (both, currently acquired and historical images)
are typically distributed in digital format. A digital image is a numeric translation
of the original radiances received by the sensor, forming a 2D matrix of numbers.
Those values represent the optical properties of the area sampled, where the pixel
represents the minimum spatial unit of measurement within the sensor coverage [2].
 The following are the file formats most generally accepted as standards for
encoding and transferring the remote sensing images:

- *HDF*[1] is a self-describing and portable, platform-independent data format for
 sharing science data, as it can store many different kinds of data objects, including
 multi-dimensional arrays, metadata, raster images, colour palettes and tables in a

---

[1] https://www.hdfgroup.org/.

single file. There is no limit on the number or size of data objects in the collection, giving great flexibility for big data.

- *NetCDF*[2] is also a self-describing, portable and scalable format that is currently widely used by climate modellers.
- *JPEG 2000*[3] is an image coding system that uses state-of-the-art compression techniques based on wavelet technology and offers an extremely high level of scalability and accessibility. Content can be coded once at any quality, up to lossless, but accessed and decoded at a potentially very large number of other qualities and resolutions and/or by region of interest, with no significant penalty in coding efficiency. Typically used for distributing Sentinel-2 images.
- *GeoTIFF*[4] is a public domain metadata standard which allows georeferencing information to be embedded within a TIFF file. The potential additional information includes map projection, coordinate systems, ellipsoids, datums and everything else necessary to establish the exact spatial reference for the file. More interestingly, *"Cloud Optimized GeoTIFF" (COG)*—a standard based on GeoTIFF—is designed to make it straightforward to use GeoTIFFs hosted on HTTP web servers, so that users/software can make use of partial data within the file without having to download the entire file. It is designed to work with HTTP range requests and specifies a particular layout of data and metadata within the GeoTIFF file, so that clients can predict which range of bytes they need to download.

These specially designed data formats work quite well when the amount of data is not very large. However, issues start to arise when data volumes increase. The most obvious problem is that it is not easy to find, retrieve and query the information needed.

A lot of effort has been spent during the last years for standardising many of the EO ground segment interfaces in the context of HMA (OGC)[5] [4] and CEOS[6] [5]. The interfaces for which widely accepted standards exist and are deployed include:

- EO dataset/product metadata [6].
- EO dataset/product discovery [7–9].
- Online data access [10–12].
- Viewing.
- Processing.

Further details concerning standards for EO metadata and discovery interfaces can be found in Chap. 2 "Standardized EO data platforms".

---

[2] https://www.unidata.ucar.edu/software/netcdf.

[3] https://jpeg.org/jpeg2000/.

[4] https://en.wikipedia.org/wiki/GeoTIFF.

[5] Heterogeneous Missions Accessibility (HMA), https://wiki.services.eoportal.org/tiki-index.php?page=HMA+AWG.

[6] https://ceos.org/.

Table 4.1 Sentinel missions[7]

SENTINEL-1	With the objectives of land and ocean monitoring, SENTINEL-1 is composed of two polar-orbiting satellites operating day and night and will perform radar imaging, enabling them to acquire imagery regardless of the weather
SENTINEL-2	Its main objective is land monitoring, and the mission is composed of two polar-orbiting satellites providing high-resolution optical imagery. Vegetation, soil and coastal areas are among the monitoring objectives
SENTINEL-3	Its primary objective is marine observation, with focus on studying sea surface topography, sea and land surface temperature, ocean and land colour. Composed of three satellites, the mission's primary instrument is a radar altimeter, but the polar-orbiting satellites will carry multiple instruments, including optical imagers
SENTINEL-4	It is dedicated to air quality monitoring. Its UVN instrument is a spectrometer carried aboard Meteosat Third Generation satellites, operated by EUMETSAT. The mission aims to provide continuous monitoring of the composition of the Earth's atmosphere at high temporal and spatial resolution, and the data will be used to support monitoring and forecasting over Europe
SENTINEL-5	It is dedicated to air quality monitoring. The SENTINEL-5 UVNS instrument is a spectrometer carried aboard the MetOp Second Generation satellites. The mission aims to provide continuous monitoring of the composition of the Earth's atmosphere. It provides wide swath, global coverage data to monitor air quality around the world
SENTINEL-5P	A precursor satellite mission SENTINEL-5P aims to fill in the data gap and provide data continuity between the retirement of the Envisat satellite and NASA's Aura mission and the launch of SENTINEL-5. The main objective of the Sentinel-5P mission is to perform atmospheric measurements, with high spatio-temporal resolution, relating to air quality, climate forcing, ozone and UV radiation

4.3.2 Data Sources

4.3.2.1 Copernicus Programme and Sentinel Missions

The Copernicus EO programme is a cooperation of the European Union (EU) and the European Space Agency (ESA). This agency is responsible for coordinating the satellite acquisition and delivery of the EO data. Since the launch in 2014 of Sentinel-1A, the fleet of Sentinel satellites is delivering data for environmental monitoring and civil security applications.

Copernicus is served by a set of dedicated satellites (the Sentinel families) and contributing missions (existing commercial and public satellites). The Sentinel satellites are specifically designed to meet the needs of the Copernicus services and their users (Table 4.1).

Thematic Services

Besides the Sentinel satellite constellation, Copernicus also provides access to specific services, which fall into **six main thematic categories**[7]: services for land management, services for the marine environment, services relating to the atmosphere, services to aid emergency response, services associated with security and services relating to climate change.

- Land Monitoring: Monitoring the Earth's land is useful for many fields, particularly agriculture, forestry, topography and land-cover and land-change studies. The data can be used to track current trends and predict future changes.
- Marine Monitoring: Information on the state and dynamics of the ocean and coastal zones can be used to help protect and manage the marine environment and resources more effectively, as well as ensure safety at sea and monitor pollution from oil spills and other events.
- Atmospheric Monitoring: Monitoring the quality and condition of our planet's atmosphere is important in that it helps us to understand how we may be affected and is an essential tool in forecasting weather events.
- Managing Emergency: When an emergency occurs, satellite data can prove essential in forming a response. Historical data can provide perspective on a situation, while current data can help to analyse and manage the emergency.
- Security: Surveillance and security can be difficult to manage from the ground. Observations from space can make monitoring borders and sea routes much easier and track developing situations.
- Climate Change: Satellites are a vital tool in monitoring our world's changing climate, providing wide-scale views of affected areas and contributing to growing archives of data for use in long-term studies.

Most of the data and information are delivered by Copernicus, and its services are made available via a "free, full and open" policy to any citizen and any organization everywhere on Earth.

For dissemination of level 0, level 1 and level 2 products, ESA provides access via the Copernicus **Open Access Hub**[9] portal, providing access to Sentinel-1, -2, -3 and -5p data through an interactive graphical user interface. Additionally, there are the Collaborative Data Hub, International Access Hub and Copernicus Services Data Hub which are providing access to public authorities, European projects and Copernicus services.

---

[7] https://sentinels.copernicus.eu/web/sentinel/thematic-areas.

[7] https://sentinels.copernicus.eu/web/sentinel/missions.

[9] https://scihub.copernicus.eu.

Table 4.2 DIAS providers

Name	Provider	Webpage
CREODIAS	Creotech instruments, Cloudferro	https://creodias.eu
Mundi	Atos Integration, DLR, e-GEOS, EOX, GAF, Sinergise, Spacemetric, Thales Alenia Space, T-Systems	https://mundiwebservices.com
ONDA	Serco, OHV	https://www.onda-dias.eu
Sobloo	Airbus, Capgemini, Orange	https://sobloo.eu
WEkEO	Eumetsat, ECMWF, Mercator Océan	https://www.wekeo.eu

4.3.2.2 DIAS

In order to facilitate the access of Earth observation products and the development of EO-powered applications for end users, five different **Data and Information Access Services** (DIAS) are available (see Table 4.2). The DIASes provide access to product repositories in cloud storage. They primarily are not thought to be used as "dissemination" hubs (download bandwidth is even lower than at Open Access Hub, and it is generally not free). The DIAS provides platforms for hosting processing in vicinity to the cloud storage. End users can bring their algorithms and run them with free and fast access to the product data (by combining simple access to curated petabyte-size collections of Copernicus, other satellite and third-party data). Eventually, the end user only needs to download the (typically low volume) processing results and not the (high volume) satellite input products.

4.3.2.3 Other

Other data access portals are available as well:

- **Amazon Web Services** (AWS) and **Google Cloud Platform** (GCP) offering storage and processing platforms services similar like the DIAS but differing in product offers and service pricing
- **Sentinel Hub**[10] is a commercial data access and on-the-fly processing software instantiated on AWS and on two of the DIAS and exposing an application programme interface (API) to user applications for accessing Copernicus and Landsat products and derivatives.

---

[10] https://www.sentinel-hub.com/.

4.4 Selected Technologies

The present section identifies information technology domains and contains further practically relevant insights (mainly from DataBio data access components) into these for builders of applications and systems using EO data and cloud-based environments.

4.4.1 Metadata Catalogue

As per the OGC definition[11]: "Catalogue services support the ability to publish and search collections of descriptive information (metadata) for data, services and related information objects. Metadata in catalogues represent resource characteristics that can be queried and presented for evaluation and further processing by both humans and software. Catalogue services are required to support the discovery and binding to registered information resources within an information community".

In the case of Earth observation datasets, a series of specific EO metadata profiles have been defined in order to facilitate their description and findability. Chapter 2 "Standards and EO data platforms" provides further details about them. The following describes the concrete EO metadata catalogue implementations used in DataBio.

FedEO Gateway

This component [13] acts as a unique endpoint allowing clients to access metadata and data from different backend EO catalogues implementing different protocols. It supports access through OGC 10-032r8 and OGC 13-026r8 OpenSearch interfaces and provides atom responses with metadata in OGC 10-157r4 format (i.e. EO profile observations and measurements). Alternative response formats such as RDF/XML, Turtle, JSON-LD and GeoJSON (OGC 17-003) are available as well. SRU-style bindings and W3C linked data platform bindings are available as well.

FedEO Catalogue

This component [13] implements an EO catalogue server allowing to store EO (satellite) collections (series) and products (datasets) metadata. It offers an API to populate the catalogue and an API to search the catalogue.

Both components have been developed by Spacebel s.a.

---

[11] https://www.ogc.org/standards/cat.

4.4.2 Object Storage and Data Access

GeoRocket

GeoRocket[12] is a high-performance data store for geospatial files developed by Fraunhofer Institute for Computer Graphics Research IGD. It can store 3D city models (e.g. CityGML), GML files or GeoJSON data sets. It provides the following features:

- High-performance data storage with multiple back ends such as Amazon S3, MongoDB, distributed file systems (e.g. HDFS or Ceph), or your local hard drive (enabled by default)
- Support for high-speed search features based on the popular open-source framework elasticsearch. You can perform spatial queries and search for attributes, layers and tags.
- Its design and implementation (based on the open-source toolkit Vert.x), makes it perfectly suitable for being deployed in Cloud environments, making it reactive and capable of handling big files and larger numbers of parallel requests.

Rasdaman

Rasdaman[13] is an array database system, which provides flexible, fast, scalable geoservices for multi-dimensional spatio-temporal sensor, image, simulation and statistics data of unlimited volume. Data are stored in a PostgreSQL database, thereby achieving full information integration (e.g. latitudes, longitudes, time coordinates, resolutions and other ancillary annotations.). Ad-hoc access, extraction, aggregation, as well as remix and analytics are enabled through a new SQL raster query language—the Rasdaman query language (RasQL)—with highly effective serverside optimization. The core features include—truly multi-dimensional—1D, 2D, 3D, 4D, and beyond—powerful, flexible query language for visualization, classification, convolution, aggregation and many more geospatial functions spatial indexing and adaptive tiling for fast data access—parallelization and for unlimited scalability from laptop to cluster and cloud—full information integration of raster data with all geo data in the PostgreSQL database—support for the raster-relevant OGC standards, reference implementation for WCS core and WCPS.

Data Cubes

EO data cubes are an advanced way how users interact with large spatio-temporal EO data [14]. Figure 4.1 illustrates the principle. The idea is to read incoming image tiles covering an area ("Dice") and arrange these in time series pixel stacks ("Stack"). This makes access to the time series of observations ("Use") much easier.

Data cubes implementations (such as Rasdaman or ADAM[14]) allow accessing a large variety of multi-year global geospatial collections enabling data discovery,

---

[12] https://georocket.io/.

[13] https://rasdaman.com/.

[14] https://adamplatform.eu.

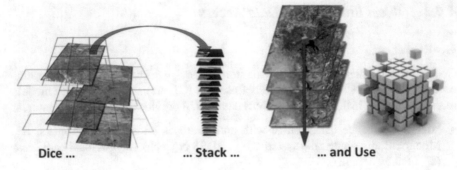

Dice ...   ... Stack ...   ... and Use

Fig. 4.1 Data cube (*Credits* Geoscience Australia)

visualization, combination, processing and download. They permit to exploit data from global to local scale (taken from distributed data sources are made accessible through the data cube layer that exposes OGC-standardized interfaces). On top of the data cube layer, platform-based interfaces (web application, mobile application, Jupyter Notebook and APIs) as well as third-party user interfaces can be deployed.

Another example is **Xcube**,[15] which is an open-source Python package for generating and exploiting data cubes. It comprises one of the core parts of the Euro Data Cube (EDC),[16] together with the Sentinel Hub. The EDC engine is able to technically serve custom raster data in addition to the freely available EO data archives like Sentinel, Modis or Landsat.

4.5   Usage of Earth Observation Data in DataBio's Pilots

A significant part of the 27 DataBio pilots uses Earth observation data as input for their specific purposes, in the context of efficient resource use and increasing productivity in agriculture [15], forestry [16] and fishery [17] (Table 4.3).

---

[15] https://xcube.readthedocs.io/en/latest/.

[16] https://eurodatacube.com.

Table 4.3 Examples of use of EO datasets in DataBio pilots

EO dataset	Pilots	Common usage
Sentinel-2	Agriculture pilots A1.1, B1.2, C1.1	Time series and multiple statistics of EO-based indicators that describe various agri-environmental conditions and are assigned to each agricultural parce (e.g. corrected products, vegetation indices like NDVI and NDWI)
	Agriculture pilots A1.2, A1.3, B1.1, B1.3, C1.2, C2.1 Forestry pilots 2.3.1, 2.3.2- AIS, 2.3.2-FH, 2.4.1	A time series of Sentinel-2 L1C images (both A and B satellites) are used to cover a growing season
	Agriculture pilot B1.4	Scenes covering vegetation period of cereals and meeting cloud cover criteria
Sentinel-3	Fishery pilots A1 and B1	Sentinel-3 SLSTR for sea surface temperature Sentinel-3 SRAL/MWR for altimetry (anomalies) Sentinel-3 OLCI for chlorophyll
Landsat	Agriculture pilot B1.1 Forestry pilot 2.3.2-FH, 2.3.2-AIS	Despite the resolution being lower than the Sentinel mission, it has been TRAGSA-TRAGSATEC reference data for years. It is used as reference, contrast and on those dates or areas with no Sentinel coverage
Landsat 8	Agriculture pilot B1.4	Scenes covering vegetation period of cereals and meeting cloud cover criteria
Proba-V	Agriculture pilots A1.2, A1.3, B1.3, C1.2	Used for long-term time series, which are not available for Sentinel-2 yet
Meteorological data	Agriculture pilots A1.2, A1.3, B1.3, C1.2	Temperature and rainfall data from national providers and/or ECMWF
CMEMS products	Fishery pilots A1 and B1	The Copernicus Marine Environment Monitoring Services of interest cover many products/variables used in the fishery pilots

References

1. https://en.wikipedia.org/wiki/Remote_sensing. Last accessed September 3, 2019.
2. Chuvieco, E. (2016). *Fundamentals of satellite remote sensing: An environmental approach* (2nd ed.). Boca Raton, FL, USA: CRC Press Inc.
3. NASA. (2010). *On-orbit satellite servicing study.* https://sspd.gsfc.nasa.gov/images/NASA_S atellite%20Servicing_Project_Report_0511.pdf
4. Heterogeneous Missions Accessibility (HMA), design methodology, architecture and use of geospatial standards for the ground segment support of earth observation missions. European Space Agency, ESA TM-21, April 2012, https://esamultimedia.esa.int/multimedia/publicati ons/TM-21/TM-21.pdf, 2012, ISBN 978-92-9221-883-6.
5. CEOS OpenSearch Best Practice. Issue 1.2, 13/06/2017. https://ceos.org/document_manage ment/Working_Groups/WGISS/Interest_Groups/OpenSearch/CEOS-OPENSEARCH-BP-V1.2.pdf.
6. OGC 10-157r4, Earth observation metadata profile of observations & measurements, Version 1.1, June 9, 2016. https://docs.opengeospatial.org/is/10-157r4/10-157r4.html.
7. OGC 10-032r8, OGC OpenSearch geo and time extensions. https://www.opengeospatial.org/ standards/opensearchgeo.
8. OGC 13-026r8, OGC OpenSearch Extension for Earth Observation Products. https://docs.ope ngeospatial.org/is/13-026r8/13-026r8.html.
9. OGC 17-047, OGC EO OpenSearch Response GeoJSON(-LD) Encoding Standard. https:// docs.opengeospatial.org/is/17-047r1/17-047r1.html.
10. OGC 13-043, Download Service for Earth Observation Products, Version 1.0, January 31, 2014.
11. OGC 14-055r2, OGC OWS Context GeoJSON Encoding, May 30, 2016. https://docs.openge ospatial.org/is/14-055r2/14-055r2.html.
12. OGC 12-084r2, OGC OWS Context Atom Encoding Standard. https://portal.opengeospatial. org/files/?artifact_id=55183
13. Coene, Y., Gilles, M. et al. (2020). "D5.1 EO component specification", Deliverable D5.1 of the European Project H2020-732064 Data-Driven Bioeconomy (DataBio), June 13, 2020 [Online].
14. Triebnig, G. (2020). D3.9 guidelines on EO-ICT support improvement to stakeholders v1. Deliverable D3.9 of the European Project H2020-821940 Bringing together the Knowledge for Better Agriculture Monitoring (EO4AGRI) July 2, 2020. [Online].
15. Estrada, J., Rogotis, S., Miliaraki, N., Mastrogiannis, K. et al. (2020). "D1.3 agriculture pilot final report", Deliverable D1.3 of the European Project H2020-732064 Data-Driven Bioeconomy (DataBio), July 15, 2020 [Online].
16. Miettinen, J., Tergujeff, R., Seitsonen, L. et al. (2020). "D2.3 Forestry Pilot Final Report" Deliverable D2.3 of the European Project H2020-732064 Data-Driven Bioeconomy (DataBio), July 15, 2020 [Online].
17. Fernandes, J. A., Quincoces, I., Reite, K.-J. et al. (2020). "D3.3 Fishery Pilot Final Report", Deliverable D3.3 of the European Project H2020-732064 Data-Driven Bioeconomy (DataBio), July 15, 2020[Online].

Chapter 5
Crowdsourced Data

Karel Charvát and Michal Kepka

Abstract Crowdsourcing together with Volunteered Geographic Information (VGI) are currently part of a broader concept – Citizens Science. The methods provide information on existing geospatial data or is a part of data collection from geolocated devices. They enable opening parts of scientific work to the general public. DataBio Crowdsourcing Solution is a combination of the SensLog server platform and HSLayers web and mobile applications. SensLog is a server system for managing sensor data, volunteered geographic information and other geospatial data. Web and mobile applications are used to collect and visualize SensLog data. SensLog data model builds on the Observations & Measurements conceptual model from ISO 19156 and includes additional sections, e.g., for user authentication or volunteered geographic information (VGI) collection. It uses PostgreSQL database with PostGIS for data storage and several API endpoints.

5.1 Introduction

Crowdsourcing is a sourcing model in which individuals or organizations obtain goods and services, including ideas, voting, micro-tasks and finances, from a large, relatively open and often rapidly evolving group of participants. Crowdsourcing can be used as a research method (Citizens Science [1]), as the involvement of the public in scientific research [2].

In the area of collection of spatial information or Earth Observation, we are often using the term Citizens' Observatories [3]. This term is usually understood as methods of community-based monitoring using novel Earth Observation applications and sensors embedded in portable or mobile personal devices [4–6].

K. Charvát (✉)
Lesprojekt-Služby, Ltd, Záryby, Czech Republic
e-mail: Charvat@lesproject.cz

M. Kepka
University of West Bohemia, Pilsen, Czech Republic

© The Author(s) 2021
C. Södergård et al. (eds.), *Big Data in Bioeconomy*,
https://doi.org/10.1007/978-3-030-71069-9_5

Another term, which is often used in this context, is Volunteered Geographic Information (VGI) [7], which is the harnessing of tools to create, assemble, and disseminate geographic data provided voluntarily by individuals [8]. Some examples of this phenomenon are WikiMapia, OpenStreetMap, and Google Map Maker. VGI can also be seen as an extension of critical and participatory approaches to geographic information systems and as a specific topic within online or web reliability. These sites provide general base map information and allow users to create their own content by marking locations where various events occurred or certain features exist. In voluntary data collection, an important part is how data is processed. An example is Neogeography (New Age Geography) focused on combining geotagged data (e.g. Keyhole Markup Language—KML) [9] with a map interface for contextualised exploration.

In this chapter we will describe two concepts developed in the DataBio project. Firstly, a solution based on SensLog [10] and a profile for VGI. The second concept is a Map composition and sharing Maps as objects among users.

5.2 SensLog VGI Profile

SensLog is a web-based solution for receiving, storing and publishing sensor data of different kinds. As VGI can be collected as sensor data, SensLog can provide a suitable operational solution. The SensLog data model was extended with new tables with emphasis on the variability of VGI. Only a few mandatory attributes characterize an VGI observation, but it can be enriched with a lot of additional attributes. The data type of an additional attribute is only limited to those that can be stored as a value in JSON format. A VGI observation can include a list of multimedia files that are also stored in a data model. The data model of a VGI module is shown on Fig. 5.1.

Added tables are following:

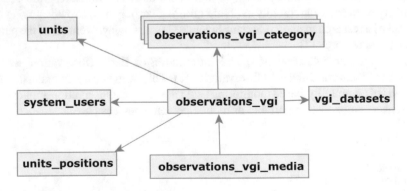

Fig. 5.1 SensLog VGI module data model

- *observations_vgi*—a main table storing VGI observation with all connected attributes
- *observations_vgi_media*—a table storing multimedia files connected to VGI observations
- *observations_vgi_category*—a table storing coded list values of categories of VGI, uses partitioning mechanism to sort categories
- *vgi_datasets*—a table storing user-defined datasets of VGI observations with metadata

This data model supports measurements and observations by users with portable devices. It is a typical way of collecting spatially referenced thematic data in the domain of the Earth Observation (Fig. 5.2).

To publish data according to Open Linked Data best practices and with a self describing data structure, we are using Virtuoso graph model engine [11]. The data are stored in so called *quads*, which consist of a graph name describing a dataset and triple mapping an attribute value (object) to Point of Interest (POI) (subject). The quad uses a property which, in the best case, is well defined in some public ontology making it easier to integrate our data into other systems [12].

Fig. 5.2 Citizens
observatories mobile Apps

5.3 Maps as Citizens Science Objects

Maps are interesting not only as visualizations of agriculture data—but also as shareable, fascinating and valuable agriculture objects in themselves. In the past, a map used to be an expensive rolled up scroll of calf skin that was drawn by a skilled artist from the manuscripts of daring sea-farers in the great age of discovery. Later, maps were produced by less picturesque but more efficient means until the advent of the Geographical Information Systems (GIS) age, when a lot of people suddenly could make professionally LOOKING maps. Nowadays, a map is not a "flat image", but a complex layered object that references data sources 'scattered' across a decentralized, democratic and, at times, volatile Internet.

Our needs are many and very different, but so are our skill sets. Thus, offering everyone sophisticated GIS tools capable of letting the users make their own maps is not the best way. It is often simpler, better and more effective to simply give them a "map".

Currently, hundreds of services offer spatial information through real-time interactive protocols such as Web Map Service (WMS) and Web Feature Service (WFS) etc. Soon, if EU member states and signatories of INSPIRE do as they are legally obliged, this number will be thousands, even ten thousands.

As a map is a composite object referring to a lot of live data sources around the internet, it requires a "Map Composition" standard that describes the map elements and how they should be combined to fit together neatly.

An early effort by the OGC was the Web Map Context specification that has not evolved since 2005. This slightly "heavy-weight" XML-based standard is limited in scope and has not kept up with the developments in standards and technology in the years that have passed since its creation. In DataBio we worked with defining a simple, lightweight specification for Map Compositions using HTML5 and bandwidth friendly JavaScript Object Notation (JSON) as a carrier of information.

The current specification of the JSON Map Composition is available on the GitHub Wiki of HSLayers NG [13].

References

1. Irwin, A. (1995). Citizen science: A study of people, expertise and sustainable development. Routledge. ISBN 9780415130103.
2. Gura, T. (2013). Citizen science: Amateur experts. *Nature, 496*(7444), 259–261. https://doi.org/10.1038/nj7444-259a.
3. Have you heard about the concept of Citizens' Observatories?, Published on 21/12/2016. Available at: https://ec.europa.eu/easme/en/news/have-you-heard-about-concept-citizens-observatories.
4. Newman, G., Wiggins, A., Crall, A., Graham, E., Newman, S., & Crowston, K. (2012). The future of citizen science: Emerging technologies and shifting paradigms. *Frontiers in Ecology and the Environment, 10*(6), 298–304.
5. Robinson, J. A., Kocman, D., Horvat, M., & Bartonova, A. (2018). End-user feedback on a low-cost portable air quality sensor system—are we there yet? *Sensors, 18*(11), 3768.

6. Charvat, K., Bye, B. L., Mildorf, T., Berre, A. J., & Jedlicka, K. (2018). Open data, VGI and citizen observatories INSPIRE hackathon. *International Journal of Spatial Data Infrastructures Research, 13,* 109–130.
7. Goodchild, M. F. (2007). Citizens as sensors: The world of volunteered geography. *GeoJournal, 69*(4), 211–221.
8. Elwood, S. (2008). Volunteered geographic information: Future research directions motivated by critical, participatory, and feminist GIS. *GeoJournal, 72*(3&4), 173–183.
9. Graham, M. (2010). Neogeography and the palimpsests of place. *Tijdschrift voor Economische en Sociale Geografie, 101*(4), 422–436.
10. SensLog. http://www.senslog.org/.
11. Virtuoso Universal Server. https://virtuoso.openlinksw.com/.
12. Kepka, M. et al. (2017). The senslog platform—a solution for sensors and citizen observatories. In J. Hřebíček, R. Denzer, G. Schimak, T. Pitner (Eds.) *Environmental software systems. Computer science for environmental protection.* ISESS 2017. IFIP Advances in Information and Communication Technology, Vol. 507. Springer, https://doi.org/10.1007/978-3-319-89935-0_31.
13. Patras DataBio Hackathon TEAM 4: Sharing maps as intelligent objects. https://www.plan4all.eu/2019/06/team-4-sharing-maps-as-intelligent-objects/.

Chapter 6
Genomics Data

Ephrem Habyarimana and Sofia Michailidou

Abstract *In silico* prediction of plant performance is gaining increasing breeders' attention. Several statistical, mathematical and machine learning methodologies for analysis of phenotypic, omics and environmental data typically use individual or a few data layers. Genomic selection is one of the applications, where heterogeneous data, such as those from omics technologies, are handled, accommodating several genetic models of inheritance. There are many new high throughput Next Generation Sequencing (NGS) platforms on the market producing whole-genome data at a low cost. Hence, large-scale genomic data can be produced and analyzed enabling intercrosses and fast-paced recurrent selection. The offspring properties can be predicted instead of manually evaluated in the field . Breeders have a short time window to make decisions by the time they receive data, which is one of the major challenges in commercial breeding. To implement genomic selection routinely as part of breeding programs, data management systems and analytics capacity have therefore to be in order. The traditional relational database management systems (RDBMS), which are designed to store, manage and analyze large-scale data, offer appealing characteristics, particularly when they are upgraded with capabilities for working with binary large objects. In addition, NoSQL systems were considered effective tools for managing high-dimensional genomic data. MongoDB system, a document-based NoSQL database, was effectively used to develop web-based tools for visualizing and exploring genotypic information. The Hierarchical Data Format (HDF5), a member of the high-performance distributed file systems family, demonstrated superior performance with high-dimensional and highly structured data such as genomic sequencing data.

E. Habyarimana (✉)
CREA Research Center for Cereal and Industrial Crops, Via di Corticella 133, 40128 Bologna, Italy
e-mail: ephrem.habyarimana@crea.gov.it

S. Michailidou
Center for Research and Technology Hellas - Institute of Applied Biosciences, 6th Km Charilaou Thermis Road, 57001 Thessaloniki, Greece

C. Södergård et al. (eds.), *Big Data in Bioeconomy*,
https://doi.org/10.1007/978-3-030-71069-9_6

6.1 Introduction

The array of techniques for probing complex biological systems such as (crop) plants is continuously expanding, providing unprecedented data on multiple phenotypic layers as well as multiple omics layers (genome, proteome, metabolome, epigenome or methylome, and more). Furthermore, new and cheap local sensor techniques as well as advances in remote sensing and geo-information systems provide extensive descriptions of the environmental conditions under which plants grow. This allows *in silico* prediction of plant performance (e.g. traits like yield, abiotic and biotic resistance) depending on genotype, environment and crop management. Several statistical, mathematical and machine learning methodologies for analysis of phenotypic, omics and environmental data typically use individual or a few of these data layers. Genomic selection is one of the applications, where heterogeneous data, such as those from genomics, metabolomics and phenomics technologies, are handled also accounting for several genetic models of inheritance [1].

Genomic selection is a new paradigm in plant breeding allowing to bypass the costly and time-consuming phenotyping step by selecting superior lines based on DNA information according to the workflow in Fig. 6.1 [2, 3].

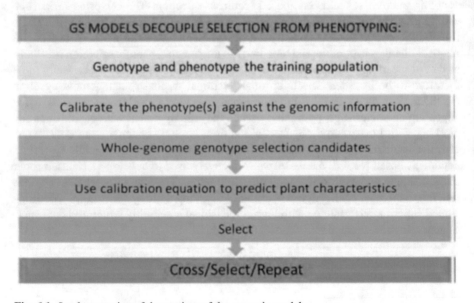

Fig. 6.1 Implementation of the routines of the genomic models

6.2 Genomic and Other Omics Data in DataBio

Genomics and other omics data were produced in sorghum (*Sorghum bicolor (L.)* Moench) and tomato (*Solanum lycopersicum* L.) crops (Fig. 6.2) evaluated in DataBio Genomics pilots; four categories of data were produced including (Tables 6.1 and 6.2): (1) in situ sensors and farm data, (2) genomic data from plant breeding efforts in greenhouses and in open field produced using Next Generation Sequencers (NGS), (3) biochemical data produced by chromatographs (LC/MS/MS, GS/MS, HPLC), wet chemistry and NIRS (near infrared spectroscopy) (Tables 6.1 and 6.2), and (4) genomics modelling output represented by integrative analytics information. In situ sensors/environmental outdoor generated wind speed and direction, evaporation, rain, light intensity, UVA and UVB data. In situ sensors/environmental indoor generated air temperature, air relative humidity, crop leaf temperature (remotely and in contact), soil/substrate water content, crop type, and several other data. Farm Data generated in situ measurements comprising soil nutritional status, farm logs (work calendar, technical practices at farm level, irrigation information), and farm profile (Static farm information, such as size).

Fig. 6.2 Tomato accessions in glasshouses (top) and sorghum pilot fields (bottom) used genomic models platform

Table 6.1 Genomic, biochemical and metabolomic data tools, description and acquisition

Data	Mission, Instrument	Data description and acquisition
Genomic data	• To characterize the genetic diversity of sorghum and tomato varieties and lines used for breeding (Fig. 2) • To identify novel variants in the sorghum and tomato genomes, associated plant characteristics of interest • To use the genomic information to guide breeding strategies (as a selection tool for higher performance) and develop a model to predict the final breeding result in order to rapidly achieve with the minimum financial burden varieties of higher performance • Data were produced using the MiSeq and NextSeq 500 sequencing platforms (Illumina Inc., San. Diego, CA, USA)	• Data were produced from plant biological samples (leaf and fruit) • Collection was conducted in two different plant stages (plantlets and mature plants) • Genomic data were produced using standard and customized protocols at CREA and CERTH facilities • Data produced from Illumina platforms were stored in compressed text files (fastq) • Genomic data, although in plain text format, are big volume data and pose challenges in their storage, handling and processing • Analysis was performed using CREA and CERTH's HPC computational facilities
Biochemistry, agronomy, metabolomics	To characterize the biochemical profile of fruits from tomato varieties used for breeding. Data were produced from different chromatographs, mass spectrometers, wet lab, NIRS	Data was mainly proprietary binary sets converted to XML or other open formats. Data were acquired from biological samples of tomato fruits
IoT, sensor, and environmental data	To characterize growing environments and crop management	Environmental indoor/outdoor, farm data/log/profile

Table 6.2 Phenomics, metabolomics, genomics and environmental datasets

Field	Value
Name of the dataset/API provider	Phenomics, metabolomics, genomics and environmental datasets
Short description	This dataset includes phenomics (sensor data), metabolomics, genomics, environmental (IoT) data, as well as genomic predictions and selection data
Data type	Raw text, CSV data
Dataset/API owner/responsible contacts	ephrem.habyarimana@crea.gov.it, argiriou@certh.gr
Data Volume	30 TB (5 TB/year/institution)
Geographical coverage	Regions of Emilia Romagna (Italy) and Thessalia (Greece)

Genomics data used in the DataBio project resulted from genomic DNA (Deoxyri-bonucleic acid) of the plant species of interest resequenced using Illumina sequencing platform consisting of high-throughput Next Generation sequencers. The genomic data included SNPs (Single Nucleotide Polymorphisms), InDels (Insertions / Dele-tions), SVs (Structure Variations), and CNVs (Copy Number Variation). A Single Nucleotide Polymorphisms is a variation caused by changing of a single nucleotide (A, T, C or G) in the genome. The SNPs, including switch and reverse of single nucleotide bases, are responsible for genome diversity between species and between individuals of the sample species. InDel refers to insertion mutation, deletion muta-tion or both, including what happened in the early stage of evolution. CNVs, a form of structural variations, are alterations of the DNA of a genome that results in the cell having an abnormal number of copies of one or more sections of the DNA. CNVs correspond to relatively large regions of the genome that have been deleted (fewer than the normal number) or duplicated (more than the normal number) on certain chromosome. Structural Variation includes deletion, insertion, duplication, inversion and transposition of long fragment (at least 50 bp) in genome.

In the process of whole-genome resequencing, genomic DNA (gDNA) libraries are prepared (Fig. 6.3) and sequenced; Images generated by sequencers are converted by base calling into nucleotide sequences, which are called raw data or raw reads and are stored in FASTQ format.

FASTQ files are text files that store both read sequences and their corresponding quality scores. Each read is described in four lines as follows [4, 5]:

@FCB068CABXX:6:1101:1403:2159#TAGGTTAT/1

GTAGAAGACTTATAGATTAAAATTCTCCAACATATAGATGTCCTTACA

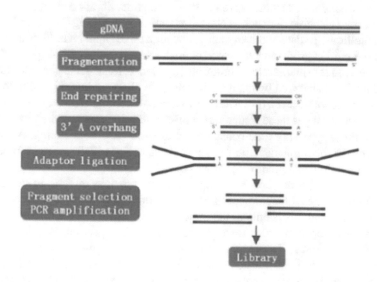

Fig. 6.3 Genomic DNA library construction workflow

CCGTTTTCCTTTGCTCAGCAGGCTCCGTGTTTGCTTGTCCTT

+

c'bcc_c^ccde_df\c_aeff'ffcfffdfedadca^'b_eed'fe\fed\babdba^
Yeebeccfdeae_eec^dbXbda']bcbebc

where line 1 is the DNA sequence identifier and description, lines 1 and 3 are sequence names generated by the sequencer; line 2 is the DNA sequence letters; line 4 is sequencing quality scores, in which every letter corresponds to a base in line 2; the base's sequencing quality is the ASCII value that the letter in line 4 refers to minus 64 (Specification). For example, the ASCII value of c is 99, so the corresponding sequencing quality value is 35. In this work, the quality value of sequencing bases ranged from 2 to 35; the higher the sequencing quality, the lower the sequencing error rate. For instance, the sequencing qualities of 13 and 30 correspond to error rates of 5% and 0.1%, respectively.

The generated raw reads were processed through bioinformatics analysis to filter the raw data and generate clean (reads) data. The filtered reads are subsequently aligned to the reference sequence, the alignment processed and the variation (SNPs, InDels, SVs, and CNVs) detected according to the standard Workflow (Fig. 6.4), which constitute the genomics data used in genomic prediction and selection models.

6.3   Genomic Data Management Systems

Generation of DNA data requires laboratories equipped with molecular biology infrastructure for basic techniques (e.g. DNA extraction, library construction), along with advanced technologies such as Next Generation Sequencing (NGS) and computational facilities. To date, there are many new high throughput NGS platforms available on the market producing sequence data at a very low cost per sequenced base, affordable even for small-scale laboratories [6]. Hence, large-scale genomic data can be produced and analyzed by many scientists, providing the breeder accurate information at the genomic level, for selection of candidates before crosses, in a short time. Among the advantages these technologies offer is accelerating breeding by genomic selection, thus, bypassing time-consuming cultivation and field testing. Additional advantages are the implementation of genomic selection to inform intercrosses and recurrent selection, and predicting instead of field evaluating the offspring.

In the real world, breeders often have a short window of time to decide and take actions on their breeding schemes by the time they receive phenotypic and genotypic data, and this is among the major challenges for many commercial agriculture applications. In order to implement genomic selection routinely as part of breeding programs, data management systems and analytics capacity have to be in order. In short, infrastructures and software that will enable scientists to design and analyse multi-phenotype and multi-omics experiments for maximal data-to-information conversion, are required. This is the major challenge in order to efficiently exploit the huge volume and complexity of the information produced.

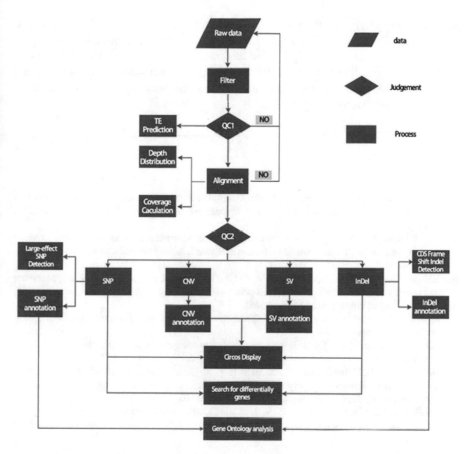

Fig. 6.4 Workflow of standard bioinformatics analysis

The genomic data management system must be able to efficiently store and retrieve huge volumes of genomic information with high complexity and provide rapid data extraction for computation. The system must be scalable and flexible for large breeding programs while being able to run effectively in situations with limited access to large computational clusters. For this purpose, traditional relational database management systems (RDBMS) offer many appealing characteristics. The RDBMS systems are designed and built to store, manage and analyze large-scale data. However, performance can be problematic, when dealing with large matrix data like those commonly encountered in genomic research. To address this performance issue, many RDBMS were upgraded with the capabilities for working with binary large objects (BLOBs). In addition, NoSQL systems have been considered more recently as effective tools for managing high dimensional genomic data [7]. NoSQL systems for distributed file storage and searching represent scalable solutions comparable to RDBMS, when dealing with semi-structured data types. MongoDB system, for instance, is a document-based NoSQL database, which has been used to develop web-based tools for visualizing and exploring genotypic information. The Hierarchical Data Format (HDF5) is a member of the high-performance distributed file systems family. It is designed for flexible, efficient I/O and for high-volume and complex data. It has demonstrated superior performance with high-dimensional and highly structured data such as genomic sequencing data making it an appealing option for a hybrid system approach.

References

1. Habyarimana, E., Lopez-Cruz, M. (2019). Genomic selection for antioxidant production in a panel of sorghum bicolor and S. bicolor × S. halepense Lines. Genes 10:841. https://doi.org/10.3390/genes10110841.
2. Habyarimana, E. (2016). Genomic prediction for yield improvement and safeguarding genetic diversity in CIMMYT spring wheat (Triticum aestivum L.). *Australian Journal of Crop Science, 10,* 127–136.
3. Habyarimana, E., Parisi, B., & Mandolino, G. (2017). Genomic prediction for yields, processing and nutritional quality traits in cultivated potato (Solanum tuberosum L.). *Plant Breeding, 136,* 245–252. https://doi.org/10.1111/pbr.12461.
4. Mount, W. D. (2004). *Bioinformatics: Sequence and genome analysis* (2nd ed.). Cold Spring Harbour Laboratory Press.
5. Gibas, C., Jambeck, P. (2001). *Developing bioinformatics computer skills* (1st ed.). O'Reilly Media, Beijing.
6. Habyarimana, E., Lopez-Cruz, M., & Baloch, F. S. (2020). Genomic selection for optimum index with dry biomass yield, dry mass fraction of fresh material, and plant height in biomass sorghum. *Genes, 11,* 61. https://doi.org/10.3390/genes11010061.
7. Nti-Addae, Y., Matthews, D., Ulat, V. J. et al. (2019). Benchmarking database systems for genomic selection implementation. Database (Oxford) 2019. https://doi.org/10.1093/database/baz096.

Part III
Data Integration and Modelling

Chapter 7
Linked Data and Metadata

**Christian Zinke-Wehlmann, Amit Kirschenbaum, Raul Palma,
Soumya Brahma, Karel Charvát, Karel Charvát Jr., and Tomas Reznik**

Abstract Data is the basis for creating information and knowledge. Having data
in a structured and machine-readable format facilitates the processing and analysis
of the data. Moreover, metadata—data about the data, can help discovering data
based on features as, e.g., by whom they were created, when, or for which purpose.
These associated features make the data more interpretable and assist in turning it
into useful information. This chapter briefly introduces the concepts of metadata
and Linked Data—highly structured and interlinked data, their standards and their
usages, with some elaboration on the role of Linked Data in bioeconomy.

C. Zinke-Wehlmann (✉) · A. Kirschenbaum
Institute for Applied Informatics (InfAI), University of Leipzig, Goerdelerring 9, 04109 Leipzig,
Germany
e-mail: zinke@infai.org

A. Kirschenbaum
e-mail: amit@informatik.uni-leipzig.de

R. Palma · S. Brahma
Poznan Supercomputing and Networking Center (PSNC), ul. Jana Pawła II 10, 61-139 Poznan,
Poland
e-mail: rpalma@man.poznan.pl

S. Brahma
e-mail: sbrahma@man.poznan.pl

K. Charvát · K. Charvát Jr. · T. Reznik
LESPROJEKT-SLUŽBY Ltd., Martinov 197, 27713 Záryby, Czech Republic
e-mail: charvat@lesprojekt.cz

K. Charvát Jr.
e-mail: charvat@lesprojekt.cz

T. Reznik
e-mail: tomas.reznik@sci.muni.cz

© The Author(s) 2021
C. Södergård et al. (eds.), *Big Data in Bioeconomy*,
https://doi.org/10.1007/978-3-030-71069-9_7

7.1   Introduction

Linked Data is a set of best practices for publishing and interlinking structured data on the Web [1]. Linked Data employs Web technologies, such as HTTP, RDF, URIs to create entities from various domains and connect them through typed links, thus building a Web of machine-readable data, rather than human-readable documents. Controlled vocabularies and ontologies are the means of organizations and communities of different disciplines to formalize entities and their relations.

The Semantic Web, called the Web of Data, is a constantly growing dataspace.[1] Besides the simple collection of data, the Semantic Web approach includes the provision of relationships between the data. "This collection of interrelated datasets on the Web can also be referred to as Linked Data".[2] Semantic Web standards, such as RDF [2], OWL [3], and SPARQL [4] have been developed to describe semantic information, including the relationship between data and concepts, on the Web, providing the basis for Linked Data.

Regarding bioeconomy, the main topic of this book, Semantic Web is a useful technology for integrating and publishing heterogeneous data—see also Section 7.6, "Enterprise Linked Data" below. This enables better querying and analyzing processes of bioeconomy.

Linked Data, which started as an initiative[3] of Tim Berners-Lee (the inventor of the World Wide Web), has been increasingly becoming one of the most popular methods for publishing data on the Web. There are different reasons for that: on the one hand, it defines simple principles for publishing and interlinking structured data that is accessible by both humans and machines, enabling interoperability and information exchange [5]. For instance, improving the data accessibility lowers the barriers to finding and reusing this data, while providing machine-readable data facilitates the integration of this data into different applications. On the other hand, Linked Data allows to discover more useful data through the connections with other datasets, and to exploit it in a more useful way through inferencing and semantic queries and rules. The term "Semantic Web" refers to W3C's vision of the Web of linked data. Semantic Web technologies enable people to create data stores on the Web, build vocabularies, and write rules for handling data. Linked data are empowered by technologies such as RDF, SPARQL, OWL, and SKOS. As a result, there is a growing number of datasets becoming available in Linked Data format, as depicted in the Linked Open Data (LOD) cloud[4] diagram (Fig. 7.1). The widespread use and interest in Linked Data has also resulted in the creation of guidelines and best practices on how to generate and publish it, as discussed later in this chapter.

Linked Data can be used and applied to virtually any application domain (as depicted in Fig. 7.1). It consists of both application data as well as data about other

---

[1] https://www.w3.org/2013/data/.

[2] https://www.w3.org/standards/semanticweb/data.html.

[3] https://www.w3.org/DesignIssues/LinkedData.html.

[4] https://lod-cloud.net/.

Fig. 7.1 The linked open data cloud diagram

data or resources (metadata). In fact, Linked Data incorporates human and machine-readable metadata along with it, making it self-describing [6]. Moreover, RDF, the underlying standard for Linked Data interchange and query, was originally developed in the 1990s with the emphasis on the representation of metadata about Web resources; however later the vision of the Semantic Web was extended to the representation of semantic information in general, beyond simple RDF descriptions and Web documents as primary subjects of such descriptions [5], which provided the ground for the creation of the Linked Data initiative later on.

In the following, we discuss more in detail metadata, with focus on agriculture and other bio-sectors, followed by more technical information on Linked Data and related best practices. Next, we present different usage scenarios and experiences of using Linked Data in DataBio.

7.2 Metadata

Metadata is, as its name implies, data about data. It describes the properties of a dataset or resource. Metadata can cover various types of information, which according to [7], can be coarsely categorized into three categories: (i) *descriptive metadata* includes elements such as the title, abstract, author, and keywords, and is mostly used to discover and identify a dataset or another resource; (ii) *structural metadata*, which indicates how compound objects are put together (logical or physical relationships between objects and their parts); and (iii) *administrative metadata* with elements such as the license, intellectual property rights, when and how the dataset was created, who can access it, etc. Datasets in agriculture are either added locally, by a user, harvested from existing data portals, or fetched from operational systems or IoT ecosystems. The definition of a set of metadata elements is necessary to allow identification of the vast amount of information resources managed for which metadata is created, its classification and identification of its geographic location and temporal reference, quality and validity, conformity with implementing rules on the interoperability of spatial data sets and services, constraints related to access and use, and organization responsible for the resource.

Metadata of datasets and dataset series (particularly relevant for agriculture are the EO products derived from satellite imagery) should adhere to the INSPIRE Metadata Regulation[5] with added theme-specific metadata elements for the agriculture, forestry and fishery domains if necessary. This approach will ensure that metadata created for the datasets, dataset series and services will be compliant with the INSPIRE requirements as well as with international standards.[6, 7, 8] In addition, INSPIRE conformant metadata may be expressed also through the *DCAT* Application Profile,[9] which defines a minimum set of metadata elements to ensure cross-domain and cross-border interoperability between metadata schemas used in European data portals. Such a mapping could support the inclusion of INSPIRE metadata[10] in the Pan-European Open Data Portal[11] for wider discovery across sectors beyond the geospatial domain.

A *Distribution* represents a way in which the data is made available. DCAT is a rather small vocabulary, which strategically leaves many details open as it welcomes "application profiles": more specific specifications built on top of DCAT[12], e.g., GeoDCAT-AP[13] as a geospatial extension. For sensors there is also SensorML[14], a standard which can be used to describe a wide range of sensors, including both

---

[5] https://inspire.ec.europa.eu/metadata/6541.

[6] https://www.iso.org/standard/39229.html.

[7] https://www.iso.org/standard/32557.html.

[8] https://docs.opengeospatial.org/is/10-157r4/10-157r4.html.

[9] https://joinup.ec.europa.eu/asset/dcat_application_profile/description.

[10] https://inspire.ec.europa.eu/metadata/6541.

[11] https://www.europeandataportal.eu/en/homepage.

[12] https://www.w3.org/TR/vocab-dcat-2/.

[13] https://inspire.ec.europa.eu/good-practice/geodcat-ap.

[14] https://opengeospatial.org/standards/sensorml.

dynamic and stationary platforms and both in situ and remote sensors. Another possibility is Semantic Sensor Network Ontology[15], which describes sensors and observations, and related concepts. It does not describe domain concepts, time, locations, etc.; these are intended to be included from other ontologies via OWL imports. This ontology is developed by the W3C Semantic Sensor Networks Incubator Group (SSN-XG).[16]

There is a need for metadata harmonization of the spatial and non-spatial datasets and services. GeoDCAT-AP is an obvious choice due to the strong focus on geographic datasets. The main advantage is that it enables users to query all geospatial datasets in a uniform way. GeoDCAT-AP is still very new, and the implementation of the new standard can provide feedback to OGC, W3C & JRC from both technical and end user point of view. Several software components are available in the DataBio architecture that have varying support for GeoDCAT-AP, being Micka[17], CKAN[18] [3], FedEO Gateway & Catalog[19], and GeoNetwork[20] [4]. For the DataBio purposes we also had to integrate Semantic Sensor Net Ontology and SensorML.

For enabling compatibility with COPERNICUS[21], INSPIRE[22], and GEOSS[23], the DataBio project made three extensions: (i) Module for extended harvesting INSPIRE metadata to DCAT, based on XSLT and easy configuration; (ii) Module for user friendly visualisation of INSPIRE metadata in CKAN; and (iii) Module to output metadata in GeoDCAT-AP respectively SensorDCAT. DataBio used Micka and CKAN systems. Micka is a complex system for metadata management used for building Spatial Data Infrastructure (SDI) and geoportal solutions. It contains tools for editing and management of spatial data, and services metadata as well as other sources (documents, websites, etc.). Micka also fully supports GeoDCAT-AP and Open Search. CKAN supports DCAT to import or export its datasets. CKAN enables harvesting data from OGC:CSW catalogues, but not all mandatory INSPIRE metadata elements are supported. Unfortunately, the DCAT output does not fulfil all INSPIRE requirements, nor is GeoDCAT-AP fully supported.

For data identification, naming, and search keywords we used the INSPIRE data registry.[24] The INSPIRE infrastructure involves a number of items, which require clear descriptions and the possibility to be referenced through unique identifiers. Examples of such items include INSPIRE themes, code lists, application schemas or discovery services. Registers provide a means to assign identifiers to items and their labels, definitions and descriptions (in different languages). The INSPIRE Registry is

[15] https://www.w3.org/TR/vocab-ssn/.

[16] https://www.w3.org/2005/Incubator/ssn/.

[17] http://micka.bnhelp.cz/.

[18] https://ckan.org/.

[19] http://ceos.org/ourwork/workinggroups/wgiss/access/fedeo/.

[20] http://geonetwork-opensource.org/.

[21] https://www.copernicus.eu/en.

[22] https://inspire.ec.europa.eu/.

[23] https://www.earthobservations.org/geoss.php.

[24] http://inspire.ec.europa.eu/registry.

a service giving access to INSPIRE semantic assets (e.g. application schemas, code-lists, themes), and assigning to each of them a persistent URI. As such, this service can be considered also as a metadata directory/catalogue for INSPIRE, as well as a registry for the INSPIRE "terminology". Starting from June 2013, when the INSPIRE Registry was first published, several versions have been released, implementing new features based on the community's feedback.

Also important is *data lineage*, which refers to the sources of information, such as entities and processes, involved in producing or delivering an artifact. Data lineage records the derivation history of a data product. The history could include the algorithms used, the process steps taken, the computing environment run, data sources input to the processes, the organization/person responsible for the product, etc. *Provenance* provides important information to data users for them to determine the usability and reliability of the product. In the science domain, the data provenance is especially important since scientists need to use the information to determine the scientific validity of a data product and to decide if such a product can be used as the basis for further scientific analysis.

7.3   Linked Data

As noted above, Linked Data refers to a set of best practices for publishing and interlinking structured data thereby enabling it to be accessed by both humans and machines. The data interchange follows the RDF family of standards and SPARQL is used for querying. In particular, the key concepts and technologies that support Linked Data are:

- Any concept or entity can be identified by assigning specific Uniform Resource Identifier (URIs) to them.
- HTTP for retrieving or description of resources.
- RDF which is a generic graph-based data model used for structuring and linking data that describes concepts or entities in the real world.
- SPARQL is the standard RDF query language.

More in detail, RDF expresses data as *triples* of the form < subject, predicate, object > . A triple encodes the relation of the object to the subject through the predicate. The subject is a URI, or more generally Internationalized Resource Identifier (IRI), which, as specified above, identifies a resource or a concept; the object may be either a literal e.g. number, string, date, or a URI which references another resource. Triples which interlink resources constitute RDF links, which construct the Web of Data.

7.4 Linked Data Best Practices

The growing popularity of Linked Data has led to the definition of more detailed guidelines for the development and delivery of (open) data as linked data. For instance, for open government (also applicable for LOD and the bioeconomic sector) data, the following best practices are recommended [8]:

- To prepare the stakeholders by explaining the process of creating and maintaining the Linked Data.
- To select a dataset which can be reused by others.
- To model the Linked Data represented as data objects and their relation in an application-independent way.
- To specify an appropriate license to ease data reuse by declaring the origin, ownership and conditions applied for the reusing of the open data.
- To use a well-considered URI naming strategy and implementation plan, based on HTTP URIs.
- To describe the objects with previously defined vocabulary so as to extend the standard vocabulary.
- To convert data to a Linked Data representation by scripting or other automated processes.
- To provide machine access to the data by providing a way for search in an engine and other automated processes using standard web mechanisms.
- To announce new datasets on authoritative domains to initiate an implicit social contact.
- To maintain the data once published.

It is important to note that although these best practices were conceived for open government data, they can be applied in most cases to many other domains.

To help prepare stakeholders, there are at least three well known life-cycle models (Hyland et al. [8], Hausenblas [9], Villazón-Terrazas et al. [10]) describing the process for publishing linked data. All of these models identify common needs of specifying, modelling and publishing data in the standard open Web format (https://www.databio.eu/wp-content/uploads/2017/05/DataBio_D4.3-Data-sets-formats-and-models_public-version.pdf, https://www.google.ca/search?q=%22standard+open+web+format%22). However even though all of the models somewhat deal with similar tasks, they have some differences between those tasks. To discuss more in detail the above mentioned tasks, we will focus on one of these models as their roles are similar and complementary. For the sake of consideration, Villazón-Terrazas et al. [10] has the following sub-tasks for each step:

- Specification:

    - Identification and analysis of the data sources by opening and publishing the data that have not yet opened up and published and by reusing or leveraging the data that had already been opened/published up by others. This may require contacting specific data owners to get access to their legacy data.

- Design URIs by using meaningful URIs rather than opaque ones whenever possible. It is important to separate TBox (ontology model) from ABox (instances) URIs.
- Definition of the license of the data sources. It is also possible to reuse and apply an existing license of the data sources.

- Modelling:

  - Ontologies ideally are expressed in OWL or RDF(S) both being based on RDF.
  - Reusing the existing and available vocabularies.
  - Reusing the available non-ontological resources like highly reliable websites, domain related sites, government catalogs etc.

- Generation:

  - Transformation of the data sources selected in the specification activity into RDF according to the vocabulary created in the modelling activity by using tools like CSV and spreadsheets, RDB or XML.
  - Data cleansing involves the finding and fixing of the possible errors specified in Hogan et al. which includes http-level issues, such as accessibility and de-referencability, reasoning issues such as namespace without vocabulary and malformed/incompatible data types.
  - Linking suitable datasets and discovering suitable relationships between the data items and validate the relationships discovered.

- Publishing:

  - Dataset publication by using tools for storing RDF (e.g. Virtuoso Universal Server, Jena, Sesame, 4Store, YARS, OWLIM etc.) using SPARQL endpoint and Linked Data front end (e.g. Pubby, Talis Platform, Fuseki)
  - Metadata publication using VoID, which allows to express metadata about RDF datasets and by OPM (Open Provenance Model).
  - Dataset discovery by registering the datasets in the CKAN registry and generating sitemap files for the dataset, by using sitemap4rdf.[25]

- Exploitation is the final step in linked data publication workflow which refers to the application and exploitation of Linked Data for various purposes and applications across different platforms.

---

[25] http://mayor2.dia.fi.upm.es/oeg-upm/index.php/en/technologies/122-sitemap4rdf/index.html.

7.5 The Linked Open Data (LOD) Cloud

The LOD cloud comprises 1,255 datasets with 16,174 links (as of May 2020). Nevertheless, although large cross-domains datasets exist (like DBpedia[26] or Wikidata[27]) and some domains are well covered, like Geography, Government, and Bioinformatics, this is still not the case for all domains. For instance, in the agriculture domain we can find relevant thesaurus like AGROVOC[28] from FAO[29], or the National Agricultural Library's Agricultural Thesaurus (NALT),[30] but there is still a lack of datasets related to agricultural facilities and farm management activities. A similar situation occurs in the fishery domain where only some taxonomies for specific types of fish or regions are available, but no catching data exists, including, for example, locations, quantities, values, equipment used, vessels used, etc. This is also true in the forestry domain, where almost no specific Open Linked Data is available. This is in part due to the lack of standardized models for the representation of such data, even though some efforts in this direction have been made in the past, as discussed below.

FOODIE project,[31] for instance, addressed this issue for the agriculture domain with the development of the FOODIE data model[32] [11], which was reused and extended in the DataBio project. To ensure the maximum degree of data interoperability, the model is based on INSPIRE generic data models, specially the data model for Agricultural and Aquaculture Facilities (AF), which is extended and specialized. In particular, the model was created based on AF version 1.0, and thus it was found that there was a lack of a concept for an entity of finer granularity than *Site* that was part of the INSPIRE AF.[33] The key motivation was to represent a continuous area of agricultural land with one type of crop species, cultivated by one user using one farming mode (conventional vs. transitional vs. organic farming). Such a concept is called *Plot* and represents the main element in the model, especially because it is the level to which the majority of agro data is related. One level lower than *Plot* is the *ManagementZone*, which enables a more precise description of the land characteristics in fine-grained areas. Additionally, the FOODIE model includes concepts for crop and soil data, treatments, interventions, agriculture machinery, etc. Furthermore, the model reuses data types defined in ISO standards (ISO 19101, ISO/TS 19103, ISO 8601 and ISO 19115) as well standardization efforts published under the INSPIRE Directive[34] (like the structure of unique identifiers). The model was consulted with several experts from various institutions like the Directorate General

---

[26] https://wiki.dbpedia.org/.

[27] https://www.wikidata.org/wiki/Wikidata:Main_Page.

[28] http://aims.fao.org/vest-registry/vocabularies/agrovoc.

[29] http://www.fao.org/home/en/.

[30] http://aims.fao.org/news/nal-thesaurus-now-available-linked-open-data.

[31] http://www.foodie-project.eu/.

[32] https://github.com/Wirelessinfo/FOODIE-data-model.

[33] http://inspire.ec.europa.eu/theme/af.

[34] https://inspire.ec.europa.eu/inspire-directive/2.

Joint Research Centre (DG JRC) of the EU Commission, the EU Global Navigation Satellite Systems Agency (GSA), Czech Ministry of Agriculture, Global Earth Observation System of Systems (GEOSS), or the German Kuratorium für Technik und Bauwesen in der Landwirtschaft (KTBL). FOODIE data model was specified in Unified Modeling Language (UML) (as the INSPIRE models), but describes the process followed to transform this model into an OWL ontology in order to enable the publication of linked agricultural data [12]. FOODIE ontology follows a modular approach. Thus, while the core ontology includes all elements common to different applications, the ontology can be further specialized with profiles for a particular application or country needs. In the DataBio project, for example, the FOODIE ontology was reused in several agriculture pilots, which resulted in the addition of several new elements in the core, and with the creation of extensions for the specific needs of the pilot.

Regarding the fishery domain, there have been also some previous efforts to fill this standard model gap. For instance, in NeOn project, FAO produced a network of fisheries ontologies[35] that included a catch record pattern, water areas (e.g. FAO division areas), species taxonomic classifications, fisheries commodities, vessels classifications, gear classifications, etc. Unfortunately, the work did not continue and many of these resources are no longer available. Nevertheless, in the DataBio project, some of these resources were reused when possible (e.g. catch pattern, species taxonomy), some others were re-created with further detail (e.g. water areas), and some new extensions were created to cover specific pilot needs in order to publish linked fishery data from them.

7.6 Enterprise Linked Data (LED)

Although Linked Data is mostly known and used to publish open data, and to link different open datasets, the underlying technologies and approach can also be applied in a (partially) closed setting, e.g. an enterprise, where potentially some data cannot be made openly available - this is especially relevant for all sectors of Bioeconomy with sensitive and geo-based data. In fact, even if the enterprise data remains closed, or accessible only via access control mechanisms to selected parties, it can still be linked with open data, and get all the benefits from that.

According to [11], Linked Enterprise Data (LED) meshes each and every enterprise data (e.g. structured records, documents or office files), wherever they come from, to create a global and unified information space from which new business information is created to solve operational needs. Hence, it federates the content of heterogeneous silos by interconnecting the data and creates a unified and coherent warehouse, called an information hub, that exposes and shares new knowledge objects [13]. Besides, as it follows the same standards, links can be established with other datasets, either internal or external (e.g. LOD).

---

[35] http://aims.fao.org/network-fisheries-ontologies.

In order to restrict the access to internal data, LED must be used in combination with access control mechanisms enabling compliance with privacy and security constraints, as described in the next section. Regarding security of the stored RDF data, one of the most typical approaches to control the access to the data is by using different RDF graphs for the restricted datasets. An RDF graph is a set of RDF triples, normally identified by an IRI, which can be assigned different access control policies.

For instance, Virtuoso, the RDF store used in the DataBio project, features SPARQL endpoints, which are Web services capable of providing more than Read-Only access to back-end graphs. So, even though they are commonly general-purpose, SPARQL endpoints can also be purpose-specific, and their privileges may, therefore, be limited to specific Create, Read, Update, and/or Delete operations. The privileges provided by a given Virtuoso SPARQL endpoint may be based simply upon the endpoint's URL, or upon sophisticated rules which associate specific user identities with specific database roles and privileges. Virtuoso offers three methods for securing SPARQL endpoints:

- Digest Authentication via SQL Accounts
- OAuth Protocol based Authentication
- WebID Protocol based authentication.

In the DataBio project, the first method was tested in order to restrict access to some of the pilot datasets. In particular, the process of setting up a secure Virtuoso SPARQL endpoint using the method of Digest Authentication via SQL Accounts is as follows:

- Step 1: Create a user for a data graph.
- Step 2: Assign the user to the specific user group assigned with a specific role. A user should become a member of an appropriate group (e.g. SPARQL_SELECT, SPARQL_SPONGE, or SPARQL_UPDATE) in order to start using its graph-level privileges.
- Step 3: Some graphs are supposed to be confidential; the whole triple store is first set to be restricted to set the overall graph store permission.
- Step 4: Set some basic privileges to some users where the specific users will not have the global access to the graphs.
- Step 5: Grant specific privileges on specific graphs to specific users:

  - User can only READ but not WRITE from the personal system data graph.
  - User can both READ and WRITE from the personal system data graph.
  - Grant specific privileges on specific graph to public where the graphs (e.g. dbpedia.org) are intended for public consumption for:

    READ but not WRITE;
    READ and WRITE.

References

1. Bizer, C., Heath, Tom, & Berners-Lee, T. (2009). Linked data—the story so far. *International Journal on Semantic Web and Information Systems., 5*(3), 1–22. https://doi.org/10.4018/jswis.2009081901.
2. Cyganiak, R., Wood, D., Lanthaler, M., Klyne, G., Carroll, J., McBride, B. (2014) RDF 1.1 Concepts and abstract syntax [W3C Recommendation]. (Technical report, W3C).
3. W3C OWL Working Group. (2012). *OWL 2 Web ontology language document overview* (2nd Ed.). W3C Recommendation 11 December 2012 World Wide Web Consortium (W3C).
4. Harris, S., Seaborne, A., Prud'hommeaux, E. (2013). *SPARQL 1.1 query language.* W3C Recommendation. W3C.
5. Hitzler, P., Krotzsch, M., Rudolph, S. (2009). *Foundations of semantic web technologies.* Chapman and Hall/CRC.
6. Hyland, B., Wood, D. (2011). *The joy of data—a cookbook for publishing linked government data on the web. linking government data.* Springer, pp. 3–26. https://doi.org/10.1007/978-1-4614-1767-5_1.
7. Hodge, G. (2001). *Metadata made simpler: A guide for libraries.* NISO Press.
8. Hyland, B., Atemezing, G., Villazón-Terrazas, B. (2014). *Best practices for publishing linked data.* W3C Working Group Note 09 January 2014. https://www.w3.org/TR/ld-bp/.
9. Hausenblas, M. (2011). Linked data life cycles. https://www.slideshare.net/mediasemanticweb/linked-data-life-cycles. Accessed 30 Sept 2020.
10. Villazón-Terrazas, B., Vilches-Blázquez, L. M., Corocho, O., Gómez-Pérez. (2011). A methodological guidelines for publishing government linked data. In: *Linking Government Data, chapter 2.* Springer, pp. 27–49. http://link.springer.com/chapter/10.1007/978-1-4614-1767-5_2.
11. Esbrí, M., Palma, R., Rodríguez, J., Noriega, A., Campos, A., Charvat, K., Reznik, T., Lukas, V., Kepka, M., Sredl, M., Ogut, G., Sumer, A. (2016). FOODIE. D2.2.3 service platform specification. https://confluence.man.poznan.pl/community/download/attachments/66584868/D2.2.3-Service-Platform-Specification.pdf?api=v2. Accessed 30 Sept 2020.
12. Palma, R., Reznik, T., Esbri, M., Charvat, K., Mazurek, C. (2015). An INSPIRE-based vocabulary for the publication of Agricultural Linked Data. In: Proceedings of the OWLED Workshop: Collocated with the ISWC-2015, pp. 124–1334.
13. Antidot. (2012). Linked enterprise data, principles, uses and benefits. https://www.antidot.net/wp-content/uploads/2012/11/LinkedEnterpriseData-WP-en-v2.2.pdf. Accessed 30 Sept 2020.

Chapter 8
Linked Data Usages in DataBio

Raul Palma, Soumya Brahma, Christian Zinke-Wehlmann,
Amit Kirschenbaum, Karel Charvát, Karel Charvat Jr., and Tomas Reznik

Abstract One of the main goals of DataBio was the provision of solutions for big data management enabling, among others, the harmonisation and integration of a large variety of data generated and collected through various applications, services and devices. The DataBio approach to deliver such capabilities was based on the use of Linked Data as a federated layer to provide an integrated view over (initially) disconnected and heterogeneous datasets. The large amount of data sources, ranging from mostly static to highly dynamic, led to the design and implementation of Linked Data Pipelines. The goal of these pipelines is to automate as much as possible the process to transform and publish different input datasets as Linked Data. In this chapter, we describe these pipelines and how they were applied to support different uses cases in the project, including the tools and methods used to implement them.

8.1 Introduction

Linked Data has been extensively used in the DataBio project as a federated layer to support large-scale harmonization and integration of a large variety of data collected from various heterogeneous sources and to provide an integrated view on them. Accordingly, as part of the project, we generated a large number of linked datasets. In fact, the triplestore populated during the course of DataBio with Linked Data has over 1 billion triples, being one of the largest semantic repositories related to agriculture. The dataset has been recognized by the EC Innovation Radar as 'arable farming data integrator for smart farming.' In addition, we have deployed different

R. Palma (✉) · S. Brahma
Poznan Supercomputing and Networking Center (PSNC), ul. Jana Pawła II 10, 61-139 Poznan, Poland
e-mail: rpalma@man.poznan.pl

C. Zinke-Wehlmann · A. Kirschenbaum
Institute for Applied Informatics (InfAI), University of Leipzig, Goerdelerring 9, 04109 Leipzig, Germany

K. Charvát · K. Charvat Jr. · T. Reznik
Lesprojekt Služby Ltd., Martinov 197, 27713 Záryby, Czech Republic

C. Södergård et al. (eds.), *Big Data in Bioeconomy*,
https://doi.org/10.1007/978-3-030-71069-9_8

endpoints providing access to some dynamic data sources in their native format as Linked Data by providing a virtual semantic layer on top of them.

Given the huge number of data sources, and data formats that were addressed during the course of DataBio, such layer has been realized in DataBio through the implementation of instantiations of a 'Generic Pipeline for the Publication and Integration of Linked Data,' which have been applied in different uses cases related to the bioeconomy sectors. The main goal of these pipeline instances is to define and deploy (semi-) automatic processes to carry out the necessary steps to transform and publish different input datasets as Linked Data. Accordingly, they connect different data processing components to carry out the transformation of data into RDF [1] format or the translation of queries to/from SPARQL [2] and the native data access interface, plus their linking, and include the mapping specifications to process the input datasets. Each pipeline instance is configured to support specific input dataset types (same format, model and delivery form), and they are created with the following general principles in mind:

- Pipelines must be directly re-executed and re-applied (e.g., extended/updated datasets).
- Pipelines must be easily reusable.
- Pipelines must be easily adapted for new input datasets.
- Pipeline execution should be as automatic as possible. The final target is to fully automated processes.
- Pipelines should support both: (mostly) static data and data streams (e.g., sensor data).

Most of the Linked Data Publication pipeline instances discussed in this chapter perform the transformation and publication of agricultural data as Linked Data; however, there are also some pipelines that are focused on fishery data or on providing access to geospatial datasets metadata as Linked Data. The ultimate target is to query and access different heterogeneous data sources via an integrated layer, in compliance with any privacy and access control needs.

A high-level view of the end-to-end flow of the generic pipeline, aligned with the top-level DataBio generic pipeline, is depicted in Fig. 8.1. Following the best practices and guidelines for Linked Data Publication [3, 4], these pipelines (i) take as input selected datasets that are collected from heterogeneous sources (shapefiles, GeoJSON, CSV, relational databases, RESTful APIs), (ii) curate and/or preprocess the datasets when needed, (iii) select and/or create/extend the vocabularies (e.g., ontologies) for the representation of data in semantic format, (iv) process and transform the datasets into RDF triples according to underlying ontologies, (v) perform any necessary post-processing operations on the RDF data, (vi) identify links with other datasets and (vii) publish the generated datasets as Linked Data and applying required access control mechanisms. The transformation process depends on different aspects of the data like format of the available input data, the purpose (target use case) of the transformation and the volatility of the data (how dynamic is the data). Based on these characteristics, there are two main approaches for making

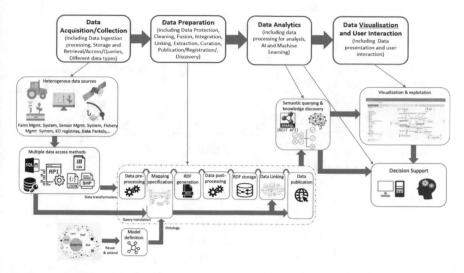

Fig. 8.1 Generic flow for Linked Data integration and publication pipeline aligned with top-level generic pipeline

the transformation for a dataset: (i) data upgrade or lifting, which consists of generating RDF data from the source dataset according to mapping descriptions and then storing it in semantic triplestore (e.g., Virtuoso) and (ii) on-the-fly query transformation, which allows evaluating SPARQL queries over a virtual RDF dataset, by rewriting those queries into source query language according to the mapping descriptions. In this former scenario, data physically stays at their source and a new layer is provided to enable access to it over the virtual RDF dataset.

In every transformation process, regardless of the method or tools chosen, a mapping specification has to be defined to specify the rules to map the source elements (e.g., table columns, JSON elements, CSV columns, etc.) into target elements (e.g., ontology terms). Generally, this specification is an RDF document itself written in RML[1]/R2RML[2] (and extensions) languages and/or nonstandard extensions of SPARQL, e.g., in the case of the Tarql CSV to RDF transformation tool.[3]

The resulting datasets can thereafter be exploited through SPARQL queries, or via a plethora of user interfaces. Some examples of these interfaces include:

- SPARQL endpoint interface, to execute queries: https://www.foodie-cloud.org/sparql
- Faceted search interface to navigate the linked datasets http://www.foodie-cloud.org/fct/
- Map visualization via HS Layer applications, e.g., http://app.hslayers.org/project-databio/land/

---

[1] https://rml.io/specs/rml/.

[2] https://www.w3.org/TR/r2rml/.

[3] https://tarql.github.io/.

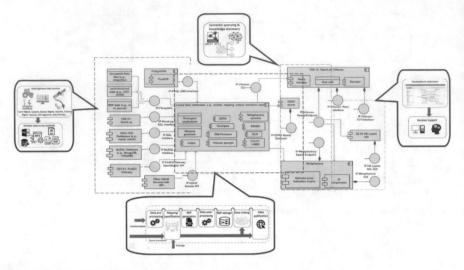

Fig. 8.2 Generic Linked Data Publication pipeline component view diagram

- Metaphactory: http://metaphactory.foodie-cloud.org/resource/Start.

The following diagram (Fig. 8.2) provides a simplified representation of the generic Linked Data Publication pipeline component view that includes the software components and interfaces involved. More information is available in [5, 6].

The URL link of the generic pipeline in the DataBioHub is https://mub.me/2f81.

8.2 Linked Data Pipeline Instantiations in DataBio

The Linked Data Pipeline, as described in the previous section, is a generalization of multiple instantiations, in particular two specific project's pilots and four additional experiments in DataBio. Thus, in order to show how this generic pipeline has been applied in each of these use cases, we present in this section for each of them the pipeline view, previously presented in [5], highlighting the specific methods and components used/applied, along with a description of the task performed and results achieved.

8.2.1 Linked Data in Agriculture Related to Cereals and Biomass Crops

This pipeline instance was focused toward publication of INSPIRE-based agricultural Linked Data from the farm data collected from cereals and biomass crop pilots, in order to query and access different heterogeneous data sources via an integrated layer. The input datasets used for this experiment include:

- Farm data (Rostenice pilot) that holds information about each field name with the associated cereal crop classifications and arranged by year.
- Data about the field boundaries and crop map yield the potential of most of the fields in the Rostenice pilot farm from Czech Republic.
- Yield records from two fields (Pivovarska and Predni) within the pilot farm that were harvested in 2017 and 2018.

The source datasets, collected as shapefiles, were transformed into RDF format and published as Linked Data, using the FOODIE ontology as the underlying model. The resulting linked datasets are available for querying and exploitation through the DataBio SPARQL endpoint deployed at PSNC' HPC facilities. More in detail, the tasks carried out are as follows:

- Definition of the data model to transform the input datasets into RDF. For this step, FOODIE ontology [7], which is based on INSPIRE schema and the ISO 19100 series standards, was used as the base vocabulary and extended as needed (with a Czech pilot extension) in order to represent all the farm and open data from the input datasets. The extension includes data elements and relations from the input datasets that were not covered by the main FOODIE ontology but that were critical for the pilot needs.
- Creation of an RDF mapping file that specifies how to map the contents of a dataset into RDF triples by matching the source dataset schema with FOODIE ontology and its extensions. A generic RML/R2RML definition of the mapping file was generated from the input shapefiles by using applications like GeoTriples and thereafter manually edited as per the data model identified to generate the final mapping definition. GeoTriples was also used to generate the RDF dump from the source data contents. FOODIE ontology and its extension were used extensively in the mapping files to match the source dataset schemas.
- The RDF datasets generated were loaded into DataBio Virtuoso triplestore. A SPARQL endpoint and a faceted search endpoint are available for querying and exploiting the Linked Data in the Virtuoso instance deployed at PSNC infrastructure.
- The final task involved providing an integrated view over the original dataset. As source datasets were particularly large (especially when considering connections with open datasets), and the connections were not of equivalence (i.e., resources are related via some properties but they are not equivalent), it was decided to use queries to access the integrated data as per need rather than using link discovery tools like SILK or LIMES. Hence, cross-querying within the datasets was done in Virtuoso SPARQL endpoint for some use cases to establish possible links between agricultural and related open datasets.
- To visualize and explore the Linked Data in a map, we have created different application/system prototypes. One such map visualization component called HS Layers NG is available at https://app.hslayers.org/project-databio/land/.

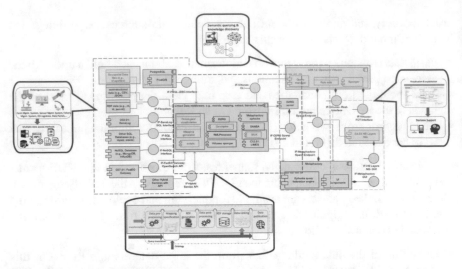

Fig. 8.3 Mapping of the generic components into cereals and biomass crop pilots in the pipeline view

Please refer to Section 'Usage and Exploitation of Linked Data' for additional information of other visualization components.

The resulting linked datasets are accessible via: https://www.foodie-cloud.org/ sparql. A figure that maps the generic components identified in this pilot is given below (Fig. 8.3). The red highlighted markings indicate the components in use in the pilot.

8.2.2 Linked Sensor Data from Machinery Management

This pipeline was performed for the machinery management DataBio pilot, where sensor data from the SensLog service (used by FarmTelemeter service) was transformed into Linked Data on the fly; i.e., data stays at the source, and only a virtual semantic layer was created on top of it to access it as Linked Data. For modeling the sensor data, the following vocabularies/ontologies were selected:

1. Semantic sensor network (SSN[4]) ontology for describing sensors and their observations, the involved procedures, the studied features of interest, the samples used to do so and the observed properties. A lightweight but self-contained core ontology called Sensor, Observation, Sample, and Actuator or SOSA was actually used in this specific case to align the SensLog data.
2. Data Cube Vocabulary and its SDMX ISO standard extensions were effective in aligning multidimensional survey data like in SensLog. The Data Cube

---

[4] https://www.w3.org/TR/vocab-ssn/.

includes well-known RDF vocabularies (SKOS,[5] SCOVO,[6] VOID, FOAF,[7] Dublin Core[8]).

The SensLog service uses a relational database (PostgreSQL) to store the data. Hence, in the mapping stage, the creation of R2RML/RML definitions required different preprocessing tasks and some on-the-fly assumptions to engineer the alignment between the SensLog database and the ontologies/vocabularies.

Once the mapping file was generated (manually), the RDF data of the dataset was published using a D2RQ server that enables accessing relational database sources as virtual RDF graphs. This on-the-fly approach allows publishing of RDF data from large and/or live databases, and thus the need for replicating the data into a dedicated RDF triple store is not required. The Linked Data from the sensor data from SensLog (version 1) was published in the PSNC infrastructure in a D2RQ server available at http://senslogrdf.foodie-cloud.org/. The associated SPARQL endpoint to query the data is available at: http://senslogrdf.foodie-cloud.org/sparql.

The figure below (Fig. 8.4) highlights the main components used in this pilot from the generic pipeline components.

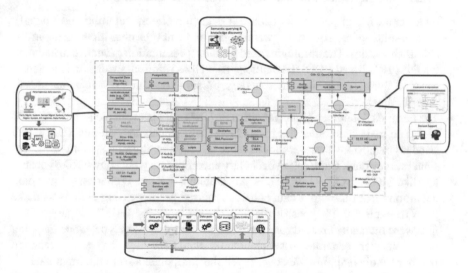

Fig. 8.4 Mapping of the generic components into machinery management pilot in the pipeline view

---

[5] https://www.w3.org/TR/skos-reference/.

[6] http://vocab.deri.ie/scovo.

[7] http://www.foaf-project.org/.

[8] https://www.dublincore.org/specifications/dublin-core/dces/.

8.2.3   Linked Open EU-Datasets Related to Agriculture and Other Bio Sectors

This pipeline focuses on EU and national open data from various heterogeneous sources from a wide range of applications in the geospatial domain. The purpose was to experiment on these datasets by transforming them into Linked Data and exploiting them on various technology platforms for integration and visualization. The sources for all of these data contents are widely heterogeneous and in various forms (e.g., in shapefiles, CSV format, JSON and in relational databases), which required extensive work to identify the most suitable mode for their transformation. This included a careful inspection of the input data contents in order to identify available ontologies/vocabularies, and any required extensions, necessary for the representation of such data in RDF format. Additionally, since the source datasets were in different formats, selecting the most suitable tools for their transformation was a key activity in order to create the correct (R2RML/RML) mapping definitions. Some of the input datasets, their formats and the ontologies/vocabularies used for the representation of data in semantic format are described below.

- Input data of land parcel and cadastral data (for Czech Republic and Poland), erosion-endangered soil zones, water buffer and soil type classification are available as shapefiles. The ontologies used for the representation of such data included the INSPIRE-based FOODIE ontology as well as different extensions created to cover all the necessary information (e.g., erosion zones and restricted areas near water bodies).
- The Farm Accountancy Data Network (FADN[9]) data is available as a set of CSV files. The main ontologies used were Data Cube Vocabulary and its SDMX ISO standard extensions that were much more effective in aligning such multidimensional survey data. Data Cube Vocabulary encompasses well-known RDF vocabulary like SKOS, SCOVO, VOID, FOAF, Dublin Core, etc. Preparing the mapping definitions from the input data sources required preprocessing actions to make them reusable for all types of the CSV data sources of FADN. Separate CSV files were manually created for each reusable common class type. Once mapping definitions were generated for each of the created CSV files, they were integrated into one whole mapping file covering all the components from the input data.
- The sample data input from Yelp is available as a set of JSON files. Different ontologies like review,[10] FOAF, schema.org, POI, etc., were used to represent the elements from the input data in semantic format during the creation of the mapping definition.
- Other ontologies from previous efforts for the representation of open geospatial datasets like Corine, Hilucs, OLU, OTM, Urban Atlas, were also used. These ontologies are available in https://github.com/FOODIE-cloud/ontology.

---

[9] https://ec.europa.eu/agriculture/rica/.

[10] https://vocab.org/review/.

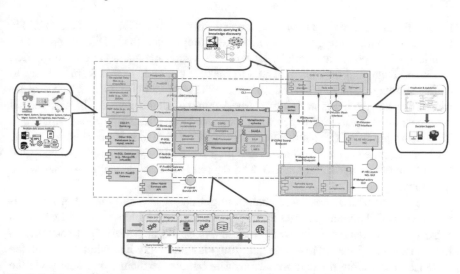

Fig. 8.5 Mapping of the components used in the use case of linked open EU-datasets in the pipeline view

The generation of RDF triples was carried out using different tools (depending on the source dataset format). For shapefiles, GeoTriples tool was used, while for the JSON and CSV data the RML processor tool was used. The resulting RDF datasets were then loaded into DataBio Virtuoso triplestore providing SPARQL and faceted search endpoints for further exploitation. Finally, for the provision of an integrated view over the original datasets in case of agricultural and open data, SPARQL queries were generated and additional links were discovered using tools like SILK. For visualization, platforms like HS Layers NG and Metaphactory were used as discussed in Chap. 13.

The resulting linked datasets are accessible via: https://www.foodie-cloud.org/sparql. The figure below (Fig. 8.5) highlights the main components used in this pilot from the generic pipeline components.

8.2.4 Linked (Meta) Data of Geospatial Datasets

This pipeline focuses on the publication of metadata from geospatial datasets as Linked Data. There were two data sources that were transformed.

The first dataset was metadata collected from the public Lesproject Micka registry,[11] which includes information of over 100 K geospatial datasets. Micka is a software for spatial data/services metadata management according to ISO, OGC and

---

[11] https://micka.lesprojekt.cz/en/.

INSPIRE standards, and it allows to retrieve the metadata in RDF using Geo-DCAT[12] for the representation of geographic metadata compliant with the DCAT application profile for European data portals. Nevertheless, such metadata cannot be queried as Linked Data, and thus the goal was to make it available in this form in order to enable its integration with other datasets, e.g., Open Land Use (OLU). The process for publication, thus, was straightforward: A dump of all the metadata in RDF format was generated from Micka, which was then loaded into DataBio Virtuoso triplestore. Some example SPARQL queries were then generated to identify connection points for integration, e.g., get OLU entries and their metadata given a municipal code and type of area (e.g., agriculture lands). The dataset is accessible via: https://www.foo die-cloud.org/sparql.

The second dataset was more challenging. The goal was to make Earth Observation (EO) Collections and EO Products metadata available as Linked Data via a SPARQL compliant endpoint which makes requests to non-SPARQL back ends on the fly. Hence, we wanted to enable querying via SPARQL without harvesting all the metadata and storing the data in a triplestore but access them dynamically via the existing online interfaces. The metadata was accessible via an OpenSearch interface provided by the FedEO[13] Clearinghouse in Spacebel (http://geo.spacebel.be/opense arch/readme.html) that enables retrieving the metadata in different formats, including atom/xml, RDF/xml, turtle, GeoJSON and LD-JSON. We used LD-JSON, which already defines the semantic properties used to represent the metadata elements. These properties comprise terms from different standard and well-known vocabularies/ontologies like Dublin Core, DCAT, SKOS, VOID and OM-Lite-lite, as well as from the OpenSearch specifications. Next, in order to enable access to a REST API via SPARQL queries that would allow linking with other Linked Datasets we used the Metaphactory platform. Metaphactory (https://www.metaphacts.com/pro duct) includes a component called Ephedra, which is a SPARQL federation engine aimed at processing hybrid queries. Ephedra provides a flexible declarative mechanism for including hybrid services into a SPARQL federation and implements a number of static and runtime query optimization techniques for improving the hybrid SPARQL query performance [8]. The RDF data is exposed via a SPARQL endpoint provided in the Metaphactory platform (http://metaphactory.foodie-cloud. org/sparql?repository=ephedra). A demo interface has also been implemented to visualize the Linked Data in Metaphactory (entry point: http://metaphactory.foodie-cloud.org/resource/:ESA-datasets).

The figure below (Fig. 8.6) highlights the main components used in this pilot from the generic pipeline components. In the figure, the components related to the first sub-case (Micka) are highlighted in green, while the components related to the second sub-case (FedEO) are highlighted in orange.

---

[12] https://ec.europa.eu/jrc/en/publication/geodcat-ap-representing-geographic-metadata-using-dcat-application-profile-data-portals-europe.

[13] http://ceos.org/ourwork/workinggroups/wgiss/access/fedeo/.

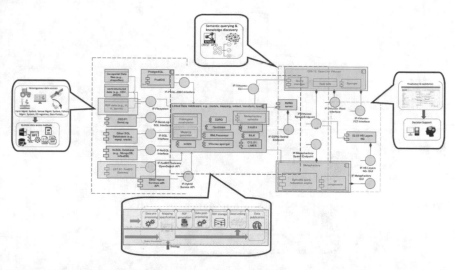

Fig. 8.6 Mapping of the components used in the use case of linked (meta) data of geospatial datasets in the pipeline view. The components related to the first sub-case (Micka) are highlighted in green, while the components related to the second sub-case (FedEO) are highlighted in orange

8.2.5   Linked Fishery Data

This pipeline focuses on the catch record data from the fisheries of Norwegian region. The purpose of this pipeline was to publish the catch record data from five years of historical data as Linked Data and perform experimentation operations to exploit and visualize them on various platforms. The input data was in the form of CSV files containing the catch record data of each year.

- The first task was to identify and map which attributes of the data are mostly in line with the transformation procedure and can be mapped with some existing ontology. Upon identifying such relevant data attributes from the main CSV file and carefully following the most relevant ontologies/vocabularies, we decided to use 'catchrecord.owl'[14] and mostly an extended version for our use of mapping.
- The CSV files were extensively preprocessed in such a way so as to generate a R2RML/RML mapping definition using a tool named GeoTriples. The mapping definitions were further analyzed and processed to settle with the final mapping definition for transformation of the CSV data. During the creation of the mapping definitions, the possibility of integration with other Linked Datasets was also considered.
- The transformation to the Linked Data was carried out using a tool named RML Processor from the final R2RML/RML mapping definitions.
- After the transformation of the Linked Data, a few post-processing steps were done to make the data ready to upload to the DataBio Virtuoso triplestore.

---

[14] http://www.ontologydesignpatterns.org/cp/owl/fsdas/catchrecord.owl.

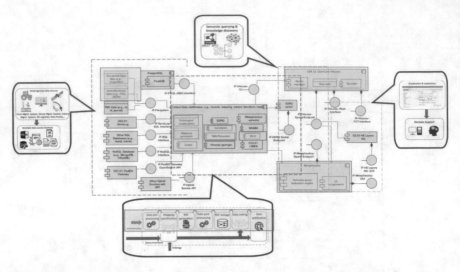

Fig. 8.7 Mapping of the components used in the fishery use case in the pipeline view

- At present, the catch data from five years was transformed and uploaded to the Virtuoso triplestore providing SPARQL and faceted search endpoints for further exploitation.

For the purpose of showcasing the integration and visualization of the dataset, a Web interface using the Metaphactory platform was created, which includes map visualizations and representation of data in the form of charts and graphs. This process is ongoing, and more experimentations are to come by. The interface is presently available at http://metaphactory.foodie-cloud.org/resource/:CatchDataNor way_v2. The resulting linked datasets are accessible via: https://www.foodie-cloud. org/sparql and https://www.foodie-cloud.org/fct.

The figure below (Fig. 8.7) highlights the main components used in this use case from the generic pipeline components.

8.3 Experiences from DataBio with Linked Data

8.3.1 Usage and Exploitation of Linked Data

The pipelines used in DataBio are part of an ongoing process and yet to be tested on other use cases and input data types. For example, as a result of the pipelines involving the LPIS and Czech field data, it was possible to perform integration experiments of the dataset for various use case scenarios of data integration.

As mentioned above, the datasets are deployed in the Virtuoso triplestore within PSNC and can be accessed via SPARQL and faceted search endpoints. The triplestore has **over 1 billion triples**, making it one of the largest semantic repositories related to agriculture.

The data in the triplestore is partitioned/organized into *named graphs*, where each named graph describes different contents and is identified by an IRI.

For example, the IRI <http://w3id.org/foodie/open/africa/GRIP> is the graph identifier of the African Roads Network dataset, which contains 27,586,675 triples.

Named graphs may be further composed of named subgraphs, as it is the case of the LPIS Poland dataset, which provides information about land-parcel identification in Poland, identified by the graph <http://w3id.org/foodie/open/pl/LPIS/>, and contains 727,517,039 triples. This graph contains, for example, the subgraph <http://w3id.org/foodie/open/pl/LPIS/lubelskie>, which refers to the data associated with the Lublin Voivodeship.

The table below shows some of the respective graphs produced by all the pipelines previously described and the number of triples contained in them.

Graph URI (note: URIs are not resolvable; they can be used to refer to the specific dataset in the triplestore)	Name of dataset	Number of RDF triples
http://w3id.org/foodie/open/pl/LPIS/{voivodeship} (where voivodeship in Poland = mazowieckie, dolnoslaskie, kujawsko pomorskie, lodzkie, lubelskie, lubuskie, malopolskie, opolskie, podkarpackie, podlaskie, pomorskie, slaskie, warminsko-mazurskie, wielkopolskie, zachodniopomorskie, swietokrzyskie)	LPIS Poland	727517039
http://w3id.org/foodie/olu agriculture-related lands (hilucs_code < 200) in CZ, PL, ES and for main cities in Czech Republic (centers of NUTS3 regions), Poland (agglomeration areas from Urban Atlas) and Spain (agglomeration areas from Urban Atlas)	Open land use	127926060
http://w3id.org/foodie/otm CZ, ES, PL; but RoadLinks only for FunctionalRoadClassValue of type: ('mainRoad,' 'firstClass,' 'secondClass,' 'thirdClass,' 'fourthClass') (see http://opentransportmap.info/OSMtoOTM.html)	Open transport map	154340785

(continued)

(continued)

Graph URI (note: URIs are not resolvable; they can be used to refer to the specific dataset in the triplestore)	Name of dataset	Number of RDF triples
http://micka.lesprojekt.cz/catalog/dataset	Open land use metadata	10456676
http://www.sdi4apps.eu/poi.rdf	Smart points of interest (SPOI)	407629170
http://w3id.org/foodie/open/cz/pLPIS_180616_WGS	LPIS Czech Republic	24491282
http://w3id.org/foodie/open/cz/lpis/code/LandUseClassificationValue	LPIS Czech Republic land use classification	83
http://w3id.org/foodie/atlas agriculture-related lands (hilucs_code < 200) and for main cities in Czech Republic (centers of NUTS3 regions), Poland (agglomeration areas from Urban Atlas) and Spain (agglomeration areas from Urban Atlas)	Urban Atlas	19606088
http://w3id.org/foodie/corine agriculture-related lands (hilucs_code < 200) and for main cities in Czech Republic (centers of NUTS3 regions), Poland (agglomeration areas from Urban Atlas) and Spain (agglomeration areas from Urban Atlas)	Corine land use	16777595
http://w3id.org/foodie/open/cz/Soil_maps_BPEJ_WGSc	Czech soil maps	8746240
http://w3id.org/foodie/open/cz/water_buffer25	Czech water buffers	3978517
http://w3id.org/foodie/core/cz/Predni_prostredni_vyfiltrovano_UTM	Yield mass in field crops (CZ Pilot)	1111852
http://w3id.org/foodie/core/cz/Pivovarka_vyfiltrovano	Yield mass in field crops (CZ Pilot)	437404
http://w3id.org/foodie/core/cz/CZpilot_fields	CZ Pilot fields and crop data	20183

(continued)

(continued)

Graph URI (note: URIs are not resolvable; they can be used to refer to the specific dataset in the triplestore)	Name of dataset	Number of RDF triples
http://ec.europa.eu/agriculture/ FADN/{FADNcategory} (Where FADN category = year-country, year-country-anc3, year-country-lfa, year-country-organic-tf8, year-country-siz6, year-country-siz6-tf14, year-country-siz6-tf8, year-country-sizc, year-country-tf14, year-country-tf8m, year-country-typology, year-region, year-region-siz6, year-region-siz6-tf8, year-region-sizc, year-region-tf14, year-region-tf8)	FADN	23520756
http://w3id.org/foodie/open/africa/ GRIP	African roads network	27586675
http://w3id.org/foodie/open/africa/ water_body	African water bodies	11330
http://w3id.org/foodie/open/gad m36/{level} where {level} = level0, level1, level2, level3, level4, level5	GADM dataset	7188715
http://w3id.org/foodie/open/kenya/ ke_crops_size	Kenya crop size	85971
http://w3id.org/foodie/open/kenya/ soil_maps	Kenya Soil Maps	10168
http://www.fao.org/aims/aos/fi/tax onomic	FAO	318359
http://www.fao.org/aims/aos/fi/ water_FAO_areas	FAO	150
http://www.fao.org/aims/aos/fi/ water_FAO_areas/inland	FAO	15779
http://www.fao.org/aims/aos/fi/ water_FAO_areas/marine	FAO	6768
http://w3id.org/foodie/open/catchr ecord/norway/	Catch record norway	192867166
http://standardgraphs.ices.dk/ stocks	ICES stocks data	1270280

(continued)

(continued)

Graph URI (note: URIs are not resolvable; they can be used to refer to the specific dataset in the triplestore)	Name of dataset	Number of RDF triples
https://www.omg.org/spec/LCC/Countries/ISO3166-1-CountryCodes/	ISO country codes	8629
https://www.omg.org/spec/LCC/Countries/Regions/ISO3166-2-SubdivisionCodes-NO/	ISO country subdivision codes	391
https://www.omg.org/spec/LCC/Countries/UN-M49-RegionCodes/	ISO region codes	569

The official SPARQL and the faceted search endpoints of the triplestore are: https://www.foodie-cloud.org/sparql (Fig. 8.8) and https://www.foodie-cloud.org/fct

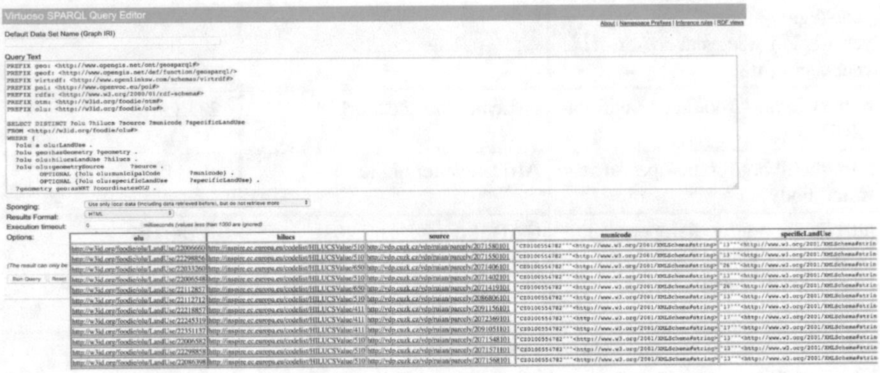

Fig. 8.8 SPARQL endpoint user interface (query and extract of result)

(Fig. 8.9).

Regarding the sensor data described in Sect. 1.3.2, it is published on the fly which serves the purpose of streaming transformation. This data can be accessed and linked through the following endpoints:

SPARQL endpoint: http://senslogrdf.foodie-cloud.org/sparql

SNORQL search endpoint: http://senslogrdf.foodie-cloud.org/snorql/

Web-based visualization: http://senslogrdf.foodie-cloud.org/ (see Fig. 8.10).

8.3.2 Experiences in the Agricultural Domain

RDF links often connect entities from two different sources, with relations which are not necessarily described in either data source. In the agricultural domain, this

Fig. 8.9 Faceted search interface

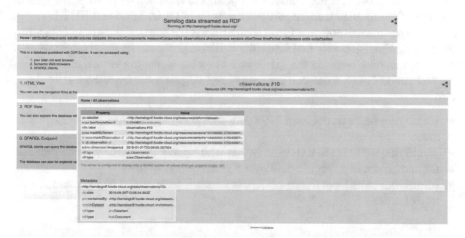

Fig. 8.10 Web interface entry page and visualization of an observation details of RDF generated on the fly

can be linking fields of specific crop type with the administrative region in which these fields reside, or find whether plots intersect with a buffer zone of water bodies in their vicinity. This is a means to control, e.g., the level and amount of pesticides used in those plots.

Creating such agricultural *knowledge graphs* is important due to environmental, economic and administrative reasons. However, constructing links manually is time and effort intensive, and links between concepts are rather to be *discovered* automatically. The basic idea of link discovery is to find data items within the target dataset which are logically connected to the source dataset. Formally, this means: Given $\mathscr{S}$ and $\mathscr{T}$, sets of RDF resources, called source and target resources, respectively, and a relation R, the aim of link discovery methods is to find a mapping $M = \{(s,t) \in \mathscr{S} \times \mathscr{T} : R(s,t)\}$. Naive computation of M requires quadratic time to test for each $s \in \mathscr{S}$ and $t \in \mathscr{T}$ whether R holds, which is infeasible for large datasets, and leads to the development of link discovery tools, which address this task.

In the agricultural domain, entities are mostly geospatial objects, and the relations are of a topological nature. Existing tools for link discovery, such as SILK and LIMES, are limited when it comes to geospatial data and therefore, as part of the DataBio project, we developed Geo-L, a system designated for discovery of RDF spatial links based on topological relations.

The system provides flexible configuration options to define to-be-linked datasets for SPARQL affine users and employs retrieval and caching mechanisms, resulting in efficient dataset management.

Geo-L uses PostgreSQL, an open-source object–relational DBMS, with PostGIS extension, as the database back end which supports geospatial data processing.

We conducted experiments to evaluate the performance of our proposed system by searching geospatial links based on topological relations between geometries of datasets of the foodie cloud, in particular subsets of OLU, SPOI and NUTS.

The experiments show that Geo-L outperforms the state-of-the-art tools in terms of mapping time, accuracy and flexibility.[15] It also proves to be more robust when it comes to handling errors in the data, as well as with managing large datasets.

We applied Geo-L to several use cases involving datasets from the foodie cloud, e.g.,

- Identifying fields from *Czech LPIS* data with specific soil type, from *Czech open data*
- Identifying all fields in a specific region which grow the same type of crops like the one grown in a specific field over a given period of time
- Identifying plots from *Czech LPIS* data which intersect with *buffer zones* around water bodies.

---

[15] As shown in a case of searching topological relations with NUTS as target dataset, where geometries are not represented as expected and were transformed on the fly to polygons by our tool.

An example for the last case is depicted in the image below (Fig. 8.11), where an overlap area between a plot and a buffer zone of a water body in its vicinity is colored with orange.

The respective dataset resulting from linking water bodies whose buffer zones are intersected by Czech LPIS plots is available on the DBpedia Databus.[16]

8.3.3   Experiences with DBpedia

DBpedia is a crowd-sourced continuous community effort to extract structured information from Wikipedia and to make this information available as a knowledge graph on the Web. DBpedia allows querying against this data and information and linking to other datasets on the Web [9, 10]. Currently, DBpedia is one of the central interlinking hubs in the Linked Open Data (LOD) cloud. With over 28 million of described and localized things, it is one of the largest and open datasets.

As part of the project, we constructed links between satellite entities, available in the European Space Association (ESA) thesaurus,[17] whose recorded images are employed in DataBio pilots and their respective DBpedia resources. These links are beneficial since the data in DBpedia is available in machine readable form for further processing, and in addition there are additional data and external links related to the satellite. We used REST API to retrieve satellite names from the ESA thesaurus and queried for DBpedia resources matching these names, which were then identified as satellites, based on their properties available in DBpedia.

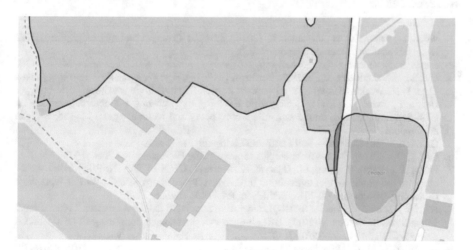

Fig. 8.11 Overlap area between a plot and a buffer zone of a water body in its vicinity, colored with orange

---

[16] https://databus.dbpedia.org/amit/geo-L/bufferzones-intersect-lpisPlots/.

[17] https://fedeo.spacebel.be/thesaurus/fr/.

```
<http://dbpedia.org/resource/Aura_(satellite)>    <http://www.w3.org/2002/07/owl#sameAs> <https://earth.esa.int/concept/aura> .
<http://dbpedia.org/resource/CASSIOPE>            <http://www.w3.org/2002/07/owl#sameAs> <https://earth.esa.int/concept/cassiope>
<http://dbpedia.org/resource/CryoSat-1>           <http://www.w3.org/2002/07/owl#sameAs> <https://earth.esa.int/concept/cryosat> .
<http://dbpedia.org/resource/EarthCARE>           <http://www.w3.org/2002/07/owl#sameAs> <https://earth.esa.int/concept/earthcare>
<http://dbpedia.org/resource/Envisat>             <http://www.w3.org/2002/07/owl#sameAs> <https://earth.esa.int/concept/envisat> .
<http://dbpedia.org/resource/Gravity_Field_
and_Steady-State_Ocean_Circulation_Explorer>      <http://www.w3.org/2002/07/owl#sameAs> <https://earth.esa.int/concept/goce>
```

Fig. 8.12 Links between ESA-platforms and their respective entities in DBpedia

The listing depicted in Fig. 8.12 presents an excerpt from the link-data result. The links allow, on the one hand, access to other properties of the respective DBpedia resources and, on the other hand, enable other DBpedia users to access the ESA set. This dataset can be found as an artifact[18] on the DBpedia Databus.

DBpedia resources which refer to geographical regions include different important properties about those areas such as temperature amplitudes and monthly precipitation. Such properties may be helpful, e.g., analysis of yields. These resources, however, do not contain the actual geometry of the regions. We used OpenStreetMap to retrieve data about regions and applied Geo-L to link between DBpedia region resources and their geometries.

These geometries can be helpful then not only for the purpose of the DataBio or for agriculture in general, but may be used for locating points of interest, which coordinates are known, within a specific region, a thing which has not been possible so far.

References

1. Wood, D., Lanthaler, M., Cyganiak, R. (2014). *RDF 1.1 Concepts and Abstract Syntax* [W3C Recommendation]. (Technical report, W3C).
2. Harris, S., Seaborne, A. (2013). *SPARQL 1.1 Query Language*. W3C Recommendation. W3C.
3. Hyland, B., Atemezing, G., Villazón-Terrazas, B. (2014). *Best Practices for Publishing Linked Data*. W3C Working Group Note 09 January 2014. URL: https://www.w3.org/TR/ld-bp/.
4. Heath, T., Bizer, C. (2011). Linked data: Evolving the web into a global data space (1st Ed.). Synthesis lectures on the semantic web: Theory and technology, 1:1, 1–136. Morgan & Claypool.
5. DATABIO. (2018). D4.i3—Full Platform Description for Trial 2.
6. DATABIO. (2019). D4.4—Technologies for supporting Bio-economy pilots.
7. Palma, R., Reznik, T., Esbri, M., Charvat, K., Mazurek, C. (2015). An INSPIRE-based vocabulary for the publication of agricultural linked data. Proceedings of the OWLED Workshop: Collocated with the ISWC-2015, Bethlehem PA, USA, October 11–15.
8. Nikolov, A. et al. "Ephedra: SPARQL federation over RDF data and services." International Semantic Web Conference (2017). URL: https://www.metaphacts.com/images/PDFs/publicati ons/ISWC2017-Ephedra-SPARQL-federation-over-RDF-data-and-services.pdf.
9. Auer et al. (2007). *"DBpedia: A nucleus for a web of open data"* https://doi.org/10.1007/978-3-540-76298-0_52.
10. Lehmann et al. (2013). *"Integrating NLP using linked data"* https://doi.org/10.1007/978-3-642-41338-4_7.

---

[18] https://databus.dbpedia.org/amit/esa/links/.

Chapter 9
Data Pipelines: Modeling and Evaluation of Models

Kaïs Chaabouni and Alessandra Bagnato

Abstract This chapter outlines the utility of data pipelines modeling in the context of a data driven project and enumerates metrics for evaluating the quality of the data modeling regarding the readability and the comprehensibility of the models. We start with explaining the challenges surrounding the DataBio project that led to the adoption of data pipelines modeling using the Enterprise Architecture language ArchiMate. Then we present the data modeling process with examples from DataBio pilot studies starting with modeling software components provided by project stakeholders and ending up with integration of components into data pipelines that achieve the data analytics lifecycle intended by the pilot study. We end the chapter with the evaluation of the quality of DataBio data pipelines models with metrics collected by a monitoring tool for ArchiMate models.

9.1 Introduction

DataBio [1] aims to develop a platform that exploits the potential of big data technologies in the domains of agriculture, fishery and forestry. Given the complexity of the task, the project decided to adopt the "Enterprise Architecture" modelling language "ArchiMate 3.0" [2, 3] as a common modelling framework for representing the requirements of the pilots and modelling the technical architecture of the components, thus facilitating communication and comprehension among partners. Most of the software components interact with data from different origins and with various formats such as satellite imagery, sensors data, geospatial data (see Chap. 4), etc. In each pilot, components are connected together through several interfaces to form a data pipeline, (see Chap. 1) in which each component has a specific function in the

K. Chaabouni (✉) · A. Bagnato (✉)
Softeam Research Department, 21 Avenue Victor Hugo, 75016 Paris, France
e-mail: kais.chaabouni@softeam.fr

A. Bagnato
e-mail: alessandra.bagnato@softeam.fr

C. Södergård et al. (eds.), *Big Data in Bioeconomy*,
https://doi.org/10.1007/978-3-030-71069-9_9

113

data value chain such as data collecting, data processing, data analytics and visualization. The modelling approach consists of representing the components and the data pipelines according to a predefined model template. The modelling environment used for this task is "Modelio" [4], which allows contributors to collaborate around a synchronized ArchiMate model. The collaboration around the models faces some challenges regarding their potential to be efficiently exploited. Hence, we define metrics for evaluating the quality of the models and we measure continuously the quality level according to these metrics using a monitoring platform.

9.2   Modelling Data Pipelines

The Enterprise Architecture language ArchiMate provides several concepts for modelling the different layers of the enterprise:

- The physical layer contains the devices and their connections, which are used in the deployment of the IT system.
- The application layer contains the software services and the data flow.
- The business layer contains business services, interfaces and actors.

The modelling of software components enabled the DataBio partners working in the various pilots to easily understand the underlying functioning of each pilot. At first, partners were asked to provide models for the software components that they have provided. In a second time round, the partners were instructed to provide data pipelines diagrams that highlight the integration of the components in each pilot study. All of the software components, pipelines and datasets can be found at the DataBio Hub [5].

9.2.1   Modelling Software Components

The project developed a naming convention, where each software component has an identifier with the pattern "Cxx.yy" where "C" refers to the word "Component", "xx" represents the number of the partner that had provided the component and "yy" represents the component number of that partner. Datasets are correspondingly expressed with the template "Dxx.yy", For example, "C16.01" denotes the first component from partner 16, which is VTT .An expanded notation is "C16.01: OpenVA (VTT)", as the component is called OpenVA, which is a platform that consists of software modules that are used as building blocks of web based visualisation and analytics applications [6]. Components are modelled with diagrams that follow a predefined template. These diagrams include deployment view, interfaces view and subordinates view.

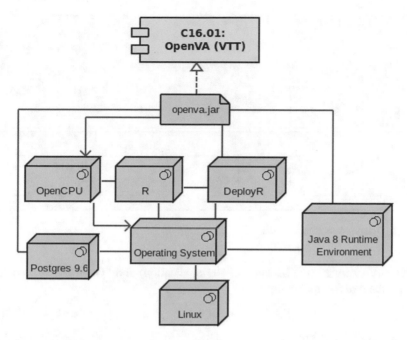

Fig. 9.1 OpenVA deployment view

9.2.1.1 Deployment View

The deployment view describes how the application is being deployed by representing the executables of the software component, the software dependencies and the physical environment required for running the application. Figure 9.1 shows an example of the deployment view of the component "C16.01: OpenVA (VTT)". As shown by the figure, OpenVA is packaged as JAR Java Package (openva.jar) which is run as a server via Java Runtime Environment (JRE). The database is handled by the Database Management System (DBMS) PostgreSQL 9.6. OpenVA server depends on two applications: DeployR and OpenCPU. DeployR is an open source application that turns R scripts into web services, so R code can be executed by applications running on a secure server. The OpenCPU server provides an HTTP API for data analysis for running R scripts on the server. OpenCPU uses standard R packaging to deploy server applications.

9.2.1.2 Subordinates View

The Subordinates view describes the subcomponents of the component such as the libraries, modules and frameworks that compose the whole application. For example, Fig. 9.2 shows the subcomponents of "C16.01: OpenVA (VTT)" which is composed

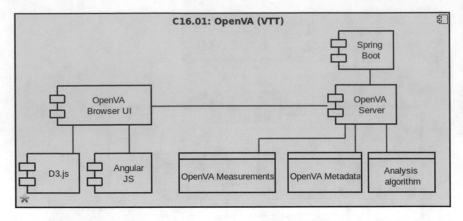

Fig. 9.2 OpenVA Subordinates view

of "OpenVA server" (the backend of the application) and "OpenVA Browser UI" (the frontend of the application).

9.2.1.3   Interfaces View

The interface view shows the provided and required interfaces of components which are designed for interactions with users or with other components through various communication protocols [7]. Figure 9.3 shows an example of the interface view of the component "C16.01: OpenVA (VTT)", which offers a web user interface for accessing OpenVA via a browser. OpenVA is also accessible via interfaces that can be provided by other components such as JDBC interface for accessing OpenVA database, Sqoop export tool for moving a set of files from HDFS (Hadoop Distributed File System) to RDBMS (Relational DataBase Management System).

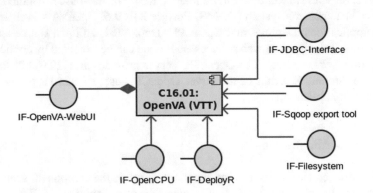

Fig. 9.3 OpenVA interface view

9.2.2 Integrating Components into Data Pipelines

Each pilot integrates in its workflow a set of software components that interact with each other in order to process huge amounts of heterogeneous data. These sets of interoperable software components are called pipelines and work as so-called white boxes showing the internal wiring and data flow between the single components of the pipeline. Hence, we model these pipelines with a "Pipeline View" that shows the different connections between components and a "LifeCycle View" that emphasizes the data value chain.

9.2.2.1 Pipeline View

Pipeline Views illustrate the connections between the different components and the interfaces that allow them to interact together. Figure 9.4 illustrates the Pipeline View of the fishery pilot "Oceanic tuna fisheries immediate operational choices" [8]. In this pilot, measurements from the ship engines are recorded continuously and are then uploaded to the ship owner server. These measurements are processed and analysed by three major components: "C16.01: OpenVA (VTT)", "C34.01: EXUS Analytics Framework (EXUS)" and "C19.01: Proton (IBM)". Each of these components offers a web interface for interacting with users and visualizing data via dashboards. "C19.01: Proton" is an event processing engine that processes events from different sources such as reading from files or from RESTful API. In this example, we receive sensor readings from the ship's monitoring and logging system which are

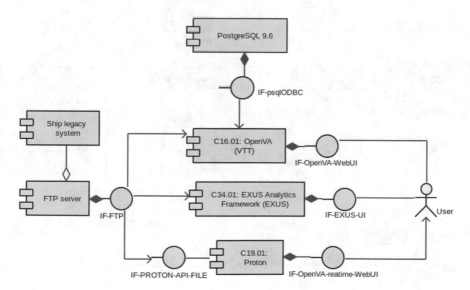

Fig. 9.4 "Oceanic tuna fisheries immediate operational choices" pilot—Pipeline view

then stored in the file system via FTP, from which it is read by Proton's file adapter and streamed into Proton engine for processing.

9.2.2.2   Lifecycle View

The lifecycle view shows the different tasks accomplished by each component along the data value chain according to the Big Data Value Reference Model [6, 9]. Figure 9.5 illustrates the Lifecycle View of the same fishery pilot as above "Oceanic tuna fisheries immediate operational choices". In this figure, we can see that the "Ship legacy system" is responsible for collecting raw sensor data. Then, custom tools and specific scripts are applied for data preparing (cleaning and transforming data) before executing the analytics tools. Finally the three major tools "C16.01: OpenVA (VTT)", "C34.01: EXUS Analytics Framework (EXUS)" and "C19.01: Proton" are used for data analytics and data visualisation.

9.3   Models Quality Metrics

The DataBio ArchiMate models are structured in five so-called projects: three projects for describing the pilots of agriculture, forestry and fishery, one project for modelling software and IoT system components and one project for modelling

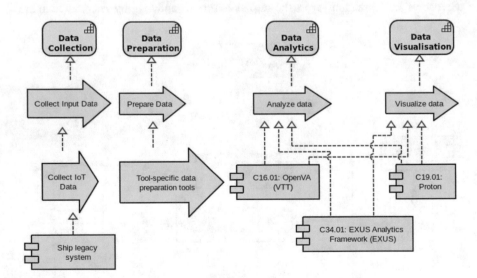

Fig. 9.5 "Oceanic tuna fisheries immediate operational choices" pilot—Lifecycle View

"Earth Observation" data services. These projects are monitored by "Measure Platform" [10], which is a monitoring platform that allows to collect periodic measurements on monitored projects. In this case, these measurements are obtained via the model indexing tool "Hawk" [11], which processes the queries of ArchiMate models. For each metric, we define a query for Hawk to interrogate from the models. After this, we store and visualize the collected measurement via the Measure Platform [12].

9.3.1   Metrics for the Quality of the Modelling with Modelio

Ensuring a better quality of the models begins with monitoring the modelling process with Modelio, which follows the creation of elements, folders, diagrams and documentation inside an ArchiMate project. We present here metrics that reflect how optimal the usage of Modelio is to guarantee a complete system design.

9.3.1.1   Percentage of Unused Elements in Diagrams

"Unused elements" are elements that have not been represented in diagrams and therefore do not bring any added value to the final generated diagrams. Each Modelio project contains a "Model Explorer" that is divided into two types of directories; one directory for managing the created elements and one directory for managing the diagrams. In the first directory we can visualize the list of all the elements that are created in the project whether they are displayed in diagrams or not. The second directory is for managing the diagrams that represent elements and their relationships. The "percentage of unused elements" metric could be an indicator of an incomplete modelling, where the element was created, but its relation with the rest of elements has not been yet specified. The unused elements could also be explained by the fact that users of Modelio sometimes create elements in diagrams and then mask them from the diagrams without deleting them from the project's Model Explorer. Moreover, this metric could also be an indicator of inefficiency, because it points to the incomplete work and to the wasted amount of work for creating useless elements. In addition, the unused elements will unnecessarily extend the list of displayed elements inside the Model Explorer, which would complicate the navigation for the user. Figure 9.6 shows that the percentage of unused elements in the monitored ArchiMate repositories in DataBio sub-projects has been between 20 and 50%.

9.3.1.2   Percentage of Duplicate Elements

The presence of duplicate elements in the models adds complexity for Modelio users as the redundancies complicate needlessly the visibility of the project and cause confusion, when choosing a suitable element. Moreover, the duplication of elements

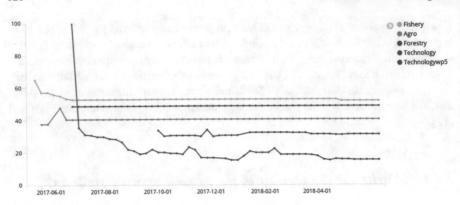

Fig. 9.6 Percentage of unused ArchiMate elements in diagrams

prevents the full exploitation of Modelio features such as identifying shared elements. Those elements are represented by several diagrams or by all the relations associated to the specified element.

9.3.1.3 Percentage of Empty Diagrams

The presence of empty diagrams is an indicator of unfinished or obsolete diagrams that need to be removed or updated.

9.3.1.4 Frequency of SVN Commits

The Modelio projects are stored as SVN (Subversion, an open-source version control system) repositories and can therefore be monitored by observing the frequency of updates of the models. The number of SVN commits per week shows the periods of time during which the work on models has been carried out. This metric does not reflect the real amount of the committed work, but rather the frequency of submitting new releases of the monitored models. Figure 9.7 shows the number of weekly SVN commits in DataBio Archimate Models. We can see from this figure that the major work on the models has been done between May 2017 and June 2018. Other SVN related measures could be conducted such as the number of contributors and the frequency of submitting new updates by each contributor.

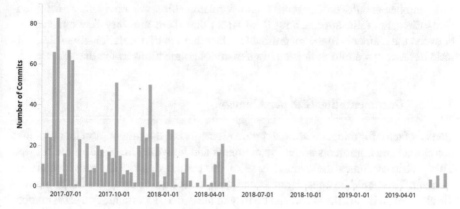

Fig. 9.7 Number of commits per week

9.3.2 ArchiMate Comprehensibility Metrics

The quality evaluation of ArchiMate views is based on several criteria that capture how well the views have fulfilled their purpose, especially their ability to help understand certain aspects in the project. Therefore, we introduce the comprehensibility metrics that evaluate how easy it is for the user to read the diagram and how easy it is to understand the model. The readability of the diagrams is impacted by how easy it is to read elements in diagrams, distinguish them from each other and find all the links between them. The understandability of the model from the provided diagrams depends on how easy it is to understand the whole organisation, the purpose of each component, service or process and the interactions between them.

9.3.2.1 Average Number of Elements per Diagram

The average number of elements per diagram shows how easy it is to read a diagram. Having a large number of elements in the same diagram will result in a dense diagram or in tiny elements inside the diagram, if it is scaled to a page or screen size. This makes it harder for users to read. On the other hand, having a very low number of elements per diagram could reflect a very fragmented model. We recommend between 8 and 25 elements per diagram, which is the case in the DataBio projects (see Fig. 9.8).

9.3.2.2 Average Number of Relationships per Element

The average number of relationships per element reflects the congestion of associations between elements and directly affects the readability of the diagram.

This number should be between 1 and 4 relationships per element. A Relationships/Elements ratio approaching 0 indicates that there are very few connections between the elements in the diagrams. On the other hand, a Relationships/Elements ratio exceeding 4 could indicate a big density of connections in the diagrams.

9.3.2.3 Documentation Size per Element

One key factor for understanding diagram elements is a documentation that provides definitions and comments about the elements and how they are used in the project. This metric evaluates the average size of the textual description provided for an element. This could be considered as an indicator of how detailed the description of the element is. Figure 9.9 shows the history of measured documentation size (number of words) per element in the monitored projects, which have an acceptable average size. However, this measure does not show the disparity of documentation, where

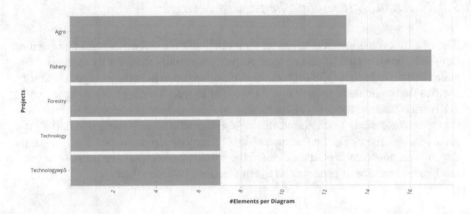

Fig. 9.8 Average number of elements per diagram

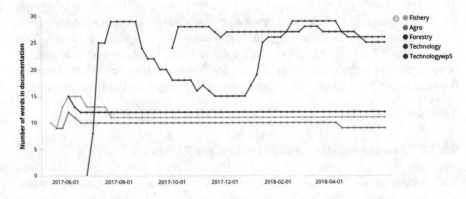

Fig. 9.9 Documentation size per element

some elements are described with big paragraphs and others have no description at all.

9.3.2.4 Documentation Size per Diagram

This metric evaluates the understandability of diagrams by measuring the documentation size diagram. It is similar to the previous one with the difference that it calculates the documentation size per diagram instead of the documentation size per element. This allows us to locate in more detail the diagrams that are lacking description.

9.3.2.5 Percentage of Documented Elements

This metric focuses on the documented part of the models. It measures the percentage of the documented elements. Apart from the self-evident elements, which are understandable just by name, it is highly recommended to describe the remaining elements, especially the elements containing abbreviations, which are not well known to everyone. Figure 9.10 shows the percentage of the documented elements in the monitored projects. The projects, which describe the agro, fishery and forestry pilots, have few documented elements (between 15% and 24%). This is explained by the clear and detailed namings of the motivation and strategy elements, which therefore do not require further explanations. On the other hand, the technology projects deal with a lot of technological components that require documentation. Hence, the documented elements represent more than 58% of the total elements in these projects.

9.3.3 Metrics for Model's Size

The model size is an indicator of the modelling progress as it reflects the number of created diagrams and elements and their relationships inside diagrams. The model size is also an indicator of the complexity of the model. The number of non-empty diagrams reflect the actual number of the models existing in the studied organisation. In our case, the ArchiMate models contain more than 500 non-empty diagrams. This makes it more complex to understand the whole project.

Fig. 9.10 Percentage of documented elements per project

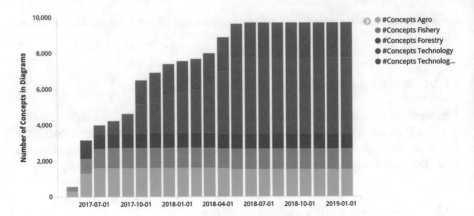

Fig. 9.11 Number of concepts represented in diagrams

9.3.3.1   Total Number of ArchiMate Concepts Used in Diagrams

Since diagrams differ in size, the number of overall ArchiMate concepts used in diagrams add information about the size of the models. The ArchiMate concepts considered here contain the elements represented in the diagrams and the relationships between the elements. Figure 9.11 shows the evolution of the total number of ArchiMate concepts and the proportion of concepts in each DataBio project. We can see that the total number of ArchiMate concepts is very close to 10000 elements, which is an indicator of the complexity of the project.

9.4   Conclusion and Future Vision

The modelling of DataBio components and data pipelines provided more clarity to the project and helped to understand the architecture of the used software components and their integration in the pilots workflows. Moreover, the created models have also contributed to the process of requirements elicitation throughout the project period and to the efficient writing of the documentation. In order to monitor the quality of the models, we have defined a metric that evaluates the efficiency of the modelling process, the comprehensibility of the models and the model size. The metric discussed here could be applied also in other projects, where the modelling tool Modelio or the modelling language ArchiMate are in use [12]. The proposed metric indicates that the quality level in DataBio is acceptable as comes to the efficiency of the modelling process and the comprehensibility of the models. However, we note that there are some areas to be improved such as the cohesion and the completeness of the models. The analysis showed that the models are lacking a more holistic view of the DataBio project, where there is a big data platform or environment offering services and components to the different pilots. Hence, we aim at finding more metrics for evaluating the cohesion of the models and expressing the interdependency between elements and diagrams inside the project. Moreover, our analysis showed that there are many incomplete and undetailed diagrams and we need therefore a metric that expresses the completeness and the maturity of the diagrams.

Acknowledgements The authors would like to thank Antonio Garcia-Dominguez from School of Engineering and Applied Science at Aston University, Birmingham, UK for his support with the Hawk tool. We would like also to acknowledge the effort of ITEA3 project 14009, MEASURE Project and thank their consortium team members for their feedback on this work. Furthermore, the research leading to these results has received funding from the European Union Horizon 2020 research and innovation program under grant agreement No. 732064, Databio Project.

References

1. DataBio Website. (2019). https://www.databio.eu/en/. Last accessed May 14, 2019.
2. Desfray, P., & Gilbert R. (2018). TOGAF, Archimate, UML et BPMN-3e éd. Dunod.
3. Fritscher, B., & Pigneur, Y. (2011). Business IT alignment from business model to enterprise architecture. In *International conference on advanced information systems engineering* (pp. 4–15). Springer.
4. Modeliosoft—Modelio BA Archimate Enterprise Architect. (2019). https://www.modeliosoft.com/en/products/modelio-ba-archimate-enterprise-architect.html (2019/04/18).
5. DataBioHub Website. (2019). https://www.databiohub.eu/. Last accessed May 14, 2019.
6. DataBio public deliverable D4.1 Platform and Interfaces. (2019). https://www.databio.eu/wp-content/uploads/2017/05/DataBio_D4.1-Platform-and-Interfaces_v1.0_2018-05-31_VTT.pdf. Last accessed Aug 13, 2019.
7. Chaabouni, K., Bagnato, A., Walderhaug, S., & Berre, A. J. Södergård, C., Sadovykh, A. (2019). *Enterprise architecture modelling with ArchiMate* (pp. 79–84). STAF (Co-Located Events).
8. DataBio public deliverable D4.2 Services for Tests (Public version). (2019). https://www.databio.eu/wp-content/uploads/2017/05/DataBio_D4.2-Services-for-Tests_public-version.pdf. Last accessed Aug 14, 2019.
9. BDVA Strategic Research and Innovation Agenda, version 4.0, October 2017. (2019). http://www.bdva.eu/sites/default/files/BDVA_SRIA_v4_Ed1.1.pdf. Last accessed Aug 14, 2019.
10. Measure Platform. (2019). http://measure-platform.org/. Last accessed May 21, 2019.
11. Hawk Tool. (2019). https://github.com/mondo-project/mondo-hawk. Last accessed May 21, 2019.
12. Chaabouni, K., Bagnato, A. (2019). *Antonio García-Domínguez: monitoring archimate models for databio project* (pp 583–589). PROFES.

Part IV
Analytics and Visualization

Chapter 10
Data Analytics and Machine Learning

Paula Järvinen, Pekka Siltanen, and Amit Kirschenbaum

Abstract In this chapter we give an introduction to data analytics and machine learning technologies, as well as some examples of technologies used in the DataBio project. We start with a short intdroduction of basic concepts. We then describe how data analytics and machine learning markets have evolved. Next, we describe some basic technologies in the area. Finally, we describe how data analytics and machine learning were used in selected pilot cases of the DataBio project.

10.1 Introduction

The goal of data analytics is to examine large quantities of data with the purpose of drawing conclusions about the data. Several techniques can be employed, each using similar methods but having a slightly different focus. The methods include, e.g., statistics, data mining, and machine learning (Fig. 10.1).

Data mining is defined as "a science of extracting useful information from large data sets or databases" [1]. Machine learning is "programming computers to optimize a performance criterion using example data or past experience" [2]. Sometimes the division between machine learning and data mining is done based on data sets. Data mining is focused on analyzing large databases, whereas in machine learning the focus is on learning patterns from data. The roots of data analysis are in statistics. The development of computers and their ability to store and manage large amounts of data has made possible large-scale statistical computation and has launched the development of new methods that would be tedious to perform manually.

A recent area of data analysis is visual data mining. Information visualization, data mining, and user interaction have evolved as separate fields in the past, but since the turn of the 2000s have become increasingly integrated as visual data mining.

P. Järvinen · P. Siltanen (✉)
VTT Technical Research Centre of Finland Ltd., Espoo, Finland
e-mail: Pekka.Siltanen@vtt.fi

A. Kirschenbaum
Institute for Applied Informatics (InfAI), University of Leipzig, Goerdelerring 9, 04109 Leipzig, Germany

© The Author(s) 2021
C. Södergård et al. (eds.), *Big Data in Bioeconomy*,
https://doi.org/10.1007/978-3-030-71069-9_10

Fig. 10.1 Data analysis
techniques [1]

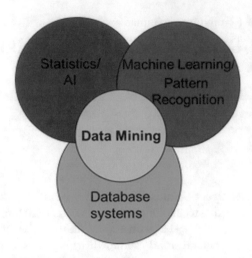

The idea of visual data mining first emerged in 1999 when Wong [3] argued that rather than using visual data exploration and analytical mining algorithms as separate tools, a stronger data mining strategy would be to couple the visualizations and analytical processes into one data mining tool. Many data mining techniques involve mathematical steps that require user intervention, and visualization could support these processes. Visual data mining is not just about using visualization to exploiting data, it is an analytical mining process in which visualizations play a major role [4].

Artificial intelligence (AI) can be defined as "a system's ability to correctly interpret external data, to learn from such data, and to use those learnings to achieve specific goals and tasks through flexible adaptation [5]."

Machine learning has been used since the 1950s by researchers in order to analyze and extract information from data. It has only been during the last decade with the rise of the generalized usage of the graphics processing units (GPUs) that enabled the true development of neural networks and in particular what is nowadays referred to as deep learning [6]. This newly found computational power gave rise to methods that are capable of solving complex, real-world problems. The capacity of modern computers not only allows for computationally intensive methods, but also facilitates the analysis of huge amounts of data, the so-called big data, in a scale that was previously intractable. In contrast to previous methods, deep learning uses multiple layers of neural networks to build architectures capable of performing a specific task, such as classification, segmentation, detection, prediction, and generation of data.

Deep learning is capable of discovering correlations in the data without the need of handcrafted features. The lack of heuristics together with the abundance of computational resources makes deep learning methods ideally suited for handling big data problems. Further to that, machine learning offers the possibility for lifelong learning where the system is capable of adapting to changing conditions. While machine learning is often portrayed as a replacement for human intelligence, it is only a tool for digitalizing human expertise into a computer model. This model is only as good as the information humans supplied it with.

10.2   Market

Data analysis has been studied intensively, and numerous algorithms exist. It has applications in different business, science, and social science domains. A wide range of tools and commercial applications is available, some of which are highly competitive in markets, such as customer relationship management (CRM). There are also several statistics programs and packages available, both for casual users and specialists (Excel, SAS, SPS, R).

Big data analysis solutions can be classified into two categories: "Data Discovery and Visualization" and "Advanced Analytics" [7]). Data discovery and visualization solutions integrate and transform big data sources using data mining algorithms to find insights into business use. Advanced analytics solutions are focused on building use case-specific predictive or descriptive solutions using advanced modeling techniques, such as deep learning or advanced statistical methods.

Frost and Sullivan estimate big data revenue at 2017 of $8.54 billion [7]. The revenue is expected to reach $40.65 billion in 2023. The market is expected to grow at a steady rate, as data discovery and visualization are expected to become more mainstream over this period and advanced analytics is expected to see more real-life use cases [7]. North America is expected to continue to be the largest market contributor, followed by Western Europe, having similar growth path.

Biggest user of data analytics techniques is business and finance, followed by governance and integrity (public sector), both over 15% of the market. In Frost and Sullivan estimations, bio-economy falls into the category of "Others," which in total covers 7.7% of the market.

According to the Zion Market Research [8], global machine learning market was valued at around USD 1.58 billion in 2017 and is expected to reach approximately USD 20.83 billion in 2024, growing at a compound annual growth rate (CAGR) of 44.06% between 2017 and 2024. Artificial intelligence experts have projected their idea that by 2050 all the intellectual tasks performed by the humans can be accomplished by the artificial intelligence technology. Some of the top applications of machine learning are financial services, virtual personal assistants, health care, government, marketing and sales, transportation, oil and gas, manufacturing, bioinformatics, computational anatomy, and more. The artificial intelligence (AI) market in agriculture is expected to register a CAGR of over 21.52%, during the forecast period of 2019–2024, offering services for the management of the crops yield, species breeding, disease detection.

Geographically, machine learning market is segmented into North America, Asia Pacific, Europe, Latin America, and Middle East and Africa. North America is predicted to govern the market in forecast period because of developed countries and their major focus on innovative technologies obtained from R&D sector. Asia-Pacific region is predicted to grow at the highest CAGR in forecast period due to increasing awareness regarding business productivity. In Asia, region vendors are offering competent machine learning proficiency due to which it is the highest potential region for the market. Moreover in Europe, the world-class research facilities, the emerging start-up culture, and the innovation and commercialization of machine intelligence technologies are stimulating the machine intelligence market. Among all regions, Europe has the largest share of intra-regional data flow. This, together with the machine learning technologies, is boosting the market in Europe.

10.3 Technology

10.3.1 Data Analysis Process

Data analysis is an iterative process starting with selecting the target data from the raw material and preprocessing and transforming it into a suitable form (Fig. 10.2). Data analysis uses several data types: database records, matrix data, documents, graphs, links, transaction data, transaction sequences, DNA sequence data, whole genome information, and spatiotemporal data. The quality of data may often cause problems. The data can contain noise, there may be missing values and duplicate data, and thus

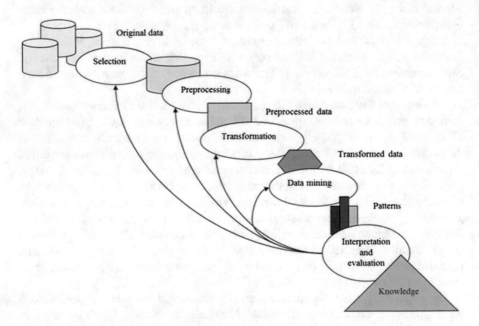

Fig. 10.2 Data mining process. Figure adapted from [1]

data cleaning phase is required before using the data. Other kinds of preprocessing may also be required, such as data aggregation, sampling, dimensionality reduction, subset selection, feature creation, and attribute transformation [1].

Next, the data is run through a data mining algorithm that creates patterns from the data. The user interprets and evaluates the results and starts a new iteration with possible modifications to the raw data, algorithm, and algorithm parameters.

10.3.2   Statistical Methods

Statistical methods are used for data exploration to gain a better understanding of the characteristics of data [1]. The central methods include, e.g., summary statistics, correlations, and visualizations. Summary statistics are numbers that summarize properties of the data. Amar et al. [9] have classified the statistical methods as

(1) computer-derived values: average, median, count, more complex values,
(2) finding extremum: finding data cases having the highest and lowest value of a defined attribute,
(3) determining range: finding a span of values of an attribute of data cases, and
(4) characterizing distributions: creating a distribution of a set of data cases with a quantitative attribute, e.g., to understand "normality." The visual methods utilize humans' ability to recognize patterns. Single variables are expressed in visual form, for instance as histograms and line charts.

Correlation is a basic statistical method of studying two variables. The prevailing method is the calculation of the Pearson correlation coefficient (r), where the correlation between two variables, x_i and y_i is calculated with the formula:

$$r = \sum_{i-0}^{n} \frac{(x_i - x)(y_i - y)}{n S_x S_y}$$

where n is the number of observation pairs, and S_x, S_y are the standard deviations, and x and y the means of the variables x_i and y_i. The correlation produces positive or negative values within the range -1 to 1. If the result is zero, there is no correlation between the variables. Values -1 and 1 indicate complete linear dependence between the variables, either negative or positive. Often the square of the correlation coefficient R^2 (also known as the coefficient of determination) is calculated. This value ranges from 0 to 1 and indicates how much one variable explains the variance of the other and is often expressed as a percentage. For instance, if R^2 is 0.32, 32% of the variance of a variable is explained by the other.

Correlations are visualized in the form of scatterplots. Exploration methods for higher dimensions use projections of data on a two-dimensional plane. These are called dimension reduction methods. They include principal component analysis

(PCA) and multidimensional scaling, as well as auto-encoders for neural networks. The result of PCA can be visualized as a two-dimensional plot.

10.3.3 Data mining

The goal of data mining is to extract useful information from large data sets [10]. Data mining can be categorized into different kind of tasks, corresponding the objectives of analysis: exploratory data analysis, descriptive modeling, predictive modeling, and discovering patterns and rules.

Exploratory data analysis (EDA) explores data without clear ideas of the findings. Visualization is effective EDA techniques, especially with relatively small and low-dimensional data sets. Bar charts, boxplots, histograms, and density plots are applicable with single variable data, scatterplots with two variable data. With multidimensional data, dimension reduction methods, such as principal component analysis, (PCA) are used. They produce informative low-dimensional projections of data that can be visualized in two-dimensional space.

The goal of descriptive methods is to describe the data. The methods include density estimation, clustering and segmentation, and models describing the relationships between variables. Clustering looks for groups of objects such that the objects in a group are similar (or related) to one another and different from (or unrelated to) the objects in other groups. The similarity of objects is defined based on similarity (or distance) measures. Euclidean distance can be used if attributes are continuous; otherwise, problem-specific measures are needed. Clustering has been an active research topic, and lots of algorithms are available. Algorithms include K-means clustering and its variants, hierarchical clustering, agglomerative clustering, and density-based clustering. Market segmentation is an application of clustering.

The purpose of predictive modeling is to build models that predict the value of one variable from the known values of other variables [10]. The predicted objects are predefined. Regression and classification are two much used predictive methods. Regression predicts a value of a continuous variable based on other variables using linear or nonlinear models [1]. Linear regression is easy to visualize, often shown as a line on a scatterplot diagram. The area is studied extensively and has its origins in statistics. It has various uses, both in commerce and science. Application examples include predicting sales based on advertising expenditure, stock markets, or wind as a function of temperature or humidity. Classification creates a model for a class attribute as a function of the values of other attributes. Unseen records are then assigned to the class. Models in both methods are developed with a learning data set, and the precision and accuracy of the models are evaluated with a test set. Several techniques have been developed including decision trees, Bayesian methods, rule-based classifiers, and neural networks. Classification is a much used method, and commercial applications are also available. Examples include classification of credit card transactions as legitimate or fraudulent, classification of e-mails as spam, or classification of news stories as finance, weather, entertainment, or sports [1].

Discovering patterns and rules involves finding combinations of items that occur frequently in databases. Sequential pattern discovery finds rules that predict strong sequential dependencies among different events. Association rule mining involves the prediction of occurrences of an item based on occurrences of other items. It produces dependency rules such as "buyers of milk and diapers are likely to buy beer." One special case of pattern discovery is anomaly detection. Anomalies are observations whose characteristics differ significantly from the normal profile. Methods of anomaly detection look for sets of data points that are considerably different from the remainder of the data. The methods build a profile of "normal" behavior and detect significant deviations from it. The profile can be patterns or summary statistics for the overall population. Types of anomaly detection schemes can be graphical-based, statistical-based, distance-based, or model-based. Credit card fraud detection, telecommunication fraud detection, network intrusion detection, and fault detection are examples of application areas [1].

10.3.4   Machine Learning

In machine learning, the idea is to learn things from data. The approach is to create mathematical models and adjust model parameters with the help of data until the model matches best the modeled phenomena. Machine learning utilizes theories from statistics combined with computer algorithms [2]. It has a strong overlap with data mining. Machine learning is focused on learning patterns from data whereas in data mining focus is on analyzing large databases. Machine learning methods can be divided into unsupervised and supervised learning. In unsupervised learning, there is only input data available, and the aim is to find patterns in data. In supervised learning, there is prior knowledge of the phenomena available in addition to the input data. Clustering belongs to unsupervised methods, whereas classification, regression, and bayesian methods are supervised. Another division is parametric and nonparametric methods. The parametric methods assume that the data is drawn from some probability distribution known before, and the model is created by estimating model parameters from data. Regression and classification methods are parametric methods. The nonparametric methods do not make such assumptions of the data but are based on finding similarities. They divide the input space into local regions, defined by a distance measure. Decision trees belong to nonparametric methods.

As in data mining, model validation is an important issue in machine learning. Input data is divided into learning part and validation part. The model is developed with the learning part and validated with validation part. Measures of the validity are model accuracy and precision.

Neural networks are a specific set of algorithms inspired by biological neural networks. The current deep neural networks (deep learning) work well in problems such as computer vision, speech recognition, and natural language processing. Currently, there are many available open-source frameworks, TensorFlow, PyTorch, Cafe, etc., that can be used for developing neural network models. These include highly optimized code that can be used for both training and using a model and thus greatly simplify the development process. Architectures for building models for specific tasks get published constantly in conferences and journals very often in an open manner. This has given rise to a variety of applications escaping the confines of academic research and reaching directly the market.

10.4 Experiences in DataBio

10.4.1 Data Analytics in Agriculture

10.4.1.1 Classification of Land Covering

This section describes the use of deep learning techniques for Earth observation data in the agriculture pilots in Part V of this book. The ongoing advancements in deep learning, and exemplary results obtained for different problems using spatiotemporal satellite images, have made deep neural networks quite popular for analyzing Earth observation data. The aim of the pilot was to design a pipeline based on deep neural networks to classify land cover using available satellite images from Sentinel-2A satellite. Initially, an investigation was done using only images and not taking advantage of the temporal nature of the signal. The results of this approach were not satisfying as the spatial information was not sufficient to differentiate crops with an adequate accuracy. For this reason, a new pipeline based on spatiotemporal data was designed. The new pipeline consisted of two steps: clean available training data, and then use this cleaned data for training crop classifiers. For the first step, instead of using traditional methods (based on data specific heuristics and handcrafted filters) to clean data, an RNN-based auto-encoder was trained to remove unreliable data. The encoder and decoder consisted of recurrent neural network (RNN) layers with long short-term memory (LSTM) cells [11]. The encoder learns the representations in latent space from the time series of pixels in crop parcels, while the decoder tries to reconstruct the time series. The representations are clustered in the latent space using K-means clustering. It is expected that most of the pixels will form one huge cluster while the outlier pixels will be away from this cluster. In this way, the parcels with clean pixels are selected and further used for training a pixel-level classifier network (inspired from [12] and [13]) for individual crops. Instead of training a neural network from scratch, the encoder part of the auto-encoder is used as initial layers of the classifier network. The pre-trained encoder network is appended with a dense layer and fine-tuned for the classification task. The classifier network produced a probability of being a particular crop for each input pixel. The details of the training for complete pipeline and obtained results can be found in [14].

The classifiers are trained for wheat, maize, and legumes for the data from regions in Greece, provided by NEUROPUBLIC for year 2016. Further, the classifiers are integrated in the DataBio online platform developed by Fraunhofer, where the probability of each pixel in the selected parcel belonging to a certain crop type can be obtained. This technology allows the user to identify the crop grown in a given area by using corresponding satellite imagery.

The presented pipeline shows the significance of data verification and provides an efficient way to create models by optimizing the efforts and the time of both engineers and experts. The data cleaning step done in an unsupervised manner increases the reliability of the data. An expert can further verify and refine these data groups by verifying only the boundary cases in the cleaning step. In this manner, the effort of the expert is optimized by focusing on targeted areas.

Additionally, the cleaning and classification done using time series of pixels (instead of parcels) are advantageous to us due to the following reasons:

- Lack of availability of fully labeled satellite images
- Due to the complexity of drawing parcel boundaries in low resolution satellite images, the pixel-level cleaning allows us to remove pixels corresponding to nearby road, lakes, etc., from the crop parcels.
- Instead of using image patches, the use of time series of individual pixel values for classification avoids influence of nearby pixels.
- The classification obtained at a pixel level enables sub-parcel level analysis which is very helpful in applications like damage assessment.

Although the presented pipeline performed well for the available data set, the results may not be as good for the following cases:

- The auto-encoder and the classifier both assume the variation of time series with in a crop type is low. In case of huge variation, we may need to subdivide the crop type for this approach to work.
- The training data corresponding to selected region in Greece creates the model depending on the temporal behavior of crops in that locality. This model may not work for the same crop having significantly different behavior in different regions across the world.
- The current model may not work well for data from other years as it may have a bias toward year 2016.

Lessons Learned

- Classification of crop using spatial data only does not have adequate performance and important information is in the temporal dimension.
- Some crops have similar varieties and can be covered with a common model. There are crops though whose varieties are very different, this approach would probably not succeed, and separate models for each sub-variety would be required.
- To develop models that can work for multiple years, implying different weather conditions throughout the season, would require further work on combining data measured on different dates each year. Similarly, multi-regional and global models would require much more data, as they would need to abstract the variation caused not only from local climates but also from a variety of different soils.

While this solution can benefit from further developments, it has the potential to form the baseline for methods targeting global scale satellite image analysis. The proposed approach for detection and classification of vegetation types operates on the sub-parcel level and is robust to noise in both the data and the labels.

10.4.1.2   Crop Detection and Monitoring

The free and open availability of Earth observation data is bringing land monitoring to a completely new level, offering a wide range of opportunities, particularly suited for agricultural purposes, from local to regional and global scale, in order to enhance the implementation of Common Agricultural Policy (CAP).

Terrasigna proposes an in-house developed fuzzy-based technique for crop detection and monitoring in Romania, based on combined free and open Sentinel-2 and Landsat-8 Earth observation data. The general methodology is based on the comparison between real crop behavior and the expected trends for each crop typology. It involves image processing, data mining, and machine learning techniques and is based on different categories of input data: Sentinel-2 and Landsat-8 SITS covering the time period of interest, farmers' declarations of intention with respect to crops types, as well as in situ/field data.

The machine learning technique used is an original one, developed taking into account the particularities of the CAP-monitoring process. The fuzzy approach allowed the use of all available scenes, provided they were not completely contaminated with clouds and shadows. The mixed time series, consisting of S2 and L8 scenes, are accompanied by relevance masks, which act as weights in the final fuzzy extraction process (i.e., drawing a firm conclusion using a series of vague and incomplete information). The strictly statistical character of the algorithm, which does not use phenology information or the intervention of a specialist with agronomic competences, makes the technique universal, being able to adapt to other regions and types of cultures, without difficulty.

The processing chain involves a series of well-defined steps:

- Image preprocessing (numerical enhancements for Sentinel-2 and Landsat-8 scenes, ingestion of external data, and clouds and shadows masking);
- Individual scene classification;
- Deriving crop probability maps at scene level;
- In the end, time series analysis allows the generation of overall crop probability maps and derived products.

The main goal of the approach within the DataBio project framework was to provide services in support to the National and Local Paying Agencies and the authorized collection offices for a more accurate and complete farm compliance evaluation—control of the farmers' declarations related to the obligation introduced by the current Common Agriculture Policy (CAP). The system produced three main types of results, all provided at a 10-meters spatial resolution as follows:

- Crop mask maps, which are pixel level maps, identifying some of the most important crop types;
- Parcel use maps, which are object-based maps, showing the most probable type of crop at plot level;
- Crop inadvertencies maps, which can be both pixel-based and object-based maps, revealing the areas for which the declared type of crop included in the LPIS appears to be different from the identified one. The pixel-based analysis states whether pixel values correspond or are different from typical spectral values of the declared crop types, whereas the object-based analysis reveals the plots for which the declared type of crop appears to be different from the one identified based on satellite imagery, based on a specific threshold.

Lessons Learned

The technology developed by Terrasigna is able to recognize a large number of crops families, of the order of tens. For Romania, it addressed the first most cultivated 32 crops families, which together cover more than 97% of the agricultural land. In 2018, the validation of results for a full agricultural season (full phonological cycle) against independent sources revealed promising results, with an accuracy higher than 95% for more than 10 crop types. The performance is quite uniform reported to parcels size and remains high even for parcels smaller than 1 ha. The highly automated proposed approach allows the performing of big data analytics to various crop indicators, being reliable, cost-, and time-saving. It leads to a more complete and efficient management of EU subsidies, strongly enhancing their procedure for combating non-compliant behaviors.

The most serious problems that had to be solved and that served as lessons were as follows:

- The use of data S2 and L8 together—which have a different format and resolution;
- Correction of the geographical positioning (georeferencing) automatically— which deeply affects the quality of the classification for small or narrow plots;
- Selecting the areas of interest from each image—which are not, as it might seem, the areas uncontaminated by clouds and shadows, but the areas where there is vegetal "activity";
- The construction of an algorithm that takes into account the matrix of semantic confusion between cultures—which required finding the natural classes of cultures that can be followed simultaneously, without serious mutual confusion.

Geospatial services together with Copernicus data can provide a really powerful tool for monitoring agricultural dynamics. The end users, the National Paying Agencies, are able to benefit from the modern and effective near real-time service, based on the principles of sustainable agriculture and saving effort both in terms of costs and time. A continuous agricultural monitoring service based on the processing and analysis of Copernicus satellite imagery time series is not just a CAP compliance tool, but can also offer a great range of supplementary information for both public authorities and citizens.

The developed technique is replicable at any scale level and can be implemented for any other area of interest.

10.4.1.3 Farm Weather Insurance Assessment

Trying to identify the parameters (weather or soil related) with the dominant impact on the crop yield such as normalized difference vegetation index (NDVI) measurements, the following approach is considered. For the first phase of this analysis k-prototypes, clustering algorithm was applied for the profile building of the parcels. Using satellite, meteorological measurements and soil characteristics are aggregated on the level of one or two months considering a full growing season. The k-prototypes algorithm is based on the k-means paradigm but removes the numeric data limitation while preserving its efficiency [15]. After this phase of analysis, each one of the parcel linear regression models [16] is trained considering only the data that belongs to this cluster. In that way after the clustering procedure, we can use historical data of a parcel in order to identify in which cluster it belongs and make predictions for the NDVI values of an upcoming period using the corresponding linear regression model.

Lessons Learned

The main challenge of this approach is that the clustering analysis cannot work with missing values, so each one of the parcels is required measurements for the same months, otherwise the parcel must be excluded from the analysis. Another challenge is the sparsity of satellite data due to weather issues (e.g., cloudy days) making it difficult to create a "complete" or usable by the machine learning algorithms data set in terms of meteorological, satellite, and soil information for the same dates. In order to deal with that issue, interpolation and aggregation of the data were applied.

10.4.1.4 Crop Disease Detection Using Satellite Images

Automated crop monitoring is an essential aspect of smart agriculture, as it allows to improve yield estimation while reducing costs and environmental imprint. We conducted a study to forecast diseases in sorghum using remote sensing via satellite imagery as a proxy for crop health. Our method uses images from Sentinel-2 satellites, which regularly provide multispectral images for land monitoring. Images of a sorghum field with infected parts taken under different weather conditions, as captured by Sentinel-2 satellites, served as our training data [17]. We use the observation that there is a strong correlation between the physiological status of a plant and its chlorophyll content, i.e., diseases have a negative influence on the chlorophyll level [18] and derive NDVI from the recorded satellite images. NDVI is an indicator for vegetation vitality which measures the difference between near-infrared light that vegetation strongly reflects, and red, which vegetation absorbs. Healthy plants, that is, with a higher level of chlorophyll, reflect more near-infrared and green light compared to other wavelengths and absorb more red light.

Lessons Learned

Accurate data on disease outbreaks in the agricultural sector is usually not publicly available, e.g., due to data protection. This posed us the challenge of a small training set, which may lead to overfitting, a general problem in training machine learning methods. To overcome this, we perform data augmentation, i.e., artificially expand the training data set, to improve the ability of the learned model to generalize. Data augmentation is performed by small changes in data, in this case—image manipulation. Such operators include rotations, reflections, random excerpts, image zooming, or combinations thereof [19].

Mask region-based convolutional network (R-CNN) [20] is then used to train a model that determines which areas are infected. Mask R-CNN is a convolutional neural network that performs instance segmentation, i.e., identifies outlines of objects on a pixel level. In our case, the segmentation would be according to the NDVI values. This method showed great potential for the task at hand, and the model achieved mean average precision very close to 1.

10.4.2   Data Analytics in Fishery

10.4.2.1   Reducing Energy Consumption of Vessels in the Fishery Domain

This study aims at reducing the ecological and economical costs of fishery vessels, by optimizing their route and speed and thus decreasing fuel oil consumption. This process requires analysis of many observations collected over time. We collected thousands of observations per day from two boats for three years, where each observation involves dozens of features, e.g., speed and angle of wind, engine load, speed of the vessel, and, of course, and fuel consumption. The first step was creating predictive models for consumption of fuel oil per nautical mile. To this end, we compared two modeling techniques: the extreme gradient boosting framework XGBoost [21] and polynomial regression [22] and opted for the latter as it provides better results. We then explore two use cases: One considers calculating an optimal route (TSP) connecting several points; and the other, selecting a single sailing destination which optimizes the energy consumption. In both use cases, varying travel distances and weather conditions were taken into account. The locations are assumed to be known and in GPS format. The weather conditions are extracted in near real time from the Sentinel-3 mission API [23].

To determine the optimal speed and corresponding fuel oil consumption, we employ a gradient descent algorithm for each possible route segment. The algorithm uses wind data, speed exploration values, as well as some control variables to estimate internal machinery values, which in turn are employed to estimate the consumption, and the minimal value gets selected. By applying this optimization method, a reduction of about 3% of the fuel oil consumption was obtained.

Lesson Learned

A general lesson learned in this study is the importance of data preparation to control input data quality. An observation considering the reliability of different sensors which varied across the ships that lead to many outliers, which negatively impacted the model accuracy, and the creation of a unified model. In addition, the timelines of wind data provided by the Sentinel mission posed limits on the methods, as they were provided once or twice a day, depending on the region of interest, and accurate forecast of wind for a period of over six hours turns out to be better by classic meteorological methods than by statistical approach.

10.4.2.2   Analyzing Historical Measurement Data

Different data analysis methods were used, e.g., in a fishery pilot (see Part VII of this book) where measurements from several fishing ship motors were analyzed using VTT OpenVA application. The main goal was to analyze ship fuel consumption, but since we imported all the measurements available, the system can be used to analyze other variables as well. VTT OpenVA is an advanced analytics solution that was tailored to create an application where a data scientist can select different measurements and get different visualizations based on the user selection. In a DataBio fishery pilot case, there are 115 measurements from the motors four different ships that were analyzed. Measurements were stored in a 10s interval from a four-year period. More three billion measurement values were stored in a standard relational database (PostgreSQL, https://www.postgresql.org/).

An analysis application using the data was implemented. Users of the application can select measurements from different ships on the selected time period, and the VTT OpenVA proposes available analysis methods, based on the measurement type. In the pilot, 18 different analysis results were shown as visualizations, and 55 single value performance indicators were calculated.

Lessons Learned

VTT OpenVA is designed to be an interactive application, but DataBio experiences show that when the amount of measurements becomes large—billions of measurements—it is hard to achieve real-time interaction, because database query response times grow to at least several seconds. This could be mitigated by more powerful database servers and specialized commercial databases, but the goal of the pilot was to use standard servers and databases, allowing easy transfer of the system to the server maintained by the system users.

To make queries faster, VTT OpenVA automatically distributes measurements into a large number of database tables instead of making queries to one huge table. Naturally, the way users use this kind of big data analysis tool, e.g., in the fishery pilot the normal time period of the data that is in practice analyzed is about three weeks, which is the average time that a fishing ship is out from harbor. Even though querying the data takes some time, actual time taken to analyze and visualize these relative small data sets is quite short.

10.4.2.3 Oceanic Tuna Fisheries Immediate Operational Choices

Exus analytics framework was integrated in the pipeline of pilot fishery to predict main engine performance and faults in advance. For the prediction of the main engine performance, a neural network was used to perform multivariate regression in order to estimate a regression model for multiple variables taking as input a considerably lower number of values. The main benefits of the neural networks are their ability to capture complex relationships between the inputs and their requirement of high number of data. The choice of this machine learning algorithm was also based on related work for fault diagnosis in engines used in vessels [24, 25].

Based on historical vessel data sets, a preprocessing of two stages is applied. First only data that corresponds to the steady state of the engine is considered. After extracting the steady-state engine data, the min-max normalization is applied for all features.

Various architectures for the number of hidden layers and units have been tested, and for the best model selection, the data set has been split into training and validation sets. The model that performed the lowest validation error is selected as the best one.

For the prediction of engine faults, the predicted variables (based on historical data) are compared to the actual values of the vessel measurements. When the variance of these differences is higher than a threshold, it is considered as an engine fault.

Lessons Learned

In this pilot, we have partial lack of knowledge about when actual faults happened or when variance on the values is due to the wear and tear. For that, only the steady data is considered for fitting the best line, so trends can be identified and when actual measurements appear different behavior from the normal trends give an early warning, even though the setting of thresholds for the identification of abnormal behavior of the vessel is challenging due to the variation of the historical data sets (e.g., different periods/years might report different statistical measurements) (Fig. 10.3).

10.4.2.4 Real-Time Data Classification for Automatic Fish Detection

The main goal of the experiment was to deploy an effective classification approach, relying only on acoustic data, that can form the basis of a real-time fish detection tool.

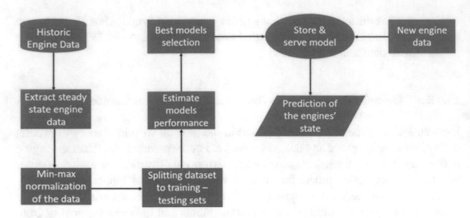

Fig. 10.3 Workflow of the Oceanic tuna fisheries immediate operational choices—pilot

For the study, echosounder sample output was appropriately preprocessed in order to produce the *mean volume backscattering strength* (MVBS) values for five frequencies: 18, 38, 70, 120, and 200 kHz. The problem with echosounder data is that the data set is quite unbalanced with respect to the presence of fish or not. In the samples that we used, about 5% of the measurements correspond to fish presence, while 95% measurements not. As a result, a random classifier can appear falsely effective.

To tackle this problem, the acoustic data set was resampled before being fed to the classifiers of the study. The comparison was made based on the *kappa coefficient*, which is more reliable in cases of unbalanced data sets. The methods tested were Naïve Bayes, K-nearest neighbors (K-NN), and SVM, both with linear and radial kernels. PCA was also examined as a preprocessing method. All classification approaches were tested on MVBS values for different combinations of the five frequencies measured.

Lessons Learned on the DataBio Use

From the process and the analyses carried out within DataBio and with respect to the specific pilot, the main conclusion and lesson learned are that many different classification algorithms should be tested, in order to identify the most efficient ones for the specific data set types. Because of the nature of the acoustic data sets, it was really challenging to identify the proper training subsets for the machine learning algorithms. This resulted in the need for a number of iterations with the pilot owner (SINTEF) in order to ensure that the algorithms are accurate enough.

References

1. Tan, P. N., Steinbach, M., & Kumar, V. (2006). *Introduction to data mining*, First edition, Addison Wesley.

2. Alpaydin, E. (2020). *Introduction to machine learning*. MIT press.
3. Wong, P. C. (1999). Guest editor's introduction: Visual data mining. *IEEE Computer Graphics and Applications, 19*(5), 20–21.
4. Ferreira de Oliveira, M. C., & Levkowitz, H. (2003). From visual data exploration to visual data mining: A survey. *Visualization and Computer Graphics, IEEE Transactions, 9*(3), 378–394.
5. Kaplan, A., & Haenlein, M. (2019). Siri, Siri, in my hand: Who's the fairest in the land? On the interpretations, illustrations, and implications of artificial intelligence. *Business Horizons, 62*(1), 15–25.
6. Arel, I., Rose, D. C., & Karnowski, T. P. (2010). Deep machine learning-a new frontier in artificial intelligence research. *IEEE Computational Intelligence Magazine, 5*(4), 13–18.
7. Frost & Sullivan. (2013). Global big data analytics market, forecast to 2023. RESEARCH CODE: K2AF-01-00-00-00, Frost & Sullivan.
8. "Machine Learning Market by Service (Professional Services, and Managed Services), for BFSI, Healthcare and Life Science, Retail, Telecommunication, Government and Defense, Manufacturing, Energy and Utilities, Others: Global Industry Perspective, Comprehensive Analysis, and Forecast, 2017–2024" (2019), Zion Market Research.
9. Amar, R., Eagan, J. & Stasko, J. (2005). Low-level components of analytic activity in information visualization. In J. T. Stasko & M. O. Ward (eds) *IEEE Symposium of Information Visualization (INFOVIS) 2005, IEEE Computer Society*, 23–25 Oct., p. 111.
10. Hand, D. J., Mannila, H., & Smyth, P. (2001) *Principles of data mining*, First edition, MIT press.
11. Hochreiter, S., & Schmidhuber, J. (1997). Long short-term memory. *Neural Computation, 9*(8), 1735–1780.
12. Russwurm, M., & Koerner, M. (2017). Multi-temporal land cover classification with long short-term memory neural networks. *ISPRS—International Archives of the Photogrammetry, Remote Sensing and Spatial Information Sciences, 42*, 551–558.
13. Mou, L., Ghamisi, P., & Zhu, X. X. (2017). Deep recurrent neural networks for hyperspectral image classification. *IEEE Transactions on Geoscience and Remote Sensing, 55*, 3639–3655.
14. Purwar, P., Rogotis, S., Chatzipapadopoulus, F., Kastanis, I. (2019). "A reliable approach for pixel-level classification of land usage from spatio-temporal images". In *2019 6th swiss conference on data science (SDS)* (pp. 93–94).
15. Huang, Z. (1997). Clustering large data sets with mixed numeric and categorical values. In *Proceedings of the 1st pacific-asia conference on knowledge discovery and data mining, (PAKDD)*, pp 21–34.
16. Draper, N., Smith, H. (1981). *Applied regression analysis*. Wiley.
17. Habyarimana, E., Piccard, I., Zinke-Wehlmann, C., De Franceschi, P., Catellani, M., Dall'Agata, M. (2019). Early within season yield prediction and disease detection using sentinel satellite imageries and machine learning technologies in biomass sorghum. *Lecture Notes in Computer Science, 11771*, 227–234. https://doi.org/10.1007/978-3-030-29852-4_19.
18. George, A. F. H., Houghton, J. D, & Brown, S. B. (1987). Tansley review no. 11. the degradation of chlorophyll—a biological enigma. *The New Phytologist, 107*(2), 255–302.
19. Mikołajczyk, A., Michał, G. (2018). Data augmentation for improving deep learning in image classification problem. In *2018 international interdisciplinary Ph.D. workshop (IIPhDW)* (pp. 117–122). IEEE.
20. He, K., Gkioxari, G., Dollár, P., Girshick, R. (2017). Mask R-CNN. In *Proceedings of the IEEE international conference on computer vision* (pp. 2961–2969).
21. Xgboost. https://xgboost.readthedocs.io/en/latest/get_started.html. Accessed: 2019.
22. Hastie, T., Gareth, J., Witten, D., Tibshirani, R. (2014). An introduction to statistical learning.
23. Sentinel-3 api. https://coda.eumetsat.int/#/home. Accessed: 2019.
24. Antory, D., et al. (2005). Fault diagnosis in internal combustion engines using non-linear multivariate statistics. *Proceedings of the Institution of Mechanical Engineers, Part I: Journal of Systems and Control Engineering, 219*(4), 243–258.
25. Basurko, O. C., & Uriondo, Z. (2015). Condition-based maintenance for medium speed diesel engines used in vessels in operation. *Applied Thermal Engineering, 80*, 404–412.

Chapter 11
Real-Time Data Processing

Fabiana Fournier and Inna Skarbovsky

Abstract To remain competitive, organizations are increasingly taking advantage of the high volumes of data produced in real time for actionable insights and operational decision-making. In this chapter, we present basic concepts in real-time analytics, their importance in today's organizations, and their applicability to the bioeconomy domains investigated in the DataBio project. We begin by introducing key terminology for event processing, and motivation for the growing use of event processing systems, followed by a market analysis synopsis. Thereafter, we provide a high-level overview of event processing system architectures, with its main characteristics and components, followed by a survey of some of the most prominent commercial and open source tools. We then describe how we applied this technology in two of the DataBio project domains: agriculture and fishery. The devised generic pipeline for IoT data real-time processing and decision-making was successfully applied to three pilots in the project from the agriculture and fishery domains. This event processing pipeline can be generalized to any use case in which data is collected from IoT sensors and analyzed in real-time to provide real-time alerts for operational decision-making.

11.1 Introduction and Motivation

To stay relevant and competitive, modern enterprises must continuously monitor events of interest, assess changing conditions, and make fast decisions. The continuous flow of event streams, such as customer orders, bank deposits, invoices, social media updates, market data, Global Positioning System (GPS)-based location information, signals from Supervisory Control and Data Acquisition (SCADA) systems, and temperature from sensors and IoT devices, are analysed to help enterprises respond in real-time to changing market and environmental conditions. Furthermore, with the emergence of the Internet of Things (IoT), organisations are taking advantage of the high volumes of data produced by sensors for real-time situational

F. Fournier (✉) · I. Skarbovsky
IBM Research—Haifa, University of Haifa Campus, Mount Carmel, 3498825 Haifa, Israel
e-mail: fabiana@il.ibm.com

© The Author(s) 2021
C. Södergård et al. (eds.), *Big Data in Bioeconomy*,
https://doi.org/10.1007/978-3-030-71069-9_11

147

awareness and real-time insights. IoT generates a huge amount of high-speed real-time data in different formats from a vast number of sources that must be analysed quickly for timely responses. IoT sensors enable decision-makers to continuously monitor and track various parameters that help them in their day-to-day operations.

Traditionally, organisations used to store data in databases and then process and analyse it after storage using batch processing. As mentioned above, the unexpected growth in the number of events due to advanced operations, massive sensor adoption, mobile devices, and high-speed networks has resulted into an exponential increase in data volume. Moreover, organisations need to be increasingly capable of extracting insights from real-time business events, because data loses value with the passage of time. Many of today's common applications such as fraud detection, algorithmic trading, network monitoring, predictive maintenance, and sales and marketing require the processing of data in real time. *Event Stream Processing* (ESP) has evolved to cope with the analysis of real-time streaming data.

To understand the essence of ESP, let's decompose the name to its three basic terms: event + stream + processing. An *event* is an occurrence within a particular system or domain; it is something that has actually happened or is contemplated as having happened in that domain. The word *event* is also used to refer to a programming entity that represents such an occurrence in a computing system [1]. A *stream* is a constant and continuous flow of events that navigate into and around companies from thousands of connected devices, IoT, and any other sensors. An *event stream* is a sequence of events arranged in some order, typically by time. Enterprises generally have three different kinds of event streams: business transactions, such as customer orders, bank deposits, and invoices; information reports, such as social media updates, market data, and weather reports; and IoT data, such as GPS-based location information, signals from SCADA systems, and temperature measurements from sensors [2]. *Processing* is the final act of analysing all this data in real-time.

ESP is the processing of continuous event data streams in real time. It helps identify the patterns and anomalies within these data streams that are important to an enterprise, such as event correlation, causality, and timing. ESP also enables organisations to respond quickly to critical events, thus saving time, money, and resources. It is also known as real-time streaming analytics, streaming analytics, and (complex) event processing [3].

Specifically, stream analytics provided by ESP platforms [4]:

- Support situation awareness through dashboards and alerts by analysing multiple kinds of events in real-time.
- Benefit decision-makers of different verticals to make data-driven decision and take proactive action before the occurrence of an event.
- Enable smarter anomaly detection and faster responses to threats and opportunities.
- Help shield business people from data overload by eliminating irrelevant information and presenting only alerts and distilled versions of the most important information.

Event Processing (EP) is a paradigm where streams of events are analysed to extract useful insights of real-world events [5]. EP systems associate precise semantics with the information items being processed: these are notifications of events that happened in the external world and were observed by sources, also called event producers [6]. The EP engine is responsible for filtering and combining such notifications to understand what is happening in terms of higher-level events (aka complex events, composite events, or situations) to be notified to sinks, called event consumers. EP systems detect complex patterns of incoming items involving sequencing and ordering relationships. An example of such a situation is the flagging of a *suspicious account* that is detected whenever there are at least three events of large cash deposits within 10 days to the same account. Event processing is in essence a paradigm of reactive computing: a system observes the world and reacts to events as they occur. It is an evolutionary step from the paradigm of responsive computing, in which a system responds only to explicit service requests.

A vast number of recent applications of EP can be found in health informatics, astronomy, telecommunications, electric grids and energy, geography, and transportation [5]. In the DataBio project, event processing applications have been developed and deployed for the domains of agriculture and fisheries, as described in the pilots section. [See Parts V and VII of this book].

11.2 Market

The massive surge in data generation and the increasing demand for real-time analysis of streaming data are expected to boost the growth of the ESP market. According to the Event Stream Processing Market—Global Forecast to 2023 report from December 2018 [3], the global ESP market size is projected to reach USD 1.838 billion by 2023, growing at a compound annual growth rate (CAGR) of 21.6% during the forecast period. The market analysis by application in Europe shows that the predictive maintenance segment is expected to grow from USD 29.2 million in 2018 to USD 81.0 million by 2023, at the highest CAGR of 22.7% during the forecast period. The market analysis by verticals in Europe shows that the ESP market by vertical is expected to grow from USD 689.9 million in 2018 to USD 1838.0 million by 2023, at a CAGR of 21.6% during the forecast period. Furthermore, the market size of the banking, financial services, and insurance (BFSI) vertical is expected to have the largest market size and projected to grow from USD 37.6 million in 2018 to USD 95.8 million by 2023, at a CAGR of 20.6% during the forecast period. This can be attributed to the growing adoption of IoT-based connected devices. All the verticals are undergoing digital transformation, which has created the need for analysing real-time data to achieve a competitive advantage in the market.

Gartner [4] characterises ESP systems as transformational, meaning they have the potential to change the way organisations interact with information to such a degree that they have a demonstrable impact on organisations' business models. Three factors are driving the expansion of ESP:

- The growth of IoT and digital interactions is making event streams ubiquitous.
- Business is demanding continuous intelligence for better situation awareness and faster, more personalised decisions.
- Vendors are launching new products, many of them open source or partly open source, giving the impression of lower acquisition costs.

From the analysts' reports covered, it's clear that ESP solutions have the potential to enable new ways of doing business; companies who have not yet adopted such systems should consider doing so in the near future. Furthermore, the bioeconomy domains investigated in DataBio (i.e., agriculture, forestry, and fishery), and not mentioned in the reports so far, have a unique opportunity to be innovative by embracing this technology.

11.3 Technical Characteristics

Event processing systems are a departure from traditional computing architectures that employ synchronous, request-response interactions between client and servers. In reactive applications, decisions are driven by events. Conventional architectures are not fast or efficient enough for some applications, because they use a "save-and-process" paradigm in which incoming data is stored in databases in memory or on disk, and then queries are applied. When fast responses are critical, or the volume of incoming information is extremely high, application architects instead use a "process-first" EP paradigm; here, logic is applied continuously and immediately to the "data in motion" as it arrives. EP is more efficient because it computes incrementally, in contrast to conventional architectures that reprocess large datasets, often repeating the same retrievals and calculations as each new query is submitted.

As mentioned above, the goal of an EP engine is to notify its users immediately upon the detection of a pattern of interest. Data flows are seen as streams of events, some of which may be irrelevant for the user's purposes. Therefore, the main focus is on the efficient filtering out of irrelevant data and processing of the relevant. Obviously, for such systems to be acceptable, they must satisfy certain efficiency, fault tolerance, and accuracy constraints, such as low latency and robustness.

As previously stated, EP is a technique in which incoming data about what is happening (event data) is processed more or less as it arrives to generate higher-level, very useful summary information, known as complex events. Event processing platforms have built-in capabilities for filtering incoming data, storing windows of event data, computing aggregates, and detecting patterns. In essence, EP software is any computer program that can generate, read, discard, and perform calculations on events. A complex event is an abstraction of one or more raw input events. One complex event may be the result of calculations performed on a few or on millions of events from one or more event sources. A situation may be triggered by the observation of a single raw event but is more typically obtained by detecting a pattern over the flow of events. Many of these patterns are temporal in nature [7], but they can also be spatial, spatio-temporal, or modal [1]. Event processing deals with these functions: get events from sources (event producers), route these events, filter them, normalise or otherwise transform them, aggregate them, detect patterns over multiple events, and transfer them as alerts to a human or as a trigger to an autonomous adaptation system (event consumers). An application or a complete definition set made up of these functions is also known as an Event Processing Network (EPN) [1].

Generally speaking, complex event processing (CEP) software offers two major components: a high-level language for programmers to easily describe how to process the incoming events and an infrastructure engine for the processing of the data streams in real-time. Events of different formats are gathered from different event producers. The event producers can be of different types, including financial feeds, news feeds, weather sensors, application logs, video streams collected from surveillance cameras, etc. The EP engine is the brain that carries out multiple types of processing on event streams, based on predefined rules. The processing includes simple filtering, counting, averaging, aggregating, of simple event processing operations, as well as more complex processing, such as pattern matching and event prediction (forecasting). Event consumers are parties that are interested in mining valuable information from the event streams, e.g., software agents, users of web/mobile applications, etc. [5].

The design of event processing applications includes the design of both the functional properties and the non-functional properties. While functional requirements define what an event processing system should do, non-functional requirements place constraints on how the system will do so. The design of requirements is implementation-specific and is carried out in either hand-coded fashion or using modern dedicated event processing tools by IT developers familiar with the event processing engine and the particular way to bypass the engine's limitations.

The event logic necessary to specify the event-driven application is typically provided by domain experts who know the domain and can express the event rules. However, the task of defining the event definitions can be tedious and difficult even for experts. To alleviate this task, in some engines the event definitions can be learned in an automated way using machine learning techniques (e.g., [8] and [9]).

Non-functional requirements include scalability, usability, availability, security, and performance objectives. Not all of these requirements apply equally to all applications, so when designing an event processing application, one needs to consider which of them are important for the case in hand. A survey in the area of non-functional requirements can be found in [10].

There is no standard for event processing languages and programming models. As a result, each event processing tool uses its own terminology and semantics. For example, the IBM PROactive Technology Online (PROTON) open source tool[1] applied in the DataBio project follows the semantics presented in Etzion's and Niblet's book [1].

---

[1] https://github.com/ishkin/Proton/.

11.4   Event Processing Tools

CEP has already built up significant momentum, manifested in a steady research community and a variety of commercial and open source products [6]. Today, a large variety of commercial and open source event processing tools are available to architects and developers who are building event processing applications. These are sometimes called event processing platforms, streaming analytics platforms, (complex) event processing systems, event stream processing systems, or distributed stream computing platforms (DSCPs). DSCPs such as Amazon Web Services Kinesis[2] and open source offerings including Apache Samza,[3] Spark,[4] and Storm[5] were introduced in recent years. In particular, Apache open source projects (Storm, Spark, and Samza) have gained a fair amount of attention and interest [11, 12].

Event processing systems are general purpose development and runtime tools that are used by developers to build custom, event-processing applications. The tools allow this to be done without having to re-implement the core algorithms for handling event streams, as they provide the necessary building blocks to build the event-driven applications. In comparison, DSCPs are general-purpose platforms without full native EP analytic functions and associated accessories. However, they are highly scalable and extensible, and usually offer an open programming model so developers can add the logic to address many kinds of stream processing applications, including some EP solutions. Today, there are already some implementations that take advantage of the pattern recognition capability of EP systems along with the scalability capabilities that DSCPs offer and provide a holistic architecture. For example, the PROTON open source event processing tool applied in the DataBio project has a Storm version (ProtonOnStorm), which allows PROTON's engine to run in a distributed manner on multiple machines using the Storm infrastructure.

A recent Gartner report from 2019 [4] states that more than 40 ESP products are available on the market.

Sample vendors include EsperTech, EVAM, IBM, Microsoft, Oracle, SAP, SAS, Software AG, the Apache Software Foundation, and TIBCO Software.

11.5   Experiences in DataBio

As mentioned previously, event-driven applications were developed for the agriculture and fisheries sectors in the DataBio project. More specifically, two agricultural implementations were developed. One focuses on monitoring temperature and air pressure measurements from sensors in the field (using SensLog) and sending warnings concerning a possible upcoming freeze. This application is designed to alert

---

[2] https://aws.amazon.com/kinesis/.

[3] https://samza.incubator.apache.org/.

[4] https://spark.apache.org/streaming/.

[5] https://storm.apache.org/.

farmers before the occurrence of freezing temperatures that can destroy crops. The second application monitors different crop parameters to predict disease and pest infestation in various types of crops and sends alerts and warnings if these are found. Crop parameters are gathered by GAIATrons and pushed to PROTON for further analysis of this data, where temporal analysis of trends is carried out to allow proactive measures.

In fisheries, PROTON monitors engine parameters to send alerts in real-time regarding potential engine problems before damage will be caused to the engine and therefore to the tuna fishing vessel. An event-driven application informs crew members to act in advance to avoid critical machinery faults prior to their occurrence. PROTON has been deployed on board the vessel and is integrated with the VTT OpenVA tool to visualise the alarms and warnings in real-time as they are detected by the event processing engine.

For detailed information on these implementations, refer to the relevant pilots sections in Part V and VII of this book.

These applications follow the event-driven paradigm and fit into the "Generic pipeline for IoT data real-time processing and decision making" articulated in the course of the project and presented in Deliverable 4.4 of the project [13]. This generic pipeline is an example of a pattern that fits the two aspects of generalisation. The main characteristic of this generic pipeline is the collection of real-time data coming from IoT devices to generate insights for operational decision-making, by applying real-time data analytics on the collected data.

Figure 11.1 depicts the common data flow among three pilots of the DataBio project: two in agriculture ("Prediction and real-time alerts of diseases and pests

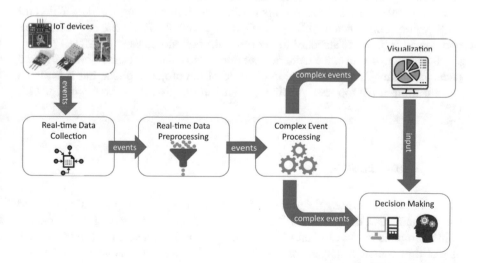

Fig. 11.1 Data flow for real-time IoT data processing and a decision-making generic pipeline

breakouts in crops" and "Cereals and biomass crop") and one in fisheries ("Monitoring, real-time alerts, and visualization for operation efficiency in tuna fishery vessels").

Streaming data from IoT sensors are collected in real-time, from sources such as agricultural sensors, machinery sensors, and fishing vessels' monitoring equipment. These streaming data (aka events) can then be pre-processed to lower the amount of data to be further analysed. Pre-processing can include filtering the data (filtering out irrelevant data and filtering in only relevant events); performing simple aggregation of the data; and storing the data (e.g., on the cloud or using other storage models, or even simply on a computer's file system) such that conditional notification on data updates to subscribers can be done. After being pre-processed, data enters the CEP component for further analysis, which generally means finding patterns in time windows (temporal reasoning) over the incoming data to form new, more complex events (aka as situations or alerts/warnings). These complex events are emitted to assist in the decision-making process that is carried out by humans ("human in the loop") or automatically by actuators (e.g., sensors that starts irrigation in a greenhouse following a certain alert). The situations can also be displayed using visualisation tools to assist humans in the decision-making process. The idea is that the detected situations can provide useful real-time insights for operational management, such as preventing possible pest infestations in crops or machinery failure.

Figure 11.1 shows the end-to-end flow. In essence, all components except the data producers (i.e., sensors) and a data consumer (either human or automatic) can be optional. The level of analysis of the data and its level of abstraction is driven by the specific use case. Sometimes, some filtering on the data is enough, while in other cases, the CEP component performs all types of analysis in a central manner. Communication between the software components is performed using standard RESTful APIs, while communication between IoT devices and the *Real-time data collection* component is based on standard IoT communication protocols (e.g., MQTT).

As mentioned above, the Generic pipeline for IoT data real-time processing and decision making is a generalization of three of the project's pilots, but it is also a specification of the top-level pipeline devised in the project as shown in Fig. 11.2 [13].

11.6 Conclusions

The major factors driving the growth of the ESP market are the increasing demand for IoT and smart devices, and the growing focus on drawing real-time insights to gain a competitive edge. IoT provides numerous opportunities for ESP vendors, such as real-time remote management, monitoring, and insights from connected devices, such as mobile phones or connected cars.

ESP is one of the key enablers of continuous intelligence and other aspects of digital business. It has transformed financial markets and become essential to smart electrical grids, location-based marketing, supply chain, fleet management, and other

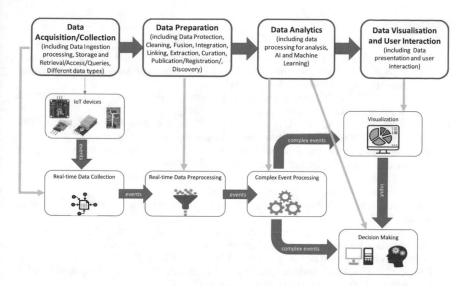

Fig. 11.2 Mapping of the steps of the top-level pipeline to the steps of the generic pipeline for data flow for real-time IoT data processing and decision-making

transportation operations. From the analysts' reports covered, we can conclude that ESP solutions can enable new ways of doing business; thus, companies who have not yet done so should consider adopting ESP systems. Furthermore, the bioeconomy domains investigated in DataBio (i.e., agriculture, forestry, and fishery) that are not mentioned in reports so far, have a unique opportunity to be innovative by embracing this technology. In DataBio, we have already paved the way for such applications by applying event-driven solutions in pilots in both the agriculture and fishery domains.

The generic pipeline for IoT data real-time processing and decision making has been applied to three pilots in the project from the agriculture and fishery domains and, as such, can be seen as a "pipeline design pattern". Conceptually, it can also be applied to other domains beyond fisheries and agriculture. Basically, use cases from any domain in which data is collected from IoT sensors and analysed in real-time to provide real-time alerts for operational decision-making can be adapted to this generic pipeline.

For example, sensor readings from a supply chain scenario in which objects are monitored for tracking and tracing can be collected for further processing by a CEP engine to detect potential delays. The detected situations can be displayed to operators so they can take action if such delays are detected (e.g., reschedule trajectory). Another use case can be found in a classical manufacturing process, in which machinery sensors are monitored to detect potential failures. The sensor data in the factory can be collected and transmitted to a CEP engine, which can detect potential failure situations and emit alerts to aid in decision-making (e.g., stop the machine, replace a part, etc.).

References

1. Etzion, O., & Niblett, P. (2010). *Event processing in action*. Manning Publications Company.
2. Watts S. (2018). *What is stream processing? Event processing explained*, [Online]. At: https://www.bmc.com/blogs/event-stream-processing
3. Sreedhar, B., & Naim, S. (2018). *Event stream processing market—Global forecast to 2023*. A Markets and Markets Report (published December 2018).
4. Hare, J., & Schlegel, K. (2019). *Hype cycle for analytics and business intelligence, 2019*. Gartner report # G00369713 (published on 18 July 2019).
5. Dayarathna, M., & Perera, S. (2017). Recent advancements in event processing. *ACM Computing Surveys, 1*(1). https://doi.acm.org/10.1145/3170432
6. Cugola, G., & Margara, A. (2012). Processing flows of information: From data stream to complex event processing. *ACM Computing Surveys, 44*(3).
7. Etzion, O. (2010). Temporal aspects of event processing. Handbook of distributed event based system.
8. Margara, A., Cugola, G., & Tamburrelli, G (2014). Learning from the past: Automated rule generation for complex event processing. In *Proceedings of the 8th ACM international conference on distributed event-based systems (DEBS2014)*, (pp 47–58).
9. Artikis, A., Sergot, M., & Paliouras, G. (2014). An event calculus for event recognition. *IEEE Transactions on Knowledge and Data Engineering (TKDE)*.
10. Etzion, O., Rabinovich, E, & Skarbovsky, I. (2011). Non-functional properties of event processing. In *Proceedings of the Fifth ACM international conference on distributed event-based systems (DEBS 2011)* (pp. 365–366).
11. Biscotti, F. et al. (2014). *Market guide for event stream processing*. Gartner report #G00263080 (published on August 14, 2014).
12. Vincent, P. (2014). *CEP tooling market survey*. December 3, 2014, https://www.complexevents.com/2014/12/03/cep-tooling-market-survey-2014/. Accessed September 25, 2020.
13. Plakia, M. et al. (2020). *D4.4—Service documentation*, [Online]. At: file:///C:/Data/DataBio/WP4/D4.4/DataBio_D4_4ServiceDocumentation_v1_2_2020_03_13_EXUS.pdf.

Chapter 12
Privacy-Preserving Analytics, Processing and Data Management

Kalmer Keerup, Dan Bogdanov, Baldur Kubo, and Per Gunnar Auran

Abstract Typically, data cannot be shared among competing organizations due to confidentiality or regulatory restrictions. We present several technological alternatives to solve the problem: secure multi party computation (MPC), trusted execution environments (TEE) and multi-key fully homomorphic encryption (MKFHE). We compare these privacy-enhancing technologies from deployment and performance point of view and explain how we selected technology and machine learning methods. We introduce a demonstrator built in the DataBio project for securely combining private and public data for planning of fisheries. The secure machine learning of best catch locations is a web solution utilizing Intel® Software Guard Extensions (Intel® SGX)-based TEE and built with the Sharemind HI (Hardware Isolation) development tools. Knowing where to go fishing is a competitive advantage that a fishery is not interested to share with competitors. Therefore, joint intelligence from public and private sector data while protecting secrets of each contributing organization is an important enabler. Finally, we discuss the wider business impact of secure machine learning in situations where data confidentiality is a concern.

K. Keerup · D. Bogdanov · B. Kubo (✉)
Cybernetica AS, Narva mnt 20, 51009 Tartu, Estonia
e-mail: baldur.kubo@cyber.ee

K. Keerup
e-mail: kalmer.keerup@cyber.ee

D. Bogdanov
e-mail: dan.bogdanov@cyber.ee

P. G. Auran
SINTEF Digital, Strandvejen 4, Trondheim, Norway
e-mail: per.gunnar.auran@sintef.no

12.1 Privacy-Preserving Analytics, Processing and Data Management

Data analysis and machine learning methods can provide great value in different areas of governance and business. By recognizing patterns in data, visualizing the patterns and developing predictive models, we can optimize farming, forestry and fishing operations.

Well-known data analysis and machine learning tools and frameworks can be used when the data originates from public sources such as Copernicus satellite images or from private sources when an agricultural business collects their own data. When data is confidential, current computers and software can protect data only while it is not being used or when data is being transferred. Typically, encryption and access restrictions are used. Traditional computers and software need to remove the technical protection to analyze data. Thus, the only protection of the owner of confidential data when using traditional software is limiting access to data to select few trusted persons and using contractual obligations.

One of the reasons for combining data from different companies and public sources is to improve the accuracy of machine learning and data analysis methods as data from different entities might capture different patterns or provide increased statistical power due to larger sample size. Learning from combined data can thus provide increased value for an industry. However, companies might be reluctant to share their data to protect the confidentiality of their operations.

Recently, secure computation technologies have been developed which enable processing confidential data without leaking individual values. By using these technologies, we are able to develop data analysis and machine learning software that retains the confidentiality of individual data providers but allows them to collectively gain improved insights from sharing their data.

When using secure computation, data is encrypted by the data owner and only then sent to a service processing the data. The host of the service will not have access to the unencrypted data nor the encryption keys. Data protection is not removed even while the data is being processed.

Secure computation technology can be used to develop solutions which are otherwise not possible due to confidentiality restrictions. There are some general types of problems where secure computation technology may be required:

- Outsourcing computations. Secure computation is a solution if one wishes to provide an analysis service to clients without learning the clients' data.
- Analyzing data governed by data protection laws. Secure statistical analysis can be used for decision-making when databases are governed by data protection laws and remain inaccessible for standard statistics software.
- Analyzing data from multiple sources. If data originates from a single provider, the provider can run analysis using their own infrastructure without giving data access to a third party. If we wish to analyze data from multiple sources without revealing the data to the party running the analysis, we can use secure computation technology.

In this chapter, we will describe two technologies for privacy-preserving data analysis and a demonstrator developed in the DataBio project which uses such technology to predict catch location and expected catch size for fisheries. The business impact of privacy-preserving data analysis and its applicability are also discussed.

12.2 Technology

Secure computation approaches can be categorized into software-based cryptographic techniques and hardware-based techniques. We bring examples from both categories.

12.2.1 Secure Multi-Party Computation

Secure multi-party computation (MPC) is a cryptographic technique for processing private data while preserving privacy. Sharemind MPC is a technology leveraging MPC which provides a framework for programming secure client–server applications. The roles of different parties involved in a Sharemind MPC process are as follows:

- Input parties who convert their public data into secret data and import it to servers hosted by computation parties.
- Computation parties who perform operations on the secret data without learning the input values or the results.
- Output parties who can retrieve the secret results from computation parties and construct the public result values.

Sharemind MPC uses an approach for MPC called additive secret sharing where private values are split into random values before being imported into an MPC system. This means that given a private 32-bit value x, two random values x_1, x_2 are generated and x_3 is computed so that $x \equiv x_1 + x_2 + x_3 \pmod{2^{32}}$. The three values are sent to three independent servers.

The servers can perform arithmetic on secret-shared values. For example, to add two values, each server adds their respective shares of the values. After the local additions, each server holds one share of the sum. More complicated operations require network communication between the servers. Figure 12.1 illustrates how two private values can be added using MPC.

As long as at most one of the servers is compromised, privacy remains protected. All three server hosts verify the analysis program before installing it. This ensures that only agreed upon results will be published to output parties. Shared responsibility also means that privacy remains protected if one of the servers is compromised. Sharemind MPC includes an auditing tool to detect tampering.

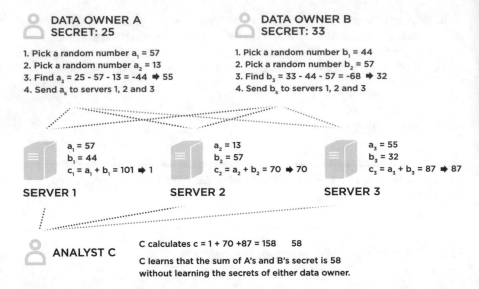

Fig. 12.1 Illustration of adding secret-shared values

MPC is a general-purpose programmable technique and has been successfully used to implement practical applications [1]. The Sharemind MPC technology has been used for tax fraud detection [2], statistical analysis of government databases for a social study [3] and a report on the state of the Estonian IT industry by combining data from companies in the IT sector [4].

The main benefit of MPC is the high security guarantees. A party hosting an MPC server cannot learn anything about the values sent to it. There are no side-channel attacks which sometimes plague cryptographic techniques. Sharemind protects data in transit, in memory, at rest and during computations.

The main downsides of MPC are its complicated deployment requirements and decreased performance when compared to conventional software. Since the three server hosts must be independent, the organizations using MPC must decide on three parties who will be managing the servers. This involves more contracts between parties participating in the process when compared to a single organization providing an analysis service, but data will be protected technically, not just by the contracts as with usual data analysis tools.

12.2.2   *Trusted Execution Environments*

An alternative to software-based techniques is using a trusted execution environment such as Intel Software Guard Extensions (SGX).[1] SGX is an extension of the instruction set of Intel processors which enables developing secure applications when even the host operating system is not trusted. SGX relies on three concepts to protect data: enclaves, attestation, and data sealing.

SGX is a set of CPU instructions for creating and operating with memory partitions called enclaves. When an application creates an enclave, it provides a protected memory area with confidentiality and integrity guarantees. These guarantees hold even if privileged malware is present in the system, meaning that the enclave is protected even from the operating system that is running the enclave. With enclaves, it is possible to significantly reduce the attack surface of an application.

Remote attestation is used to prove to an external party that the expected enclave was created on a remote machine. During remote attestation, the enclave generates a report that can be remotely verified with the help of the *Intel attestation service.*[2] Using remote attestation, an application can verify that a server is running trusted software before private information is uploaded.

Data sealing allows enclaves to store data outside of the enclave without compromising confidentiality and integrity of the data. The sealing is achieved by encrypting the data before it exits the enclave. The encryption key is derived in a way that only the specific enclave on that platform can later decrypt it.

Sharemind Hardware Isolation (HI) is a technology using Intel SGX which provides the ability to process confidential data. Sharemind HI is built as a client–server service similar to Sharemind MPC. The client is an application that calls operations on the server, encrypts data and performs remote attestation on the server. The Sharemind HI server does the bulk of the work and is responsible for the following: checking if a user has the right to access the system; checking if a user has the correct roles to perform an operation; managing the encrypted user data and the encryption keys of the data; managing task descriptions of how a data analysis process is carried out; storing a log of the operations performed in the server and scheduling the tasks to run.

Figure 12.2 illustrates the security model of Sharemind HI applications. The input data, shown in red, is encrypted at the client side and sent to the server. The input data encryption keys of the data are securely transferred to the SGX protected enclaves. Likewise, the output data, shown in green, is encrypted inside of the enclave and stored on the server. When requested, the enclave securely transfers the output data encryption keys to the authorized clients.

At any point during the deployment, a client can request a cryptographic proof of what analysis code is running in the server, shown in blue on the figure. This proof can be compared against a previously generated proof by an auditor who has validated the code to be secure.

---

[1] Intel® Software Guard Extensions | Intel® Software.

[2] https://software.intel.com/en-us/sgx/attestation-services.

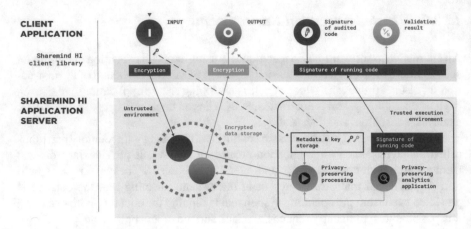

Fig. 12.2 Sharemind HI security model

The main benefits of Sharemind HI over Sharemind MPC are performance and simpler deployment. There is only one computational party, and unlike Sharemind MPC network communication is not required while the enclave is running.

Another benefit of Sharemind HI is that enclaves are programmed in the C++ programming language, whereas Sharemind MPC programs are written in a domain-specific language called SecreC which resembles C. This allows Sharemind HI programmers to adapt libraries and other existing code written in C or C++ .

The main downside of Sharemind HI is that it requires users to trust Intel. Details of how SGX-enabled processors are produced are undisclosed information, and Intel cannot prove that SGX is secure. It is also possible that side-channel attacks against SGX will be developed which would require more careful design of the enclave software. Practical applications should consider the security and performance trade-offs between cryptographic and hardware-based techniques.

12.2.3 *Homomorphic Encryption*

Another alternative for privacy-preserving computation is fully homomorphic encryption (FHE). FHE allows arbitrary computations on encrypted data. Privacy is ensured by encryption and is thus independent of the trustworthiness or security of the server that is executing the computation. See the UN Handbook on Privacy-Preserving Computation Techniques[3] for a summary of this family of encryption schemes.

---

[3] https://publications.officialstatistics.org/handbooks/privacy-preserving-techniques-handbook/ UN%20Handbook%20for%20Privacy-Preserving%20Techniques.pdf.

12.2.4   On-The-Fly MPC by Multi-Key Homomorphic Encryption

One major disadvantage of classical MPC schemes (such as secret sharing) is that they need to be planned out in advance. The number of participants needs to be known and fixed before the calculation starts. In contrast, there is the concept of *on-the-fly MPC*, which is much more flexible in those regards. The main criteria an on-the-fly MPC scheme should meet are as follows:

1. The cloud can perform arbitrary, dynamically chosen computations.
2. It can use data from an arbitrary, non-pre-fixed set of participants (on-the-fly).
3. The computations are non-interactive, i.e., they do not require communication with all the participants (like with secret sharing).

On-the-fly MPC can be achieved by using multi-key fully homomorphic encryption (MKFHE). While most FHE schemes allow only one encryption key to be used, MKFHE schemes allow for multiple keys to be used for one computation.

Figure 12.3 illustrates how an MKFHE scheme can facilitate on-the-fly MPC. In this case, we have four different Alices with their secret message m_1, m_2, m_3 and m_4. Each of them encrypts their message using a different key (k_1, k_2, k_3 and k_4) and sends it to Bob. Out of these four encrypted messages, Bob can choose any subset (say $Enc(m_1, k_1)$, $Enc(m_2, k_2)$, $Enc(m_3, k_3)$) and any function that he wishes to perform on it (say f). Note that these choices can be made *after* the messages have been encrypted and sent to Bob.

He then calculates $f(Enc(m_1, k_1), Enc(m_2, k_2), Enc(m_3, k_3))$ and sends the result back to Alice1, Alice2 and Alice3, who agree to approve or disapprove the calculation. If approved, they can decrypt the result together and obtain $f(m_1, m_2, m_3)$. The decryption is only possible if the three of them work together. Note that there is no need for any communication with Alice4, since her message is not involved in the calculation. Also note that the other three Alices need not communicate until after Bob has finished his calculation. This gives MKFHE a huge advantage over classical MPC in terms of scalability and flexibility. However, like for other FHE schemes, the computation of f is very costly.

12.2.5   Comparison of Methods

All the methods discussed above have their advantages and disadvantages. The following table gives a rough overview.

Method	Advantages	Disadvantages
MPC by secret sharing	– Relatively efficient – Easy to handle – Already mature technology	– Requires coordinating multiple servers – Requires planning and setup

(continued)

(continued)

Method	Advantages	Disadvantages
Trusted execution environments	– High efficiency – Secure even if OS is not	– Vendor (Intel) proprietary technology that is not disclosed
Single-key homomorphic encryption	– Very flexible – Security independent of software and hardware – Needs only one server	– High computational cost – Difficult to understand/use – Allows for one key only
Multi-key homomorphic encryption	– Full flexibility – Security independent of software and hardware – On-the-fly execution	– High computational cost – Difficult to understand/use

For most practical use cases, computational cost (and thereby scalability) is by far the most important factor. The better flexibility that homomorphic encryption schemes offer may be crucial for some applications, but is generally less relevant. It was therefore decided that MPC and trusted execution environments would be feasible for the project.

12.3 Secure Machine Learning of Best Catch Locations

In order to demonstrate how secure computation technologies could be used in agriculture, forestry and fisheries, a demonstrator which predicts the best fish catch location and expected catch size on a given day was developed in the frame of the DataBio project.

Catch data with geographical positions was retrieved from the Norwegian Directorate of Fisheries [5]. Although we used public data for experimentation, our approach demonstrates that secure machine learning models can be trained on data from multiple fisheries and enables combining private data with public data.

12.4 Pipeline

In the pilot, we implemented the model using both Sharemind MPC and Sharemind HI [6]. Due to better performance, we chose the Sharemind HI solution as the backend for a web-based tool. The Sharemind MPC version is efficient enough to train models that can be reused for estimation afterward even if the model is kept private. As there are fishery-specific parameters, a model would need to be trained for each fishery. The Sharemind HI version trains a model in the order of a minute instead of hours it takes with Sharemind MPC.

Figure 12.4 illustrates the prediction pipeline using secure machine learning.

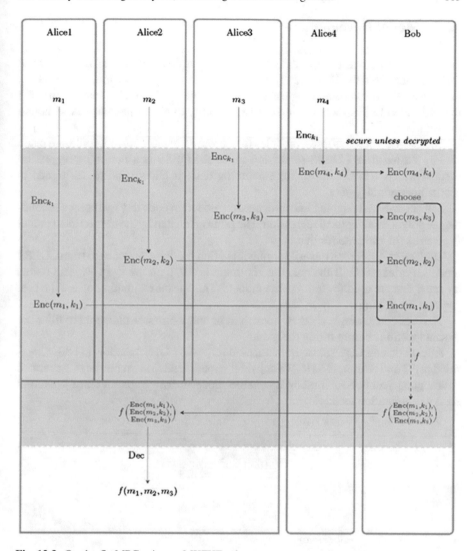

Fig. 12.3 On-the-fly MPC using an MKFHE scheme

The analysis takes into account the following parameters: harbor location, distance threshold, quantile of best catch, size of the ship and whether to maximize a single species catch or all species (total biomass output).

12.5   Model Development

Public catch data was used in the R[4] statistical analysis software to find a method for modeling the data. Since catch size and position vary by season, we could not use linear regression or autoregression for accurate prediction. A local regression method called LOESS was chosen due to its ability to model phenomena without a known function.

The program predicts three variables on a given date: latitude, longitude and catch size by fitting three LOESS regression models. LOESS is a nonlinear regression method which was developed for smoothing data. It allows one to see trends in scatterplots of noisy data.

LOESS trains a weighted linear regression model for each day by fitting a second-degree polynomial for local regression. The point estimated by the trained local model is given as the estimate for that day.

The user can specify a quantile argument to find the "best" catches to train LOESS models. For example, if the quantile argument is 0.9, then the top 10% data points by catch size are used for training the models. This means estimating where the best captains are fishing.

The user can also specify their home harbor and a distance threshold to filter out distant locations before fitting the model.

After choosing LOESS, we implemented fitting of LOESS models in both Sharemind MPC and Sharemind HI. We consider experimentation on public or generated data a good practice for finding a suitable model before implementing it using a secure computation technology.

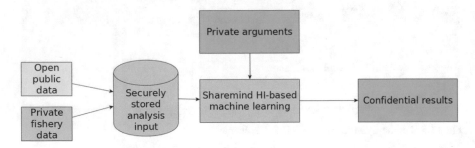

Fig. 12.4 Abstract overview of the proposed Sharemind HI-based solution

---

[4] R: The R Project for Statistical Computing.

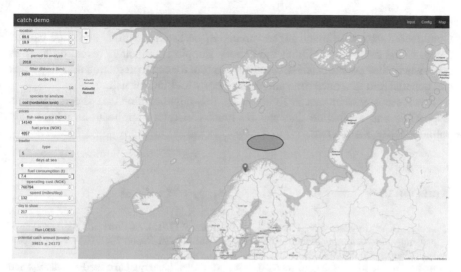

Fig. 12.5 Catch location prediction demonstrator user interface

12.6 User Interface

A web-based interface was developed for the tool. It allows input parties to encrypt
and import their data. Fisheries can use the tool to train the predictive model using
their parameters.

The user can select the fish species, home harbor, distance threshold, vessel type
and top catch quantile. After training the models, the enclave returns three vectors
to the client application: latitude curve, longitude curve and catch size curve. The
interface will display a map with the estimated position on a given day. The user
can change the day with a slider to see how the position changes. The enclave also
calculates prediction intervals for the fitted curves which allows the catch area to be
displayed as an ellipse on Fig. 12.5.

12.7 Conclusions and Business Impact

The ability to handle confidential data in privacy-preserving analytics opens up for
a number of new applications opportunities, not only in the fishery domain, but also
in agriculture and forestry.

There are many situations where sensitive data is not made available because of
concerns that the data becomes accessible by competitors or by others that might
misuse the data.

The purpose of this demonstrator is to show that it is possible to handle confidential data as part of data analytics, potentially combining open data and confidential

data in analytics that both provide business value and preserve data confidentiality. Confidential data with much higher precision on catch locations and time can be analyzed the same way, without the fishery shipping companies revealing to each other where they got the catches, resulting in a tool for catch prediction that all parties can benefit from to reduce time and energy costs looking for fish.

A wide business impact is foreseen by this demonstrator that shows that this is possible and a pipeline that can be reused in future applications where data confidentiality is a concern.

References

1. Archer, D. W., Bogdanov, D., Pinkas, B., & Pullonen, P. (2015). *Maturity and performance of programmable secure computation.* Cryptology ePrint Archive, Report 2015/1039, 2015. https://eprint.iacr.org/2015/1039
2. Bogdanov, D., Jõemets, M., Siim, S., & Vaht, M. (2015, January). How the Estonian tax and customs board evaluated a tax fraud detection system based on secure multi-party computation. In *International conference on financial cryptography and data security* (pp. 227–234). Springer.
3. Bogdanov, D., Kamm, L., Kubo, B., Rebane, R., Sokk, V., & Talviste, R. (2016). Students and taxes: A privacy-preserving study using secure computation. *Proceedings on Privacy Enhancing Technologies, 2016*(3), 117–135.
4. Talviste, R. (2011). Deploying secure multiparty computation for joint data analysis—a case study. *Master's thesis. University of Tartu. 2011.* https://sharemind.cyber.ee/files/papers/itl_app_talviste_2011.pdf
5. Fiskeridirektoratet. "Åpne data: fangstdata koblet med fartøydata." *Fiskeridirektoratet*, 2 Oct. 2018. https://www.fiskeridir.no/Tall-og-analyse/Aapne-data/Aapne-datasett/Fangstdata-koblet-med-fartoeydata
6. DataBio Hub Privacy-aware analytics. https://www.databiohub.eu/registry/#service-view/Privacy-aware%20analytics/0.0.1

Chapter 13
Big Data Visualisation

Miguel Ángel Esbrí, Eva Klien, Karel Charvát, Christian Zinke-Wehlmann, Javier Hitado, and Caj Södergård

Abstract In this chapter, we introduce the topic of big data visualization with a focus on the challenges related to geospatial data. We present several efficient techniques to address these challenges. We then provide examples from the DataBio project of visualisation solutions. These examples show that there are many technologies and software components available for big data visualisation, but they also point to limitations and the need for further research and development.

13.1 Advanced Big Data Visualisation

Data visualisation is the graphical representation of information and data. By using visual elements like charts, graphs and maps, data visualisation tools provide an accessible way to see and understand trends, outliers and patterns in data [1]. More

M. Á. Esbrí (✉) · J. Hitado
Atos Spain, Albarracín 25, 28004 Madrid, Spain
e-mail: miguel.esbri@atos.net

J. Hitado
e-mail: javier.hitadosimarro@atos.net

E. Klien
Fraunhofer-Institute for Computer Graphics Research IGD, Fraunhoferstr. 5, Darmstadt, Germany
e-mail: eva.klien@igd.fraunhofer.de

K. Charvát
LESPROJEKT-SLUŽBY Ltd, Martinov 197, 27713 Zaryby, Czech Republic
e-mail: charvat@lesprojekt.cz

C. Zinke-Wehlmann
Institut for Applied Informatics e.V. at the University of Leipzig, Goerdelering 9, Leipzig, Germany
e-mail: zinke@infai.org

C. Södergård
VTT Technical Research Centre of Finland, Espoo, Finland
e-mail: Caj.Sodergard@vtt.fi

particularly, the defining feature of big data visualisation is scale, in terms of the vast amounts of data to be dealt with.

In that sense, the amount of data created by the private and public sectors around the world is growing every year, skyrocketing with the emergence and popularisation of the Internet of Things and the many open data initiatives that have made available a wealth of datasets (typically owned by the public sector) to the public. The Copernicus programme and the data provided by its Sentinel satellite constellation are a paradigmatic example of this (see Chap. 4).

The underlying problem for decision-makers is that all this data is only useful if valuable insights can be extracted (sometimes in near real-time) from it, and decisions can be made based on them. Big data visualisation is not the only way for decision-makers to analyse data, but big data visualisation techniques offer a fast and effective way to [2]:

- Review large amounts of data—Data presented in graphical form enables decision-makers to take in large amounts of data and gain an understanding of what it means very quickly.
- Spot trends—Time-sequence data often captures trends, but spotting trends hidden in data is notoriously hard to do—especially when the sources are diverse, and the quantity of data is large. The use of appropriate big data visualisation techniques can make it easy to spot these trends and take decisions.
- Identify correlations and unexpected relationships—One of the huge strengths of big data visualisation is that it enables users to explore datasets—not to find answers to specific questions, but to discover what unexpected insights the data can reveal. This can be done by adding or removing datasets, changing scales, removing outliers and changing visualisation types.
- Present the data to others—An often-overlooked feature of big data visualisation is that it provides a highly effective way to communicate any insights that it surfaces to others by conveying meaning very quickly and in an easily understandable way.

Besides, an important aspect of big data visualisation is choosing the most effective way to visualise the data to surface any insights it may contain. In some circumstances, simple graphic tools such as pie charts or histograms may be enough, but with large, numerous and diverse datasets more advanced visualisation techniques may be more appropriate. Various big data visualisation graphics examples include:

- Linear: Lists of items, items sorted by a single feature, text tables, highlight table
- 2D/Planar/geospatial: Cartograms, dot distribution maps, proportional symbol maps, contour maps.
- 3D/Volumetric: 3D computer models, computer simulations.
- Temporal: Timelines, time series charts, connected scatter plots, arc diagrams, circumplex charts.
- Multidimensional: Pie charts, histograms, histogram, matrix, tag clouds, bar charts, tree maps, heat maps, spider charts, area chart, Box-and-whisker Plots, bubble cloud, bullet graph, circle view, Gantt chart, network, polar area, scatter plot (2D or 3D), streamgraph, wedge stack graph.

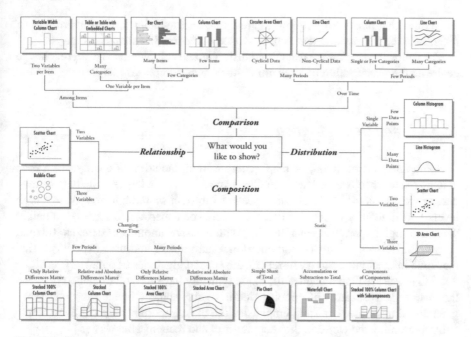

Fig. 13.1 Chart selector guide [3]

- Tree/hierarchical: Dendrograms, radial tree charts, hyperbolic tree charts.
- Any mix-and-match combination in a dashboard.

The following chart selection guide (Fig. 13.1) summarises the selection of the most appropriate chart types depending on what it is intended to be shown:

The variations in the visualisation of geoinformation (GI) are more limited because it is fundamentally linked to spatial context and geographical maps. The first priority of GI visualisation tends to be more geographical than to be informational or graphical. Maps allow us to communicate spatial information effectively. Big data visualisation opens the possibilities of GI visualisation in terms of spatial extent, spatial resolution and density of content. New techniques help mastering the vast amount of information, thus strengthening the spatial context and facilitating the exploration of new meanings and insights through map and other kinds of representations.

13.2   Techniques for Visualising Very Large Amounts of Geospatial Data

Different visualisation charts were presented in the previous section, the selection of which is dependent on the type of information and the goals of the target audience. However, in many occasions, the resulting visualisation requires the use of different techniques that allow simplifying, aggregating and reducing in various orders of magnitude the information that is finally used in the graphic charts and maps.

The following section presents three different and complementary approaches to deal with the visualisation of large amounts of geospatial data.

13.2.1 Map Generalisation

Cartographic generalisation, or map generalisation, includes all changes in a map that are made when one derives a smaller-scale map from a larger-scale map or map data or vice-versa [4]. Generalisation seeks to abstract spatial information at a high level of detail to information that can be rendered on a map at a lower level of detail. This is of high importance when dealing with massive amounts of data, as it would be prohibitively—in terms of computation, data transfer and user experience (i.e. real-time interactivity)—to try to render the several gigabytes of data "as it is".

In that sense, suitable and useful maps typically have the right balance between the map's purpose and the precise detail of the subject being mapped. Well-generalised maps are those that emphasise the most important map elements while still representing the world in the most faithful and recognisable way [5].

There are many cartographic techniques that may fall into the broad category of generalisation [4, 6]. Among the most commonly used methods, we can find:

- **Simplification**—allowing to reduce the complexity of the geometries (i.e. lines and polygons) by eliminating or merging some of their vertices
- **Aggregation**—allowing to combine or merge some of the geometries (e.g. using the distance between polygons or by common attribute values) and thus resulting in a more reduced set of geometries.
- **Selection/Elimination**—allowing to reduce the number of features in the map by filtering or retaining them according to certain criteria (e.g. attribute values, spatial relations such as overlaps and distance between them).
- **Typification**—This method can be seen as an extreme case of simplification, where a detailed geometry is replaced by a simpler one to represent the feature in the map (e.g. a polygon defining the boundaries of a city is replaced by a point).
- **Exaggeration**—allowing to visually make more prominent some aspects we are interested in presenting in the map (e.g. represent cities with larger or smaller point sizes depending on the number of inhabitants).
- **Classification**—allowing to group into the same category and present in the map features with similar values.
- **Resampling**—which allows to reduce the amount of information provided in a map by changing its spatial resolution (e.g. changing the resolution of a raster dataset where the original pixel size is resampled from 100 m^2 to 1 km^2). This can be seen as a particular case of the aggregation method, involving interpolation techniques for determining the pixel values of the new resulting raster.

13.2.2   Rendered Images Versus the "Real" Data

In general, the process of rendering geospatially enabled information into maps is quite costly. Usually, the information, either raster- or vector-based, is stored in files or databases, which must be searched, queried, filtered and then transformed into a georeferenced map that can be integrated in a desktop or web client. This process can take longer the more information we have in our repositories, which can be very inefficient when several concurrent users make requests to the web mapping service asking for different areas or zoom levels. This can lead to unresponsive services due to the large workload imposed to the server.

In order to alleviate this issue, web mapping services offer the possibility to send the maps in the form of a tiled map, which is displayed in the client by joining dozens of individually requested image or vector data files (tiles) over the Internet. The advantage of this approach is that instead of loading the whole map instantly, for each zoom level, the web mapping service divides the imagery into a set of map tiles, which are logically put in an order which the application can understand. When the user scrolls the map to a new location, or to a new zoom level or location, the service decides which tiles are necessary and translates those values into a set of tiles to retrieve.

Concerning the tiling formats, there are two possibilities, each of them with their advantages and drawbacks:

Raster tiles are used to divide raster data into small, manageable areas that are stored as individual files in the filesystem (or BLOBs in a database). The tile-based approach is fundamental for efficient and improved performance for data loading, querying, visualisation and transfer of information over the networks. Thus, for instance, if a user zooms in a map into a small two tile area in a single band image, the underlying management service (e.g. OGC WMS) needs to fetch only two raster tile files from the filesystem instead of the entire raster dataset in order to compose the final image sent to the client.

Raster tiles of 256×256 pixel images are a de facto standard; however, 512×512 pixel seems to be the usual size of high-resolution tiles. Other sizes are possible depending on the purpose (e.g. 64×64 pixel images for mobile use), and in fact, a common approach is to generate a pyramid of different tile sizes that are used depending on the zoom level requested on the client side (Fig. 13.2).

Vector tiles are similar to raster tiles, but instead of raster images, the data returned is a vector representation of the features in the tile [7].

At the client side, it is possible to mix raster tiles with vector tiles and make the best usage of both, e.g. satellite map (raster tiles) with an overlay of streets with labels available in many languages (vector tiles) (Table 13.1).

As it can be seen, it could be possible to mix raster tiles with vector tiles and make the best usage of both, e.g. satellite map (raster tiles) with an overlay of different cartography and thematic layers (vector tiles).

---

[1] https://wiki.osgeo.org/wiki/Vistsos.

13.2.3 Use of Graphics Processing Units (GPUs)

Large-scale visualisation is an ideal application for graphics processing unit (GPU) computing for several reasons [10]:

- Visualisation is a data-intensive application, particularly as the problem size increases into the petascale. GPUs are well suited for data-intensive tasks.
- Visualisation computations exhibit substantial parallelism, typically both object parallelism (many objects or parts of objects can be computed/viewed in parallel) and image parallelism (visualisations produce large images, and image parts can be computed/viewed in parallel). Parallel computations are necessary for GPUs to be effective.
- Visualisation tasks should be closely coupled to the graphics system; even though much of overall visualisation computation may not be graphics centric, the final stage typically is, and so moving computation closer to the graphics device offers potential benefits in terms of interactivity and computational steering.
- GPUs can offload computation from CPUs, permitting the entire application to run faster when GPUs are involved.

More particularly, the many different functions used to manipulate geospatial data create additional processing workloads ideally fitted to GPU-accelerated solutions. Examples of these functions include:

- Filtering by area, attribute, series, geometry, etc.
- Aggregation, potentially in histograms.
- Geo-fencing based on triggers.
- Generating videos of events.
- Creation of heat maps.

Nowadays, there are big data solutions and frameworks leveraging in GPU capabilities for improving the data processing and rendering (both at server and client side), among others:

- Server side:

Fig. 13.2 Pyramid tile structure[1]

Table 13.1 Comparison raster and vector tiles use (pros/contras) [8, 9]

	Raster tiles	Vector tiles
Pros	• Tiles are generally rendered in advance on the server and streamed to the destination • Detailed tiles can be generated and served • More suitable for the display of imagery and shaded terrain • Lower requirements for end users hardware • Still a bit better support in web JavaScript libraries and desktop GIS software	• Tiles are rendered quickly and are only 20–50 per cent the file size of raster tiles • More tiles can be produced per second • Less bandwidth is needed due to the smaller size of tile packages—making vector tiles a better choice when streaming to devices • Map styles (colour, grey, night mode, etc.) can be changed without needing to download more information or other tile sets • Dynamic labelling allows size and font types to be changed on the fly • Better user experience —smooth zooming • No need for zoom levels—- users zoom and pan throughout all scales • De facto mobile standard
Contras	• Each map style must be created in a separate raster tile set • Labelling is preset and cannot be changed • A bigger size of each tile and data on servers • Takes more time to generate—can be CPU and memory consuming • Not the greatest for real-time rendering. Slower loading disrupts the user experience when moving around the map	• Rendering occurs on the client side, where limited resources can hamper speed • Compromises clarity by reducing display detail • Requires OpenGL/WebGL/DirectX support, which is an issue for some mobile devices • Not suitable for imagery or other raster maps • Vectors are generalised (i.e. not raw data) so they may not be suitable for editing

- Rasdaman array database (only available in the enterprise version)[2]
- OmniSci database (formerly MapD)
- AresDB
- Apache Spark
- PostgreSQL and PG-Strom extension[3]

• Client side:

- Cesium
- MapD-charting
- Kepler.gl.

---

[2] https://rasdaman.com/commercial-free.php.

[3] https://heterodb.github.io/pg-strom/.

Another example of visualisation leveraging on graphical cards is exploratory visualisation. Exploratory visualisation is the process that involves an expert creating maps and other graphics while dealing with relatively unknown geographic data. Generally, these maps serve a single purpose and function as an expedient in the expert's attempt to solve a particular (geo) problem. While working with the data, the expert should be able to rely on cartographic expertise to be able to view data from different perspectives. As such, the resulting maps and graphics are available in an interactive viewing environment that stimulates visual thinking and promotes informed decision-making. WebGLayer[4] is a JavaScript library focused on fast interactive visualisation of big multidimensional spatial data through linked views. The library is based on WebGL and uses GPU for fast rendering and filtering. Using commodity hardware, the library can visualise hundreds of thousands of features with several attributes through heatmap or point symbol map. The library can render data on the map provided by third party libraries (e.g. OpenLayers, Leaflet, GoogleMap API). Figure 13.3 shows an example for the analysis of yield potential [11].

13.3 Examples from DataBio Project

13.3.1 Linked Data Visualisation

Linked data visualisation is about providing graphical representations of interesting aspects within the Semantic web. The high variety of linked data and its types is huge. An example of agriculture linked open data is the FOODIE data model, which

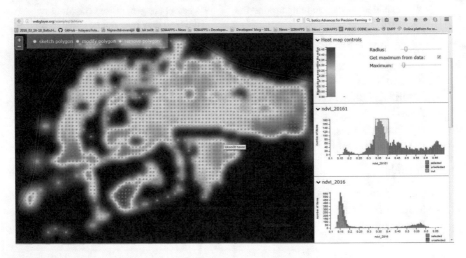

Fig. 13.3 WebGLayer showing yield potential

---

[4] https://webglayer.org.

was originally developed as part of the FOODIE project and later extended in the
DataBio project. The FOODIE data model is based on the generic data models of
INSPIRE, especially the data models for agricultural and aquaculture facilities and
Land-Parcel information system. The key motivation was to represent a continuous
area of agricultural land with one type of crop species, cultivated by one user in one
farming mode (conventional vs. transitional vs. organic farming). Additionally, the
FOODIE data model includes concepts for crop and soil data, treatments, interventions, agriculture machinery and others. Finally, the model reuses data types defined
in ISO standards (ISO 19101, ISO/TS 19103, ISO 8601 and ISO 19115) as well stan-
dardisation efforts published under the INSPIRE directive (like structure of unique
identifiers). The FOODIE data model was specified in UML (as the INSPIRE models)
but can be transformed into an OWL ontology in order to enable the publication of
linked data compliant with FOODIE data model [12].

As mentioned in Chap. 8 "Linked Data Usages in DataBio" the triplestore with
linked data has over 1 billion triples—which is organised into named graphs (IRI)
and sub-graphs. For example, the LPIS-Poland dataset (Land-Parcel identification
in Poland) can be identified by the graph <https://w3id.org/foodie/open/pl/LPIS/>
and contains 727,517,039 triples with a subgraph <https://w3id.org/foodie/open/pl/
LPIS/lubelskie#> , referring to the data with the Lublin Voivodeship. Thus, querying
and pre-processing, including link discovery, are very important for an efficient way
to visualise linked data. Depending on the size of linked datasets (amount, distributed
etc.) and the linkages between the data, there are different ways to visualise them. In
DataBio, metaphactory, a linked data exploitation platform, has been used to query,
browse and navigate linked data—for example, the catch records data from Norway
(see Fig. 13.4).

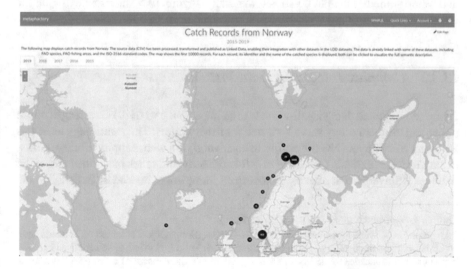

Fig. 13.4 DataBio metaphactory custom view (map with catch records from Norway)

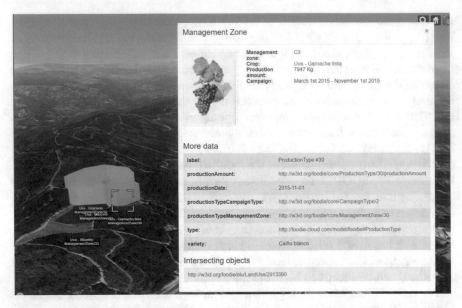

Fig. 13.5 Screenshot of the application showing result of use case crops types based on linked data

The second way to visualise is to query the SPARQL endpoint(s) (using GeoSPARQL[5]) and get RDF or JSON-LD.[6] There is also the possibility to discover more data (types and links) and put them together. Finally, the results can be transformed into the form of JSON resp. GeoJSON, which are easily processed by most visualisation clients. Leading technology providers are aware of this need and plan to develop some features to do so automatically. Figure 13.5 shows an example for visualising different crop types based on information from linked data.[7]

13.3.2 Complex Integrated Data Visualisation

Complex integrated data visualisation was an important part of the Czech agriculture pilots, and the technology was also tested for fishery pilots. The technology used was HSlayers NG. Hlayers NG (https://ng.hslayers.org/) is a web mapping library written in JavaScript. It extends OpenLayers 4 functionality and takes basic ideas from the previous HSlayers library but uses modern JS frameworks instead of ExtSJS 3 at the

---

[5] https://www.opengeospatial.org/standards/geosparql.

[6] An extension of JSON for Linked Data is JSON-LD (JavaScript Object Notation for Linked Data), which is a method of encoding Linked Data using JSON. This allows data to be serialised in a way that is like traditional JSON. JSON-LD is designed around the concept of a "context" to provide additional mappings from JSON to an RDF model.

[7] Further examples for integrated data visualisation on maps from DataBio can be explored under the following link: https://app.hslayers.org/project-databio/land/.

frontend and provides better adaptability. That is why the NG ("Next Generation") is added to its name. It is still under development and provided as open-source software. HSLayers is built in a modular way which enables the modules to be freely attached and removed as far as the dependencies for each of them are satisfied. The dependency checking is done automatically. The core of the framework is developed using AngularJS, requireJS and Bootstrap. This combination of frameworks was chosen mainly for providing fast and scalable development and for providing a modern responsive layout for the application. Figure 13.6 gives an example for a complex integrated data visualisation.

The most important modules are:

- The map functionality is provided by OpenLayers4 and extended by some controls.
-

Fig. 13.6 Integration of yield potential data (3D maps) with meteorological data (time series) [11–13]

Layer manager is used for listing all the map layers, displaying or hiding them and setting the transparency.

- OGC web services parser is used for GetCapabilities requests to different map servers and parsing the response.
- Linked Open Data explorer: Eurostat explorer is a demo application (module) which queries Semantic web data sources via SPARQL endpoints.
- HSlayers visualises geographical data in a 3D environment.
- Support for visualisation of sensors and agrometeorological data for farmers can help with forecast of weather and better planning of operations.

13.3.3   Web-Based Visualisation of Big Geospatial Vector Data

Chapter 15 introduces various pilots on smart farming for sustainable agricultural production in Greece. In these applications, information about growing crops in millions of parcels spread over the country needs to be visualised. The information about the growing plants, trees and grain types is updated periodically, which makes the data dynamic. Providing a map interface that supports end users to explore this amount of dense data using a vector-based approach is a big challenge to the implementation.

In order to address this challenge, an approach to visualise huge sets of geospatial data in modern web browsers along with maintaining a dynamic tile tree was developed in the DataBio project and successfully applied to the pilot application [14]. The approach makes it possible to render over one million polygons integrated in a modern web application by using 2D vector tiles (see Sect. 13.2.2). Figure 13.7 shows an example for an in-depth parcel assessment with vegetation index colour coding for Greece.

This novel approach to build and maintain the tile tree database provides an interface to import new data and a more flexible and responsive way to request vector tiles. There are three essential steps involved [14]:

1. Data storage is re-organised in a way to have efficient access to geospatial vector tiles. This is achieved by using a geospatial index along with the fast and scalable distributed file system GeoRocket.[8] GeoRocket uses MongoDB to persist data and Elasticsearch to build a spatial index for data query and aggregation tasks. GeoJSON can be imported directly without conversion.
2. Secondly, it is essential to speed up the vector tile creation process, which is important for both, the initial creation of the tile tree and serving tiles. For this, a new tiling algorithm was implemented. The tiling implementation is a server component itself and provides a REST interface. It can be configured using different file storage backend technologies for persisting the tiles. The configuration includes a range of zoom levels in which the tiles are created,

---

[8] GeoRocket—https://georocket.io.

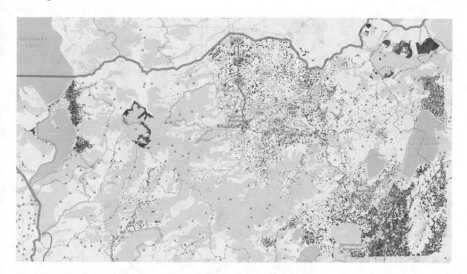

Fig. 13.7 In-depth parcel assessment with vegetation index colour coding

which is 2 to 15 by default. These are enough for most users' map interface experience, but for a more detailed view, it is also possible to build tiles on higher zoom levels.

3. Finally, data must be transmitted to a web application running in modern web browsers. The geometries are rendered using a WebGL map application framework. It is possible to add interaction concepts such as filters and user-defined styling. The most common and stable frameworks are OpenLayers and MapboxGL JS. The young vector tile implementation in OpenLayers has many issues, most critical a memory leak, no data-driven styling and no WebGL support for vector tiles. Therefore, Mapbox GL JS was used in the pilot application and evaluation.

13.3.4 Visualisation of Historical Earth Observation

Earth observation measurements provided by satellites from the Sentinel and Landsat programmes are one of the largest sources of big geospatial data, which are not only challenging in terms of data storage and access management (as presented in Chap. 4 Remote Sensing) but also for filtering, processing and visualising due to the large size of the files. Figure 13.8 shows an example from the DataBio fisheries pilot, where a web client is used for 3D visualisation of oceanic historical datasets, such as ocean salinity, temperature, concentration of chlorophyll.), in the whole Indian Ocean region where the fishery vessels operate.

The satellite imagery time series is served through the Rasdaman service via the OGC WMS-T and WMST interfaces and integrated with the HSLayers and Cesium JS library, which allow to display geospatial data available in various raster and vector

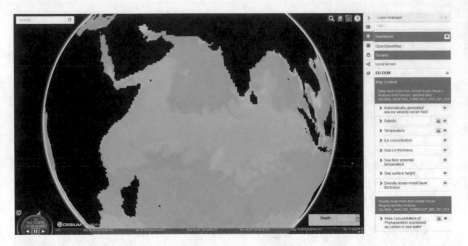

Fig. 13.8 3D web visualisation of historical oceanic measurements using HS layers and Rasdaman

formats. The web client component allows to control the visualisation by additional (non-spatial) dimensions of the data. In this specific case, the web client enables the user of the application to choose the time and depth level parameters, which are then used to query the Rasdaman service, returning the rendered map in the form of a series of raster tiled images.

13.3.5 Dashboard for Machinery Maintenance

Visualisation is important when informing the user about the status of technical processes, e.g. in machine maintenance. Especially, it is central to show alerts about critical events, like too high temperature, pressure and so on. Use of colours and visual effects, like blinking, must be considered with great care. In Fig. 13.9 is an example from DataBio, where we designed a visual dashboard for showing information about the status of the engines of fishing vessels.

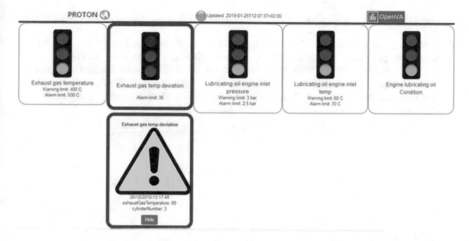

Fig. 13.9 Visual dashboard from a DataBio pilot on fishery. The dashboard shows information and alerts about the status of the fishing vessel's engines

References

1. https://www.tableau.com/learn/articles/data-visualization (Accessed September 2019).
2. Rubens, P. (2017). *Big data visualization.* Available at: https://www.datamation.com/big-data/big-data-visualization.html (Accessed September 2019).
3. Abela, A. V. (2013). *Advanced presentations by design—creating communication that drives action* (2nd ed.). Wiley.
4. Li, Z. (2007). Digital map generalization at the age of the enlightenment: A review of the first forty years. *The Cartographic Journal, 44*(1), 80–93. https://doi.org/10.1179/000870407x173913.
5. 'Cartographic generalization'. (2020). *Wikipedia.* Available at: .https://en.wikipedia.org/wiki/Cartographic_generalization (Accessed: 24 September 2020).
6. Raposo, P. (2017). Scale and Generalization. In J. P. Wilson (Ed.), *The geographic information science & technology body of knowledge* (4th Quarter 2017 Edn.), https://doi.org/10.22224/gistbok/2017.4.3. Available at: https://gistbok.ucgis.org/bok-topics/scale-and-generalization-1
7. 'Vector Tiles'. (2020). Openstreetmap Wiki. Available at: https://wiki.openstreetmap.org/wiki/Vector_tiles (Accessed: 24 September 2020).
8. 'Vector tiles vs raster tiles—the pros and cons'. (2017). From TechBlog of MapData Services. Available at: https://mapdataservices.wordpress.com/2017/02/22/vector-tiles-vs-raster-tiles-the-pros-and-cons/ (Accessed: 24 September 2020).
9. Janak, D. (2019). 'What are vector tiles and why you should care'. Available at: https://www.maptiler.com/blog/2019/02/what-are-vector-tiles-and-why-you-should-care.html
10. Owens, J. D. (2007). Towards multi-GPU support for visualisation. *Journal of Physics Conference Series (Online), 78*(1), 012055. https://doi.org/10.1088/1742-6596/78/1/012055.
11. Řezník, Tomáš et al. (2016). MONITORING OF IN-FIELD VARIABILITY FOR SITE SPECIFIC CROP MANAGEMENT THROUGH OPEN GEOSPATIAL INFORMATION. ISPRS - International Archives of the Photogrammetry, Remote Sensing and Spatial Information Sciences. XLI-B8. 1023–1028. https://doi.org/10.5194/isprsarchives-XLI-B8-1023-2016.
12. Charvat, K., Junior, K. C., Reznik, T., Lukas, V., Jedlicka, K., Palma, R., & Berzins, R. (2018, July). Advanced visualisation of big data for agriculture as part of databio development. In

IGARSS 2018–2018 IEEE International Geoscience and Remote Sensing Symposium (pp. 415–418). IEEE.

13. Jedlička, K., & Charvát, K. (2018, May). Visualisation of Big Data in Agriculture and Rural Development. In *2018 IST-Africa Week Conference (IST-Africa)* (pp. Page-1). IEEE.

14. Zouhar F., Senner I. (2020) Web-Based visualisation of Big Geospatial Vector Data. In: Kyriakidis P., Hadjimitsis D., Skarlatos D., Mansourian A. (eds) Geospatial Technologies for Local and Regional Development. AGILE 2019. Lecture Notes in Geoinformation and Cartography. Springer, Cham. https://doi.org/10.1007/978-3-030-14745-7

Part V
Applications in Agriculture

Chapter 14
Introduction of Smart Agriculture

Christian Zinke-Wehlmann and Karel Charvát

Abstract Smart agriculture is a rising area bringing the benefits of digitalization through big data, artificial intelligence and linked data into the agricultural domain. This chapter motivates the use and describes the rise of smart agriculture.

14.1 Situation

Agriculture is a central sector for all of us, but there are significant challenges that this sector and the whole society face:

- A growing populationraises the demand for food "by roughly 50 percent compared to 2013 agricultural output" [1].
- Globalisation is mixing food cultures.
- Healthy living and the elderly population are requiring different diets than before.
- Urbanization with an increasing demand for processed and high-quality food.
- Land abandonment due to growing urbanization.
- Limited and highly stressed natural resources—overused farmland becomes degraded (e.g., soil erosion, unbalanced fertilizer usage), water resources are threatened.
- Climate change affects crop growth negatively due to higher temperatures and poses higher risks for yield loss by droughts and floods.
- New policies influence agriculture production, and changes in the subsidies system can rapidly influence agriculture production [2].

Thus, the supply chain security of high-quality food products becomes very relevant, while at the same time, the global demand for food is growing [3]. To

C. Zinke-Wehlmann (✉)
Institute for Applied Informatics e.V. at the University of Leipzig, Goerdelering 9, Leipzig, Germany
e-mail: Zinke@infai.org

K. Charvát
Lesprojekt-Sluzby, Ltd, Zaryby, Czech Republic

address these challenges, digitalization and data-driven approaches for agriculture have emerged [4]. However, digitalization of agriculture is not only economically driven, but also advanced by legal requirements, fertilizer ordinances and sustainable management of natural resources [5].

Crop modeling, yield monitoring, satellite navigation, earth observation, and cheap and high precision sensors are well-known examples of digitalization, or to be more precise, of precision farming/agriculture and smart agriculture [6].

14.2 Precision Agriculture

"Precision farming makes use of information technologies in agriculture. With the satellite positioning system and electronic communication standards, position and time may be integrated into all procedures connected to farming" [7].

The goal of precision farming is to do the right things at the right places with the right intensity—e.g., fertilizing [8]. However, it is an information-driven approach to support the farmer's decisions, mostly resulting in farm-management-systems. The forecasted market value of these technologies in 2023 is 9.53 billion US dollars. With the growth of technological possibilities and development (more sensors, the expansion of Internet of Things, more data sources, e.g., earth observation and weather forecasts), cyber-physical systems became relevant for agriculture [9]. The growth of information came along with the demand for intelligent solutions.

14.3 Smart Agriculture

Smart agriculture is not only about bringing information technology in agriculture, but rather more about creating and using knowledge through technology. Agricultural machines and devices should be enabled by information technology to process and analyze data—and finally, make some decisions, or prepare them semi-automatically [4, 10]. It is based upon the rise of big data technologies [11], the Internet of Things [12], satellite observation [13], linked data [14], and artificial intelligence [15] in all the agriculture supply chain stages [5]. The forecasted market value of smart farming worldwide is 23.1 billion dollars (including precision farming). The following chapters in Part V underlines the importance of smart farming in terms of agricultural productivity, environmental impact, food security, and sustainability, with applications in the areas of crops, soil, biodiversity, farmer's decision-making, and many more—in line with works like [6]. Concretely, the following chapters demonstrate how smart agriculture can be applied.

- Chapter 15 demonstrates smart farming services based on IoT, EO data and big data analytics. They are able to provide advice for fertilization, irrigation, and crop protection in a flexible way to the farmers. The services promote sustainable farming practices for better control and management of the resources.

- Chapter 16 presents an approach for genomic prediction and selection of biomass. The data came from several sources, such as phenomics, genomics and sensors. The presented approach of smart agriculture provides the enabling technologies and knowledge to support crop breeding companies.
- Chapter 17 introduces yield prediction models for sorghum and potatoes. High-resolution satellite images were used to predict yields. Through the presented smart farming approach, farmers can improve their business operations through informed decision-making in planning field work, logistics and supply chains.
- Chapter 18 demonstrates the variable application of nitrogen fertilizers on farm fields based on satellite monitoring..
- Smart agriculture is not only about the primary supply chain; it is also about services to protect farmers. Considering the current challenges related to climate change effects and the increasing world population, insurance assessments may ensure a higher resilience of agriculture. Chapter 19 presents a first step towards data-based insurance for smart agriculture.
- To set up more environmentally friendly and efficient agricultural practices, tools and services to support compliance management, e.g., CAP, is needed. Chapter 20 demonstrates how the processing and analysis of Copernicus satellite imagery can offer compliance checking and a great range of supplementary information for public authorities and farmers.
- The concluding chapter in this Part V summarizes the presented work and gives a brief outlook on smart agriculture in the near future.

References

1. De Clercq, M., Vats, A., & Biel, A. (2018). Agriculture 4.0: The future of farming technology. Proceedings of the World Government Summit, Dubai, UAE, (p. 1113).
2. Charvat, K., Gnip, P., Vohnout, P., & Charvat jr, K. (2011). *Vision for a farm of tomorrow*. IST Africa.
3. Tilman, D., Balzer, C., Hill, J., & Befort, B. L. (2011). Global food demand and the sustainable intensification of agriculture. *Proceedings of the National Academy of Sciences, 108*(50), 20260–20264.
4. Weltzien, C. (2016). Digital agriculture or why agriculture 4.0 still offers only modest returns. *Landtechnik, 71*(2), 66–68.
5. Bovensiepen, G. & Hombach, R. (2016). Quo vadis, agricola? Smart farming: Nachhaltigkeit und Effizienz durch den Einsatz digitaler Technologien (pwc).
6. Kamilaris, A., Kartakoullis, A., & Prenafeta-Boldú, F. X. (2017). A review on the practice of big data analysis in agriculture. *Computers and Electronics in Agriculture, 143*, 2337.
7. Auernhammer, H. (2001). Precision farming—the environmental challenge. *Computers and Electronics in Agriculture, 30*(1–3), 31–43.
8. Lal, R., & Stewart, B. A. (Eds.). (2015). Soil-specific farming: precision agriculture (Vol. 22). CRC Press.

9. Rad, C. R., Hancu, O., Takacs, I. A., & Olteanu, G. (2015). Smart monitoring of potato crop: A cyber-physical system architecture model in the field of precision agriculture. *Agriculture and Agricultural Science Procedia, 6,* 73–79.
10. Ozdogan, B., Gacar, A., & Aktas, H. (2017). Digital agriculture practices in the context of agriculture 4.0. *Journal of Economics Finance and Accounting, 4*(2), 186–193.
11. Sun, Z., Zheng, F., & Yin, S. (2013). Perspectives of research and application of Big Data on smart agriculture. *Journal of Agricultural Science and Technology (Beijing), 15*(6), 63–71.
12. Gondchawar, N., & Kawitkar, R. S. (2016). IoT based smart agriculture. *International Journal of Advanced Research in Computer and Communication Engineering, 5*(6), 838–842.
13. Corbari, C., Salerno, R., Ceppi, A., Telesca, V., & Mancini, M. (2019). Smart irrigation forecast using satellite LANDSAT data and meteo-hydrological modeling. *Agricultural Water Management, 212,* 283–294.
14. Kamilaris, A., Gao, F., Prenafeta-Boldu, F. X., & Ali, M. I. (2016, December). Agri-IoT: A semantic framework for internet of things-enabled smart farming applications. In *2016 IEEE 3rd World Forum on Internet of Things (WF-IoT)* (pp. 442–447). IEEE.
15. Zhu, N., Liu, X., Liu, Z., Hu, K., Wang, Y., Tan, J., ... & Guo, Y. (2018). Deep learning for smart agriculture: Concepts, tools, applications, and opportunities. *International Journal of Agricultural and Biological Engineering, 11*(4), 32–44.

Chapter 15
Smart Farming for Sustainable Agricultural Production

Savvas Rogotis and Nikolaos Marianos

Abstract The chapter describes DataBio's pilot applications, led by NEUROPUBLIC S.A., for sustainable agricultural production in Greece. Initially, it introduces the main aspects that drive and motivate the execution of the pilot. The pilot set up consisted of four (4) different locations, four (4) different crop types and three (3) different types of offered services. The technology pipeline was based on the exploitation of heterogeneous data and their transformation into facts and actionable advice fostering sustainable agricultural growth. The results of the pilot activities effectively showcased how smart farming methodologies can lead to a positive impact from an economical, environmental and societal perspective and achieve the ambitious goal to "produce more with less". The chapter concludes with "how-to" guidelines and the pilot's key findings.

15.1 Introduction, Motivation and Goals

The global population is expected to reach 9 billion by 2050 and feeding that population will require a 70% increase in food production (FAO 2009[1]). At the same time, farmers are facing a series of challenges in their businesses that affect their farm production, such as crop pests and diseases, with increased resistance along with drastic changes due to the effects of climate change. These factors lead to rising food prices that have pushed over 40 million people into poverty since 2010, a fact that highlights the need for more effective interventions in agriculture (World Bank 2011[2]). In this context, agri-food researchers are working on approaches that aim to maximize agricultural production and reduce yield risk. The benefits of the

---

[1] https://www.fao.org/fileadmin/user_upload/lon/HLEF2050_Global_Agriculture.pdf.

[2] https://www.worldbank.org/en/topic/agriculture/overview.

S. Rogotis (✉) · N. Marianos
NEUROPUBLIC SA, Methonis 6 & Spiliotopoulou, 18545 Piraeus, Greece
e-mail: s_rogotis@neuropublic.gr

N. Marianos
e-mail: n_marianos@neuropublic.gr

© The Author(s) 2021
C. Södergård et al. (eds.), *Big Data in Bioeconomy*,
https://doi.org/10.1007/978-3-030-71069-9_15

ICT-based revolution have already significantly improved agricultural productivity; however, there is a demonstrable need for a new revolution that will contribute to "smart" farming and help to address all the aforementioned problems (World Bank 2011). There is a need for services that are powered by scientific knowledge, driven by facts and offer inexpensive yet valuable advice to farmers. In this context, smart farming is expected to reduce production costs, increase production (quantitatively) and improve its quality, protect the environment and minimize farmers' risks.

The main focus of the pilot activities is to offer smart farming advisory services referring to the cultivation of olives, peaches, grapes (pilot application scenario (1) and cotton (pilot application scenario (2) based on a unique combination of technologies such as earth observation (EO), big data analytics and Internet of Things (IoT).

The pilot activities exploit heterogeneous data, facts and scientific knowledge to facilitate decisions and field applications. They promote the adoption of big data-enabled technologies and the collaboration with certified professionals helps to manage the natural resources better, optimize the use of agricultural inputs (i.e. agrochemicals such as fertilisers) and lead to increased product quality and farm productivity.

Smart farming services provide advices for fertilization, irrigation and crop protection, adapted to the specific needs of each pilot parcel and offered through flexible mechanisms to the farmers or the agricultural advisors.

The main aspects that motivate and drive this pilot are:

- to raise the awareness of the farmers, agronomists, agricultural advisors, farmer cooperatives and organizations (e.g. group of producers) on how new technological tools could optimize farm profitability and offer a significant advantage on a highly competitive sector,
- to promote sustainable farming practises over a better control and management of the resources (fresh water, fertilizers, etc.),
- to increase the technological capacity of the involved partners through a set of pilot activities involving big data management data for high-value crops.

15.2 Pilot Set-Up

This section contains pilot set-up descriptions for the two (2) distinct pilot application scenarios that are considered together as they are provided by the same team of partners and are based on the same big data pipeline that has been adjusted to address their distinct needs. More specifically, pilot application scenario 1 worked with three (3) different crop types in three (3) different pilot areas offering a set of advisory services for irrigation, fertilization and crop protection:

- Chalkidiki (Northern Greece), where the pilot worked with olive groves of 600 ha for the production of table olives,

- Stimagka (Southern Greece), where the pilot worked with vineyards of 3.000 ha for the production of table grapes,
- Veria (Northern Greece), where the pilot worked with peach orchards covering an area of 10.000 ha.

At the same time, pilot application scenario 2 worked with one (1) crop type in one (1) site offering irrigation advisory services in the context of arable farming:

- Kileler (Thessaly), where the pilot worked with cotton of 5000 ha (Fig. 15.1).

The underlying reason for selecting these particular crop types is the great economic impact they share in the Greek farming landscape. As an example, olive tree cultivation accounts for nearly 2 billion euros in annual net income, while peach and grape cultivations reach close to 460 million and 390 million annual net income, respectively (Table 15.1).

In the pilot sites, NP was leading the activities, supported by GAIA EPICHEIREIN as the primary business partner and liaison with the farming communities, IBM (only contributing in application scenario 1) and FRAUNHOFER joined the pilot activities as technology providers. By the end of the project, a set of validated fully operational smart farming services were developed, adapted at each crop type and the microclimatic conditions of each pilot area.

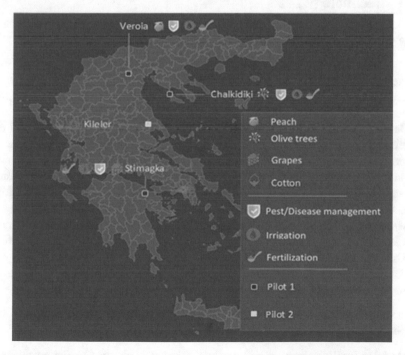

Fig. 15.1 Pilot application scenario 1 (marked as Pilot 1) and pilot application scenario 2 (marked as Pilot 2) joint high-level overview indicating pilot sites, targeted crop types and offered advisory services

Table. 15.1 Overview of the big data-driven smart services deployed at the four pilot sites

Service	Pilot application scenario 1 locations			Pilot application scenario 2 Location
	Chalkidiki (Olives)	Veria (Peaches)	Stimagka (Grapes)	Kileler (Cotton)
Irrigation	+	+	+	+
Fertilization	+	+	–	–
Crop protection	+ Exploitation of scientific models for 1 pest and 1 disease)	+ Exploitation of scientific models for 3 pests and 4 diseases)	+ Exploitation of scientific models for 2 pests and 3 diseases)	–

Goal achievement was measured by defining specific key performance indicators (KPIs). For each goal, baseline KPIs were measured and compared to achievements after the pilot activities finished (after two consecutive trial seasons).

15.3 Technology Used

15.3.1 Technology Pipeline

The technology pipeline of the solutions applied in these pilot activities (both application scenarios) consists on a high level of abstraction of data collection, data processing and data visualisation components (Fig. 15.2).

Data collection: To provide advice related to irrigation, fertilization and crop protection, a set of heterogeneous data is required, capturing critical parameters for crop status monitoring in different spatial and temporal resolutions. Weather, soil and plant-related data, crowdsourced samples, observations and information for the applied farming practices, intra-field—inter-field EO-based vegetation indices consist of different data flows that find their way into the technology pipeline.

Moreover, historical data from at least one cultivating period prior to pilot activities is required for calibrating/fine-tuning the scientific models that constitute the backbone of the advisory services.

For addressing the pilot needs in terms of data collection, the following technological modules are being exploited:

- In situ telemetric stations provided by NP, called gaiatrons, that collect field-level data related to weather, soil and plant (Fig. 15.3),

- Modules for the collection, pre-preprocessing of earth observation products, the extraction of higher level products and the assignment of EO-based vegetation indices at parcel level,

Fig. 15.2 Concept underpinning the pilot activities

- Android apps for crowdsourcing data from farmers (farm logs), agricultural advisors and agronomists about field status and the applied farming practices,
- Web-based user interfaces for collecting and updating the available farm data.

Data processing: The collected datasets are processed by several complementary data processing components provided by the pilot partners. Big data components that should be mentioned in this context are:

- GAIABus DataSmart Real-time streaming Subcomponent (offered by NP): This component allows for: the real-time data stream monitoring resulting from NP's telemetric stations installed in all pilot sites; the real-time validation of data and the real-time parsing and cross-checking.
- PROTON (offered by IBM): PROTON is an early warning system for managing pests and diseases using sophisticated temporal reasoning for olives, grapes and

Fig. 15.3 NP's IoT agro-climatic station used in the pilot activities

peaches (it is used only in pilot application scenario 1). It exploits the numerical output (risk indicator) of NP's crop and area-tailored scientific models for pest/disease breakouts. In total, NP sends one (1) pest and one (1) disease risk indicator from each pilot site (6 scientific crop protection models are sent in total), namely:

– spilocaea oleaginea and bactocera olea (for olives cultivation)
– downy mildew and lobesia botrana (for grapes cultivation)
– grapholita_molesta and curl leaf (for peaches cultivation).

PROTON conducts sophisticated complex event processing on top of the risk indicators offering even earlier alerting/warning before conditions reach critical states. The results are being sent back to NP at specified intervals (e.g. once a week) for integration.

• Georocket, Geotoolbox, SmartVis3D (offered by FRAUNHOFER): The integration of these components has a dual role: It offers a back-end system for big data preparation, handling fast querying and spatial aggregations of data, as well as a front-end application for interactive data visualization and analytics.

Data visualisation and presentation: After all data is processed, it needs to be provided in an understandable and decision-relevant way suitable for the pilot end-users (farmers, agronomists). The primary data visualization component used in the pilot is NeuroCode (offered by NP). Neurocode allows the creation of the main pilot UIs that support the provision of smart farming advisory services for optimal decision making. An additional DataBio component explored for its information visualization functionalities was Georocket (offered by FRAUNHOFER) (Fig. 15.4).

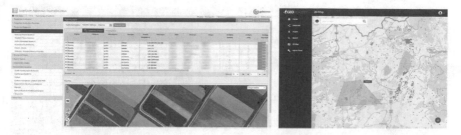

Fig. 15.4 Data visualization tools that were used in the pilot activities (Left: Neurocode, Right: Georocket)

15.3.2 Data Used in the Pilot

The specific pilot uses four (4) different data types as graphically depicted in Fig. 15.2. More specifically, the pilot exploits the following data assets:

- agro-climatic data recorded by in-situ IoT sensing units (field dimension),
- remote sensing data from satellite missions (remote dimension),
- farmer calendars and logs that capture farm profile and the applied field applications (farm dimension),
- samples, observation and field measurements offered by certified professionals (eye dimension).

However, the datasets that can be acknowledged for their big data aspects (in terms of volume, velocity, etc.) are the following:

- **Sensor measurements (numerical data) and metadata (timestamps, sensor id, etc.):** This dataset is composed of measurements from NP's telemetric IoT agro meteorological stations (gaiatrons) for the pilot sites. More than 20 gaiatrons are fully operational at all pilot sites, collecting >30MBs of data per year each with current configuration (offering measurements every 10 min).
- **EO products in raster format and metadata:** This dataset is comprised of ESA's remote sensing data from the Sentinel-2 optical products (6 tiles). High volumes of satellite data are continually being processed in order to extract the necessary information about each crop type and parcel participating in the pilot.

15.3.3 Reflection on Technology Use

The pilot has completed two rounds of trials. It conclusively demonstrated how big data-enabled technologies and smart farming advisory services can offer the means for better handling the natural resources and optimizing the use of agricultural inputs. The following figures indicate how technology can provide added value to farmers and lead to improved farm management (Figs. 15.5, 15.6 and 15.7).

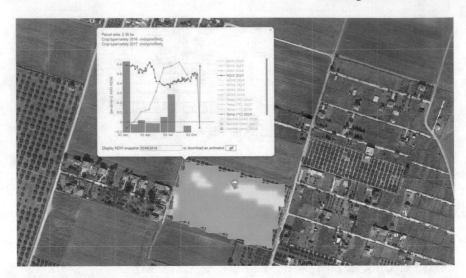

Fig. 15.5 Parcel monitoring at Chalkidiki pilot site indicating intra-field variations in terms of vegetation index (NDVI) and cross-correlations among the latter with: **a** ambient temperature (°C) and **b** rainfall (mm)

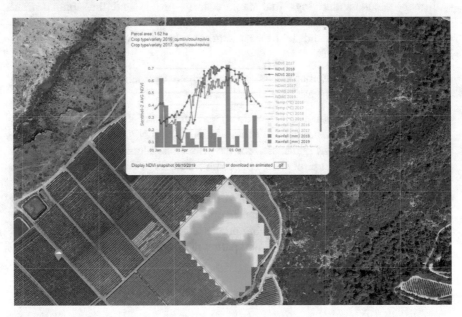

Fig. 15.6 Parcel monitoring at Stimagka pilot site indicating intra-field variations in terms of vegetation index (NDVI) and cross-correlations among the latter with **a** NDVI from 2018 cultivating period and **b** rainfall (mm) from 2018 and 2019 cultivating periods

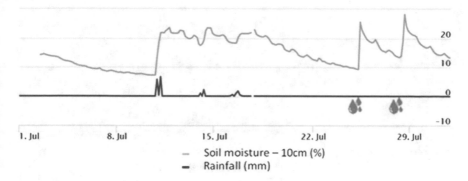

Fig. 15.7 Irrigation monitoring at a Veria pilot parcel showing two (2) correct irrigations (water drop icons) after following the advisory services during 2019 cultivating period. The impact of rainfalls in the soil water content is obvious (~10/6) and if translated correctly can prevent unnecessary irrigations

Getting more in-depth regarding irrigation advice generation, a critical factor that influences its provisioning is daily evapotranspiration. It essentially reflects the water content being lost each day from both the plant and the soil. By calculating this parameter using EO or model-based approaches, the requirement for installing a tense network of irrigation sensors for monitoring soil moisture ceases to exist. This significantly reduces infrastructure costs and leads to economy of scale, as irrigation advices can be extrapolated for a large number of parcels that share similar agro-climatic characteristics (soft facts) (Figs. 15.8, 15.9 and 15.10).

The technology pipeline can be easily used at other crop types and locations. This will require, however, an initial period of data collection (one cultivating period) to be used for the precise and complete documentation of the soil and microclimate

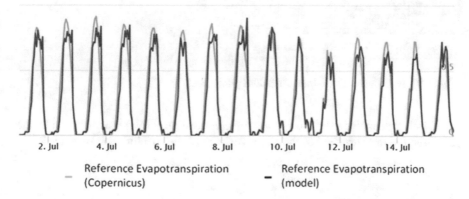

Fig. 15.8 Reference evapotranspiration monitoring at Kileler (both modelled using ML methods developed by NP and based on Copernicus EO data) for July 2019

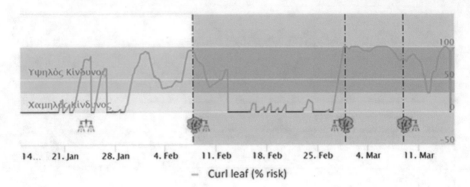

— Curl leaf (% risk)

Fig. 15.9 Crop protection monitoring at a Veria pilot parcel showing four (4) correct sprays (spraying icons) after following the advisory services and the indications for high curl leaf risk during 2019 cultivating period (high risk is when the indicator passes to the pink zone). The dashed vertical lines indicate critical crop phenological stages

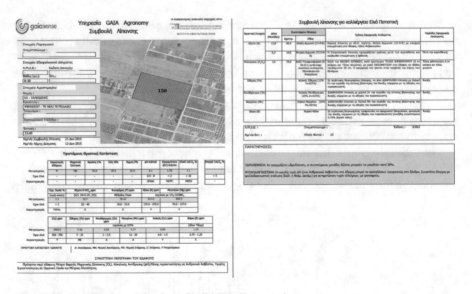

Fig. 15.10 Fertilization advice for a Chalkidiki pilot parcel

conditions that apply in the specific area, the cultivation activities undertaken by the producer, the measurement of the characteristics of the specific crop type, etc.

15.4 Business Value and Impact

15.4.1 Business Impact of the Pilot

Both pilots managed to achieve the expected results for input cost reduction, which was validated by the quantification of the results after trial stages 1 and 2. This was achieved as farmers and agricultural advisors showed a collaborative spirit and followed the advice generated by DataBio's solutions. Aggregated findings can be found at the following figures (Figs. 15.11 and 15.12).

For pilot application scenario 1, it is clear that in certain cases (irrigation), the results exceeded the initial set targets for input cost reduction. This is due to the

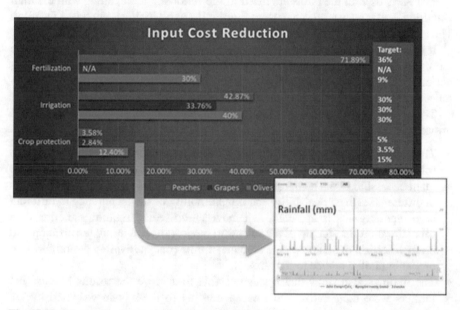

Fig. 15.11 Pilot application scenario 1 aggregated findings

Fig. 15.12 Aggregated results of pilot application scenario 2 in comparison with the target values

fact that the farmers both: (a) showed collaborative spirit and adapted their farming practices using all advice offered and (b) were benefiting from the weather conditions (rainfalls during June, July 2019) and this reduced the freshwater requirements during critical phenological stages. The aforementioned phenomenon was the underlying reason for slightly not reaching the targeted crop protection goals. The farmers chose to conduct additional proactive sprays for securing their production against threatening situations (e.g. fruit mucilage presence at the stage of swelling in Veria pilot site). In terms of fertilization, the exhibited deviation (under-fertilization) is part of the farmers' overall strategy that derives from the fact that fertilization advice is offered with a two-to-three-year application window. This allows them a window for taking fertilization measures and is expected that this deviation will be acknowledged and significantly shape the fertilization strategy over the next cultivating periods.

The KPIs used in the pilots are listed in the following table, along with the final DataBio results (measured values) that support the exploitation potential of the pilot. The following table sums the measured savings of the pilots per hectare (Table 15.2).

It is evident that the pilot's business impact would be further validated and reach more conclusive insights as KPI measurements from more (and different) cultivating periods get aggregated over the years. More trials would allow to get more business-related KPI measurements maximizing the pilot's impact.

The achieved results allow for the following conclusions regarding the business impact:

- The findings show that technology use results in real financial savings per hectare for all considered crop types and regions. As different crop types have various input necessities from an agronomical point of view, the technology used results in different savings. Scalability and transferability of the technology in different crop types/regions is apparent, as a new set-up would require gathering data for calibration/fine-tuning of the scientific models for irrigation, fertilization and crop protection of an acceptable amount of time (one cultivating period) prior to producing initial advice to the farmers.
- The findings also show that it was possible to achieve the results because the farmers were cooperative and acted according to the advice proposed by the technology.

Besides these gains, other factors can be quantified and add value to the solution:

- By reducing the number of sprays, the farmer increases the productivity of spraying and saves time that he or she can invest in other value-creating activities. This also means that the cost for labour decreases as well.
- Further gains can be achieved also by increasing the harvest from the field supported by the technology. Even though this might be difficult to measure because at the end the quality and quantity of the harvest might depend on many factors than the ones controlled by the technology. However, the more factors influencing the growth and quality of the plants can be controlled by technology, the higher the output in terms of quantity and quality should be.

Table 15.2 Quantification of business gains (baseline–achieved measured value) in both pilot application scenarios

Saving	Pilot application scenario 1			Pilot application scenario 2
	Chalkidiki (olive trees)	Stimagka (Grapes)	Veria (Peaches)	Kileler (Cotton)
Reduction of the average cost of spraying per hectare	250 − 219 = 31 Euro/Hectare	990 − 963 = 27 Euro/Hectare	810 − 781 = 29 Euro/Hectare	
Reduction of the average number of unnecessary sprays per farm	5 − 1.4 = 3.6 Number of sprays	4 − 1.8 = 2.2 Number of sprays	4 − 1.6 = 2.4 Numbers of sprays	
Reduction of the average cost of irrigation per hectare	330 − 198 = 132 Euro/Hectare	3030 − 2007 = 1023 Euro/Hectare	870 − 497 = 373 Euro/Hectare	2670 − 1881 = 789 Euro/Hectare
Reduction of the amount of fresh water used per hectare	817 − 492.4 = 324.6 m^3/Hectare	1868 − 1232 = 636 m^3/Hectare	1703 − 971.18 = 731.82 m^3/Hectare	
Reduction of the nitrogen use per hectare	230 − 161 = 69 Kg/Hectare		220 − 161 = 59 Kg/Hectare	
Quantify % divergence in the cost of the applied fertilization	−40 + (−11.27) = 51.27 %/Hectare		20 − 44 = −24 %/Hectare	
Increase in production	10,375 − 7010 = 3365 Kg/Hectare	17,117 − 18,011 = − 894 Kg/Hectare	49,825 − 52,044 = −2219 Kg/Hectare	
Decrease in inputs focused on irrigation				2670 − 1881 = 789 m^3/Hectare

As multiple parameters (climate and crop type related) affect agricultural production, it became clear that a "one-fits-all" solution is not applicable. Several factors need to be taken into consideration in translating the trial results (e.g. biennial bearing phenomenon in olive trees, heavy seasonal/regional rains, multi-year fertilization strategies, etc.).

15.4.2   Business Impact of the Technology on General Level

The pilot activities have highlighted another exploitation potential that arises from the plethora of stored heterogeneous data. The various data streams collected and stored in this pilot's context can be valuable for data scientists/researchers that could evolve their research activities and take full advantage through them.

15.5   How to Guideline for Practice When and How to Use the Technology

Farmers are constantly struggling to produce more food, to meet the increased global demand. At the same time, there is a push towards more sustainable farming practices in order to minimize the environmental impact of agriculture. In this context, the future Common Agricultural Policy (which is currently under development) focuses on digitization, inviting farmers to produce "more with less".

In order to improve farm productivity and increase their profits, farmers were traditionally asked to invest in expensive technological tools and learn how to use them—an offer usually combined with the use of specific brands of agrochemicals. This not only incurred high costs for farmers with a slow depreciation curve (in fact a high percentage of farmers—Greek farmers are in their majority smallholders—did not have the capacity to make such investments), but also required farmers to have digital skills that they lacked.

To support the business expansion of the big data-enabled technologies intro-duced within the present DataBio pilot, NP and GAIA EPICHEIREIN have already established an innovative business model that allows a swift market uptake—the "Smart-Farming-As-A-Service" model. With no upfront infrastructure investment costs and a subscription fee proportionate to a parcel's size and crop type, each smallholder farmer can now easily participate and benefit from the provisioned advi-sory services. The proposed approach takes all the complexity out of the picture and provides a simple and easy-to-use advice that both agricultural advisors and farmers can exploit.

Moreover, and as more than 70 agricultural cooperatives are shareholders of GAIA EPICHEIREIN, it is evident that there is a clear face to the market and an excel-lent liaison with end-user communities for introducing the pilot innovations and promoting the commercial adoption of the DataBio's technologies.

Finally, while the proposed data-driven solution of the pilot is appealing to small-holder farmers, it is also applicable to large farms and agricultural cooperatives. Thanks to their increased capacity (e.g. financial and technical), the application of smart farming services can multiply the benefits for these organizations, as they are applied in a larger scale.

15.6   Summary and Conclusions

NP and GAIA EPICHEIREIN have already launched in 2013 their smart farming program, called "gaiasense",[3] which aims to establish a nationwide network of tele-metric stations with agri-sensors and use the data to create a wide range of smart farming services for agricultural professionals.

Within the DataBio, the quality of the provided services greatly benefited from collaborating with leading technological partners like IBM and FRAUNHOFER, which specialize in the analysis of big data. Moreover, feedback from the end-users and lessons learnt from the pilot execution significantly fine tuned and will continue to shape the suite of dedicated tools and services, thus, facilitating the penetration of "gaiasense" in the Greek agri-food sector.

Thee pilot's success was established by high profile events[4] and online articles[5] that were promoting the pilot's findings. Consequently, the wider adoption of big data-enabled smart farming advisory services in the next years.

The sustainability of all DataBio-enhanced smart farming services, after the end of the project is achieved through: (a) the commercial launch and market growth of "gaiasense" and (b) the participation to other EU and national R&D initiatives. This will allow continuously evolving/validating the outcomes of the project, by working with both new and existing (to DataBio) user communities and applying its innovative approach to new and existing (again to DataBio) areas/crops.

---

[3] https://www.gaiasense.gr/en/gaiasense-smart-farming.

[4] https://www.gaiasense.gr/en/a-greek-innovation-gaiasense-evolves

[5] https://www.ypaithros.gr/en/yannis-olive-grove-reduction-by-30-in-production-costs-and-par allel-increase-of-sales/

Chapter 16
Genomic Prediction and Selection in Support of Sorghum Value Chains

Ephrem Habyarimana and Sofia Michailidou

Abstract Genomic prediction and selection models (GS) were deployed as part of DataBio project infrastructure and solutions. The work addressed end-user requirements, i.e., the need for cost-effectiveness of the implemented technologies, simplified breeding schemes, and shortening the time to cultivar development by selecting for genetic merit. Our solutions applied genomic modelling in order to sustainably improve productivity and profits. GS models were implemented in sorghum crop for several breeding scenarios. We fitted the best linear unbiased predictions data using Bayesian ridge regression, genomic best linear unbiased predictions, Bayesian least absolute shrinkage and selection operator, and BayesB algorithms. The performance of the models was evaluated using Monte Carlo cross-validation with 70% and 30%, respectively, as training and validation sets. Our results show that genomic models perform comparably with traditional methods under single environments. Under multiple environments, predicting non-field evaluated lines benefits from borrowing information from lines that were evaluated in other environments. Accounting for environmental noise and other factors, also this model gave comparable accuracy with traditional methods, but higher compared to the single environment model. The GS accuracy was comparable in genomic selection index, aboveground dry biomass yield and plant height, while it was lower for the dry mass fraction of the fresh weight. The genomic selection model performances obtained in our pilots are high enough to sustain sorghum breeding for several traits including antioxidants production and allow important genetic gains per unit of time and cost.

E. Habyarimana (✉)
CREA Research Centre for Cereal and Industrial Crops, via di Corticella 133, 40128 Bologna, Italy
e-mail: ephrem.habyarimana@crea.gov.it

S. Michailidou
Centre for Research and Technology Hellas – Institute of Applied Biosciences, 6th Km Charilaou Thermis Road, 57001 Thessaloniki, GR, Greece

16.1 Introduction, Motivation and Goals

Genomic selection (GS), fitting the big data generated from several sources such as phenomics, genomics, and Internet of Things (IoT), provides the enabling technologies to support crop breeding companies and research and development institutions. Genomic selection models were deployed as part of DataBio project infrastructure and solutions tailored to the end user requirements. Specific challenges, which GS addresses in agriculture, are mostly represented by the need for cost-effectiveness of the implemented technologies, simplified breeding schemes, and shortening the time to cultivar development selecting for genetic merit estimated through genomic modelling in order to sustainably improve productivity and profits. One of the interesting features of genomic selection is the possibility to customize the solutions to fit the farmer's requirements such as putting major emphasis on a single characteristic or several plant characteristics aggregated in selection index. Genomic selection allows therefore to close the gap between agricultural business planning and the responsible and sustainable maximization of the profit deriving mainly from increased crop productivity and efficiency of resource use, and reduced uncertainty of management decisions.

Another key feature of genomic selection is its ability to decouple selection from phenotyping—the assessment of expressed plant characteristics as influenced by genetic make-up and changes in the environment—in the process of crop improvement (Fig. 16.1). Genomic selection is implemented in coherent steps starting from genotyping (determining the individual's genetic constitution through Deoxyribonucleic acid sequencing) and phenotyping the training population, and then proceeding with calibrating the phenotypes against the genomic information, whole-genome genotyping the selection candidates, using calibration equation to predict plant characteristics, operating selection upon genetic merit (genomic estimated breeding

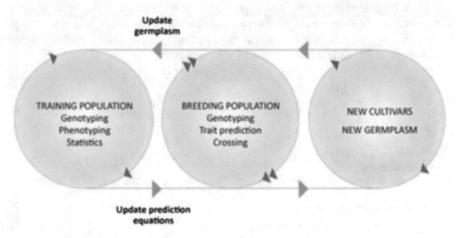

Fig. 16.1 Overall genomic prediction and selection operational steps. Refer to text for further description

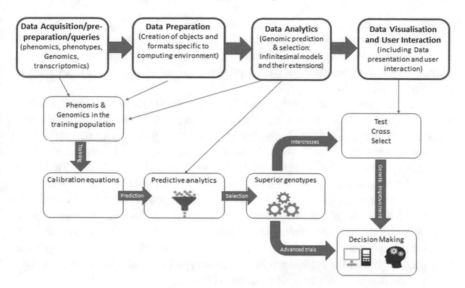

Fig. 16.2 Generic pipeline for data flow in genomic selection and prediction

values, GEBVs), and implementing repetitive cycles of crossing and selection based on GS-generated information.

The diagram below (Fig. 16.2) represents the generic pipeline for data flow of genomic selection and prediction: from data collection to data processing and decision-making, and its mapping to the steps of the top-level pipeline that is in compliance with the Reference Architecture for Big Data Application Providers [1].

One of the most compelling merits of the genomic selection technology is the possibility to integrate Marker Assisted Selection for yield into practical breeding programmes, particularly in the areas of population genetics and quantitative genetics. This has been a puzzle to breeders, geneticists and other scientists for the last 30 years of Quantitative Traits Loci (QTL, a chromosomal region that correlates with variation of a plant characteristic) breeding. Genomic selection represents the gold standard approach to expedite cultivar development, and for estimating breeding values upon which superior cultivars are identified and selected. Genomic selection allows superior response to selection, and hence superior breeding progress, due to its intrinsic attributes that expedite breeding works by shortening generation intervals through genomic prediction and selection-driven intercrosses. The genomic selection technology is therefore expected to significantly improve genetic gain by unit of time and cost, allowing farmers to grow a better variety sooner relative to conventional approaches, and hence make more income.

The pilot trials for this work were run by a collaborative effort between Council for Agricultural Research and Economics, Italy (CREA) and Centre for Research and Technology Hellas, Greece (CERTH). Genomic data (SNPs) produced in tomato was enough to run genomic models, but the size of tomato population phenotyped was too low (less than 40) and it was not therefore possible to run genomic models in tomatoes as genomic models require a big size of the training population. We therefore report herein the results obtained from CREA's sorghum pilot experiments where a sufficiently bigger population (380) had been genotyped and phenotyped, to improve yields of biomass and health-promoting compounds used to manufacture specialty foods.

In the GS approach, different assumptions of the distribution of marker effects were accommodated in order to account for different models of genetic variation including, but not limited to: (1) the infinitesimal model, (2) finite loci model, (3) algorithms extending Fisher's infinitesimal model of genetic variation to account for non-additive genetic effects. Many problems were modelled including the performance of new and unphenotyped lines, untested environments, single trait, multi-traits, single environment, and multi-environment. Models were fed several data types: open-field phenotypic data, biochemical data, phenomic and genomic data and other data sources (environmental indoor/outdoor, farm data/log/profile) collected to describe the crop management and production environment. Next, the GS equations were used to predict the breeding values of genotyped but unphenotyped candidates and the outcome was encouraging as detailed below.

16.2   Pilot Set-Up

The first stage of the sorghum pilot trials started in 2018 in several locations in Emilia Romagna Region, Northern Italy. In this year, the CREA's platform for genomic prediction and selection was specified to accommodate the requirements of the breeding programmes, particularly the upcoming genomic and phenomic/phenotypic data from sorghum field experiments. In the second stage of the trials in 2019, a second temporal replication of sorghum pilot trials was established in the same region of Emilia Romagna but in locations different from 2018 as dictated by the rules of crop rotations. Sorghum lines were genotyped using a genotyping-by-sequencing (GBS) strategy on Illumina next-generation sequencing platform.

Genotypic variability is an important precondition for genomic selection and prediction. To evaluate the genotypic variability for the evaluated traits, the Bayesian regression model was implemented in R using the probabilistic programming language Stan, implementing Hamiltonian Monte Carlo and its extension, the no-u-turn sampler (NUTS). Our choice for these algorithms was motivated by their faster convergence relative to other commonly used Markov chain Monte Carlo algorithms, like the Metropolis Hastings and Gibbs sampler. The default rules were applied to choose hyperparameters. For each trait, the models were fitted using four chains, each with 50,000 iterations of which the first 10,000 were warmup (burn-in) to calibrate the sampler, leading to a total of 160,000 posterior samples upon which our analyses were based. Genotypic variability was measured using the mean (estimate) and the standard deviation (estimate error) of the posterior distribution as well as two-sided 95% credible intervals (l–95% CI and u–95% CI) based on quantiles. Variance components and trait broad-sense heritability (repeatability) were estimated by fitting the appropriate linear mixed model equation.

16.3   Technology Used

16.3.1   Phenomics

In this work, we measured a set of phenotypes from sorghum plants (physical and biochemical traits) that were produced over the course of development and in response to environmental stimuli. The biochemical analysis was carried out both with colorimetric and chromatographic methods. Total polyphenol content was measured with the Folin-Ciocalteu method, total antioxidant activity was assessed with DPPH (2,2-diphenyl-1-picrylhydrazyl) radical assay, and total flavonoid content was measured with AlCl3 method. The phenotypic characterization of sorghum lines was carried out according to international standard operating procedures following International Board for Plant Genetic Resources (IBPGR) and International Union for the Protection of New Varieties of Plants (UPOV) as described in previous works [2, 3].

   To analyse total phenols, tannins, flavonoids and antioxidant capacity (TAC), a 10 g sample from each genotype was ground using a Cyclotec Udy Mill (sieve: 0.5 mm), the moisture in the sample was determined after they were oven dried overnight at 105 °C, and antioxidants and TAC were analysed in duplicate using 100 mg of each sample. For the phenolic compounds, the absorbance of samples was measured at 750 nm and expressed as gallic acid equivalents (gGAEkg^{-1} dry mass basis). For condensed tannins and total flavonoids assays, the absorbances were measured at 500 nm and 510 nm, respectively, and expressed as µg CE (catechin equivalents) g^{-1} dry mass basis. The TAC was determined using the 2,20-azino-bis/3-ethylbenzthiazo-line-6-sulphonic acid (ABTS) assay and expressed as mmol TE (Trolox equivalents) kg^{-1} dry basis. Internet of things (IoT) technology was implemented to collect and characterize soil, plant, and environmental properties.

16.3.2   DNA Isolation, Next-Generation Sequencing/Genotyping, and Bioinformatics

In sorghums, DNA was isolated from plantlets using the GeneJET Plant Genomic DNA Purification Kit. The methylation sensitive restriction enzyme ApeKI was used for library preparation, and genotyping-by-sequencing (GBS) was carried out on an Illumina HiSeq X Ten platform. The final working matrix consisting of 61,976 high-quality SNPs was used in this work for genomic selection and prediction analytics.

16.3.3   Genomic Predictive and Selection Analytics

To evaluate the performance of GS models, the Monte Carlo (repeated hold-out) cross-validation approach [4, 5] was applied using 70% and 30%, respectively, as training and validation sets. In a standard hold-out cross-validation, the data is randomly divided into two subsets: a training and a test (validation) set. The test set represents new, unseen data to the model. To obtain a more robust performance estimate that was less variant to how the data was split into training and test sets, the hold-out method was repeated 50 times with different random seeds and the average performance was computed over these 50 repetitions. The repeated hold-out procedure provides a better estimate of how well our model may perform on a random test set, compared to the standard hold-out validation method [5]. In addition, it provides information about the model's stability as to how the model, produced by a learning algorithm, changes with different training set splits. In the Monte Carlo method, models were implemented fitting best linear unbiased predictions (BLUP) data using Bayesian ridge regression (BRR), genomic best linear unbiased predictions (GBLUP), Bayesian least absolute shrinkage and selection operator (LASSO), and BayesB algorithms accounting for all spatial and temporal replications of the trials (Table 16.1).

In the case of multi-environment scenario, different cross-validation experiments (Table 16.2) were evaluated using GBLUP. Cross-validation CV1 reflected prediction of sorghum lines that have not been evaluated in any of the target environments, while cross-validation CV2 reflected prediction of lines that have been evaluated in some, but not all, target environments. The rationale being that prediction of non-field evaluated lines benefits from borrowing information from lines that were evaluated in other environments. This is critical in cutting costs for varietal adaptability trials

Table 16.1 Assessment of alternative genomic models accuracy fitting BLUP yield data

	[a]GBLUP	BRR	LASSO	BayesB
Mean	0.47	0.48	0.48	0.46
Standard deviation	0.049	0.050	0.049	0.048

[a]GBLUP, BRR, LASSO, BayesB, respectively, genomic best linear unbiased predictions, Bayesian ridge regression, Bayesian least absolute shrinkage and selection operator, bayes B

Table 16.2 Assessment of genomic models accuracy fitting multi-environment scenarios

	CV1				CV2			
	[a]Single Env	Across Env	M x E	RNorm	Single Env	Across Env	M $\times$ E	RNorm
Env1	0.41	0.33	0.38	0.38	0.41	0.62	0.64	0.63
Env2	0.29	0.32	0.32	0.32	0.38	0.61	0.59	0.59
Env3	0.47	0.49	0.51	0.51	0.43	0.41	0.46	0.45

[a] Env, M $\times$ E, RNorm, respectively, environment, marker x environment, reaction norm

of large numbers of lines in several target environments. The model was run on a single environment basis, across environments, marker-by-environment interaction, and using the reaction norm model.

Our findings show that genomic models perform comparably under single environments (Table 16.1, Fig. 16.3). On the other hand, under multiple environments, CV2 was superior to CV1. Under CV2 settings, single-environment model performed poorly. Accounting for environmental noise, marker information x environment or implementing the reaction norm model performed comparably and produced superior results relative to single environment model (Table 16.2).

When faced with the necessity to simultaneously improve more than one trait, a breeder can use three approaches: tandem selection, independent culling levels, and

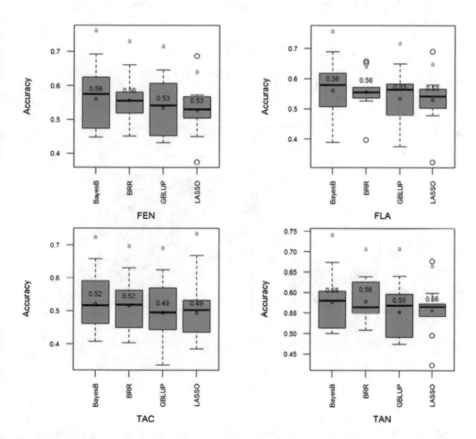

Fig. 16.3 Distribution (boxplot) of GS models validated accuracy in external sample (not used during model training) of 34 (30% of the total population) sorghum lines. FEN, FLA, TAC, TAN, respectively, polyphenols, flavonoids, total antioxidant capacity, and condensed tannins. Traits means are included within the boxplot. Trait means with same letter are not significantly different at the 5% level using the Tukey's honestly significant difference (HSD) test. Refer to text for the description of the GS models. Reprinted from Habyarimana et al. [3] under a CC BY 4.0 license (http://creativecommons.org/licenses/by/4.0/), original copyright 2019 by the authors

index selection [6]. In tandem selection, only one character is selected in each cycle; in independent culling levels, all genotypes with a phenotypic value below the culling threshold for at least one characteristic are discarded; the selection index aims at improving several traits simultaneously in such a way as to make the biggest possible improvement in overall genetic merit [7]. In this work, we implemented the optimum selection Index of Smith [2, 3, 8], the performance of which was demonstrated in previous studies [7, 9]. Our findings showed accuracy that was higher (acc = 0.52 – 0.59) and comparable in genomic selection index, aboveground dry biomass yield and plant height, while it was lower (acc = 0.36) for the dry mass fraction of the fresh weight (Fig. 16.4). In this work, the accuracy of the models was defined as the Pearson correlation coefficient (r) between observed (y) and predicted ($\hat{y}$, genomic estimated breeding values) phenotypic values as represented in the following formula:

$$ r = \frac{\sum_{i=1}^{n} (y_i - \overline{y})\left(\widehat{y}_i - \overline{\widehat{y}_i}\right)}{\sqrt{\left(\sum_{i=1}^{n} (y_i - \overline{y_i})^2 \sum_{i=1}^{n} \left(\widehat{y}_i - \overline{\widehat{y}}\right)^2\right)}} $$

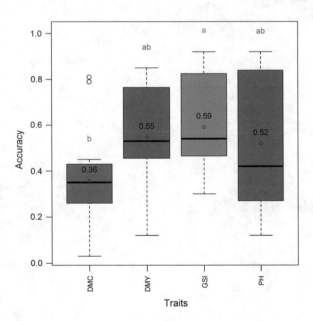

Fig. 16.4 Distribution (boxplot) of genomic selection index accuracy using single traits and all three traits of interest simultaneously in the entire panel. DMC, DMY, GSI, and PH, respectively, denote selection indices relative to dry mass fraction of fresh material, aboveground dry biomass yield, all the three traits simultaneously, and plant height. Means are indicated by open dots and are included within the boxplot. Means with same letter are not significantly different at the 5% level using the Tukey's HSD (honestly significant difference) test. Refer to text for the description of the GS models. Reprinted from Habyarimana et al. [8] under a CC BY 4.0 license (http://creati vecommons.org/licenses/by/4.0/), original copyright 2019 by the authors

where $\overline{y}$ and $\overline{\hat{y}}$ are, respectively, the means of the observed and the predicted values.

16.4   Business Value and Impact

Genomic predictive and selection (GS) modelling was developed as response to the lengthier and costlier phenotypic selection. In business, time to market is important just as the production cost. In addition, specifically for plant breeding, the longer it takes to bring the new cultivar to the market, the shorter will that cultivar stay on the market, in virtue of the naturally occurring crop degeneration. Some of the most attractive GS attributes are enabling cutting time and cost to cultivar development with high selection accuracy. The high accuracy means that the plant lines selected will breed true to type, implying diminished risks in the breeding and production processes.

In this pilot, the GS technology showed meaningful and attractive results as reflected by the key performance indices (KPIs) presented in Table 16.3. The predictive performance obtained in this pilot was encouraging. Over the two-year trial, with data integration, the four genomic selection models implemented in this pilot performed comparably across traits and are considered suitable to sustain sorghum breeding for antioxidants production and allow important genetic gains per unit of time and cost. In comparison to conventional phenotypic breeding, the genomic predictive and selection modelling allows cutting costs five times and cutting four times the time of cultivar development (Table 16.3). The results produced in this pilot are expected to contribute to genomic selection implementation and genetic improvement of sorghum for several traits including grain antioxidants for different purposes including the manufacture of health-promoting and specialty foods in Europe in particular, and in the world in general. In addition, the NGS genotyping platforms were validated and were found to be usable for sequencing and genotyping (variants calling) services in other plant species and animal husbandry.

16.5   How to Guideline for Practice When and How to Use the Technology

The method for implementing genomic prediction and selection analytics was depicted in the above diagram (Fig. 16.1), while a reusable generic pipeline for data flow genomic selection and prediction was described in Fig. 16.2. Several scenarios can be modelled including a single trait, multiple traits as index selection, a single environment, and multi-environment. A generic technological flowchart is that, in the genomic predictive and selection modelling, phenotypic and marker data are scored in the training population and fitted into appropriate algorithm to produce individuals' whole-genome marker effects. Most practically, the training set is the

Table 16.3 KPIs of the sorghum pilot trials

KPI short name	KPI description	Goal description	Base value	Target value	Measured value	Unit of value	Comment
A2.1-KPI-01	Accuracy	Increased accuracy	0.4	0.4–0.7	0.5–0.6	Pearson'r	Pilot was successful
A2.1-KPI-02	Breeding cycle (years)	Decrease the cycle relative to phenotypic breeding	–	3 times	4 times	Ratio Phenotypic/Genomic selection	Too early to assess
A2.1-KPI-03	Breeding costs (index)	Decrease costs relative to phenotypic breeding	–	2 times	5 times	Ratio Phenotypic/Genomic selection	Too early to assess

germplasm or a population that best samples the frequency of the genetic information (allele frequency) useful for the breeding programme. The marker effects are used in subsequent cycles of selection to compute the genomic estimated breeding values (GEBVs) that are used as predictors of breeding values in testing unphenotyped population. The genomic estimated breeding values are obtained as a product of the estimated marker effects in the training population and the coded marker values obtained in the testing population. To apply genomic selection, GEBVs are obtained in the selection candidates and then used to predict and rank the net genetic merit of the candidates for selection, and superior strains are selected in the process; GEBVs become the criteria for crossing block management and cultivar development. Genomic predictive and selection modelling is a gold standard for selecting for breeding values and is well poised to help breeders and seed industries to drastically cut breeding cost and time and bring new cultivar earlier on the market, thus generating higher incomes.

16.6 Summary and Conclusions

Current empirical evidence for genomic selection efficiency in plant breeding is set to $r = 0.5$ as the baseline for genomic selection prediction accuracy in plant breeding. Also, recent research works demonstrated that genomic selection accuracy as low as 0.2 can allow substantial within-generation yield improvement [10]. Therefore, the genomic selection model performances obtained in our pilots are high enough to sustain sorghum breeding for several traits including antioxidants production and allow important genetic gains per unit of time and cost. In addition to the accuracy, the importance of the genomic selection strategy is also evaluated using other criteria such as the possibility that this technology offers to shorten the breeding cycle with significant economic returns due to intercrosses driven by genetic predictions, the quick delivery of novel superior cultivars onto the market. In the case of antioxidants, genomic selection offers the possibility to select for or against this trait early (e.g. at the seed or seedling stages) without waiting for seed setting or harvest. The genomic selection algorithms developed in this work can be directly used in sorghum breeding programmes and can be adapted to other plant species and animal husbandry. The genomic selection results presented herein and the experimental designs used in this pilot can be implemented in antioxidants and other traits genetic investigations and in breeding programmes to qualitatively and quantitatively improve plant characteristics and the antioxidant production for different purposes including the manufacture of health-promoting and specialty foods.

References

1. NIST Big Data Public Working Group Reference Architecture Subgroup (2015) *NIST big data interoperability framework: Reference architecture*, (Vol. 6). National Institute of Standards and Technology.
2. Habyarimana. E., Dall'Agata, M., De Franceschi, P., Baloch, F. S. (2019). Genome-wide association mapping of total antioxidant capacity, phenols, tannins, and flavonoids in a panel of Sorghum bicolor and S. bicolor × S. halepense populations using multi-locus models. PLoS ONE 14:e0225979. https://doi.org/10.1371/journal.pone.0225979
3. Habyarimana, E., Lopez-Cruz, M. (2019). Genomic selection for antioxidant production in a panel of sorghum bicolor and S. bicolor × S. halepense Lines. Genes 10:841. https://doi.org/10.3390/genes10110841
4. Scutari, M., Mackay, I., & Balding, D. (2016). Using genetic distance to infer the accuracy of genomic prediction. *PLOS Genetics, 12,* e1006288. https://doi.org/10.1371/journal.pgen.1006288.
5. Raschka, S. (2018). Model evaluation, model selection, and algorithm selection in machine learning. arXiv:181112808 [cs, stat].
6. Wricke, G., Weber, E. (1986). Quantitative genetics and selection in plant breeding, Reprint 2010 ed. edition. De Gruyter.
7. Bradshaw, J. E. (2017). Plant breeding: Past, present and future. *Euphytica, 213,* 60. https://doi.org/10.1007/s10681-016-1815-y.
8. Habyarimana, E., Lopez-Cruz, M., & Baloch, F. S. (2020). Genomic selection for optimum index with dry biomass yield, dry mass fraction of fresh material, and plant height in biomass sorghum. *Genes, 11,* 61. https://doi.org/10.3390/genes11010061.
9. Baker, R. J. (1986). Selection indices in plant breeding. CRC Press
10. Habyarimana, E. (2016). Genomic prediction for yield improvement and safeguarding genetic diversity in CIMMYT spring wheat (Triticum aestivum L.). *Australian Journal of Crop Science, 10,* 127–136.

Chapter 17
Yield Prediction in Sorghum (*Sorghum bicolor* (L.) Moench) and Cultivated Potato (*Solanum tuberosum* L.)

Ephrem Habyarimana and Nicole Bartelds

Abstract Sorghum and potato pilots were conducted in this work to provide a solution to current limitations (dependability, cost) in crop monitoring in Europe. These limations include yield forecasting based mainly on field surveys, sampling, censuses, and the use of coarser spatial resolution satellites. We used the indexes decribing the fraction of absorbed photosynthetically active radiation as well as the leaf areas derived from Sentinel-2 satellites to predict yields and provide farmers with actionable advice in sorghum biomass and, in combination with WOFOST crop growth model, in cultivated potatoes. Overall, the Bayesian additive regression trees method modelled best sorghum biomass yields. The best explanatory variables were days 150 and 165 of the year. In potato, the use of earth observation information allowed to improve the growth model, resulting in better yield prediction with a limited number of field trials. The online platform provided the potato farmers more insight through benchmarking among themselves across cropping seasons, and observing in-field variability Site-specific management became easier based on the field production potential and its performance relative to surrounding fields. The extensive pilots run in this work showed that farming is a business with several variables which not all can be controlled by the farmer. The technologies developed herein are expected to inform about the farming operations, giving rise to well-informed farmers with the advantage to be able to adapt to the circumstances, mitigating production risks, and ultimately staying longer in the business.

E. Habyarimana (✉)
CREA Research Center for Cereal and Industrial Crops, via di Corticella 133, 40128 Bologna, Italy
e-mail: ephrem.habyarimana@crea.gov.it

N. Bartelds
NB Advies, Smart Farming Advisory Services, Zuiderdiep 222, 9571 BM 2e Exloërmond, The Netherlands

© The Author(s) 2021
C. Södergård et al. (eds.), *Big Data in Bioeconomy*,
https://doi.org/10.1007/978-3-030-71069-9_17

219

17.1   Introduction, Motivation, and Goals

Under the climate change scenarios, the rapid increase of world population and industrial development is expected to increase carbon dioxide concentration in the Earth's biosphere. At the same time, environments are predicted to be warmer and dryer, all of which will favor the cultivation of crops with a C4 photosynthetic pathway over C3 crops [1–3]. Humans will, therefore, rely heavily on C4 crops like sorghum (*Sorghum bicolor* (L.) Moench). As sorghum is becoming a world's staple food and a biofuel-dedicated biomass business, its cultivation and yields will have to be closely monitored and forecast for efficient management locally and globally.

Potato has been the major crop in the Netherlands for many years. Due to the reform of the CAP (Europe's Common Agricultural Policy), the market is changing and farmers are urged to increase their yields, but in a sustainable way. This means they need to be more conscious of the energy and other resources they use in producing their crops. AVEBE is a cooperative for the potato growing farmers and supports their growers in an innovation program called "Towards 20-15-10", to realize in 2020 an average of 15 tons of starch per ha with a variable cost price of €10 per 100 kg starch. To monitor these objectives, farmers are sharing data about their yields and farming practices in study groups. Crop yield forecasting is a key strategy in agriculture as it enables sustainable development and helps avoid famines and commodity shortages [4–7]. Crop monitoring and yield forecasting represent a good source of actionable information that can be used by governmental institutions, companies, and farmers for price predictions and adjustment and for efficient agricultural trade. They simplify business operations through better planning of harvest, delivery of the produce, deployment of machineries, logistics, and the use of resources [8].

Conventionally, crop monitoring and yield forecasting rely on field surveys, censuses, and sampling in predefined locations (e.g., potato), which are costly processes associated with high uncertainties [9]. Results are hard to relate to other fields that were not visited, making it difficult for the farmer to objectively examine the status of his crop and for the processing industry to plan logistics of transport and processing capacity at an early stage. Modern crop monitoring relying on remote and proximal sensing technologies resulted in a superior solution [9–15]. This sensor-based monitoring is dependent upon differential reflectance of light by plants [16] which generally absorb the portion of light in the wavelength range of 400–700 nm (i.e., in the blue 440–510 nm, and red 630–685 nm wavelengths), and reflect light in the green and near-infrared portions of the light spectrum. Crop monitoring technologies have been used to exploit this phenomenon, including satellites and hand-held sensors measuring light in narrow wavebands or wavelength intervals. Plant reflectance measurements have been successfully used in several instances including the quantification of canopy vigor [17–19], nutrient, and soil moisture stresses [20, 21] and to predict yields [8, 22]. However, in most studies, remote sensing-based biomass yield estimation or prediction makes use of low- or medium-resolution satellite images from sensors such as SPOT-VEGETATION or MODIS [8]. These satellite products have a coarser spatial resolution (250–1000 m) compared to the data collected from the two Sentinel-2 satellites in this work (10-m spatial resolution). With the launch of the Sentinel-2 constellation of satellites the overpass frequency (five days and locally even two to three days), the temporal resolution is nearly as good as for SPOT-VEGETATION and MODIS satellites (one to two days). The high spatial resolution of the Sentinel-2 images is a valuable asset when monitoring crops in agricultural regions characterized by many small fields like in the Mediterranean region where this study was conducted.

Deriving yield information from satellite imagery has shown promising results but this technology is not extensively applied across farmers and crop species world-wide [8, 22]. In the sorghum pilot, we developed models for in-season prediction of annual and perennial sorghum biomass yields in Emilia-Romagna, Italy, based on the fraction of absorbed photosynthetically active radiation (fAPAR) measurements from Sentinel-2A and Sentinel-2B satellite images on 42 mostly full-fledged commercial sorghum fields. Unlike other crops in which the yield is directly correlated to the aboveground biomass, potatoes follow a different pattern in the growth of the productive yield (Fig. 17.1). Crop growth models simulate both the aboveground dry matter and the tuber dry matter and can help to estimate the yield gap and yield at an early stage.

In the potato pilot, we therefore used imagery from the Sentinel-2 satellites to provide a semi-continuous flow of data about the development of the potato crop and the WOFOST [24] crop model using local weather data to provide field-specific yield information. Sample data were used to calibrate the remote sensed data.

In the pilots implemented in this work, we used machine learning algorithms to create yield prediction equations. These equations can be implemented in decision support systems to allow farmers and/or farming stakeholders to predict biomass

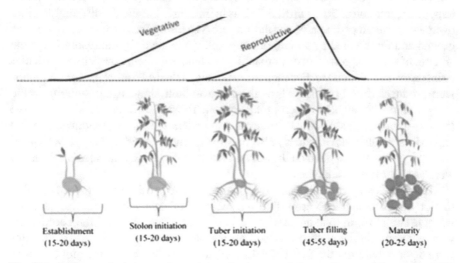

Fig. 17.1 Difference in the vegetative and reproductive growth stages of potato. Adapted from [23]

yields from sorghum fields of interest early on in the cropping season. This information is very helpful to efficiently schedule fleets of harvesting machinery, transport
vehicles, and storage facilities. The fAPAR-derived predictive models for biomass
yields can also be implemented by extension services and policy-makers for several
purposes, including the possibility to anticipate potential biomass availability and
plan ahead, to avoid specific crises such as fuel shortage. The potato pilot's goal was
an online *decision support system* (DSS) for potato farmers, which would provide
them objective information about the yield gap and yield potential of their fields given
the actual weather conditions. The developed online platform provided the farmers
more insight by benchmarking their crops during the growth period with crops in
the region and/or previous growing seasons. These new insights will improve farm
management decisions on timely and more efficient location-specific treatment of
the crops.

17.2   Pilot Set-Up

The sorghum pilot consisted of private farmers and/or farming cooperatives. During
the 2017 and 2018 cropping seasons, 43 sorghum pilots were run covering 240 ha.
The access to EO platform was made through "WatchITGrow" (VITO, Vlaamse
Instelling voor Technologisch Onderzoek N.V., Mol, Belgium), which was also the
end-to-end backbone for the technical pipeline used in this pilot. The plot sites were
geolocated and the coordinates used for site-specific monitoring the fAPAR index
throughout the cropping season. Fields were geolocalized, geolocation data saved
as kml files before they were integrated into WatchITGrow application. The fAPAR
estimates were generated at decametric spatial resolution (10 m pixel size), and a
temporal resolution of 5 days up to 2–3 days in those areas where the different satellite
overpasses overlapped. Spatial resolution refers to the surface area measured on the
ground and represented by an individual pixel, while the temporal resolution is the
amount of time, expressed in days that elapses before a satellite revisits a particular
point on the Earth's surface. For each experimental field, fAPAR or "greenness" maps
were produced (Fig. 17.2), and a growth curve was built, showing the evolution of the
fAPAR values throughout the cropping season. To correct for artifacts in the curve
(such as abnormally low fAPAR values due to undetected clouds, shadows, or haze)
and to interpolate fAPAR values between subsequent acquisition dates, a Whittaker
smoothing filter was applied on the curve. Finally, the fAPAR values from the curves
were used for further analytics.

   During the two years (2018 and 2019), groups of AVEBE farmers provided information about their potato crop, like the location of their plot, planting date, and
variety. The plots, in total an area of 111 ha, were geolocated and the coordinates
were entered into the platform. Based on the plot location, the soil characteristics
were determined from the BOFEK2012 [25] soil map. Moreover, the plot locations
were used to identify the nearest official weather station, providing a daily update of
rainfall, temperature, and solar radiation. Both soil characteristics and weather data

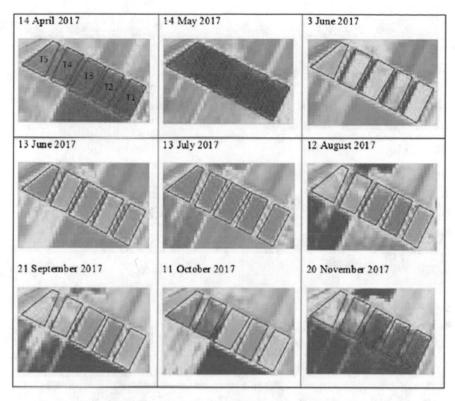

Fig. 17.2 Greenness (fAPAR) maps derived from Sentinel-2 satellite imagery for five sorghum fields in Anzola (from left to right: T5-grain sorghum, T4-dual purpose sorghum, T3-sweet sorghum, T2-forage sorghum, T1-biomass sorghum) for a selected number of dates in 2017, as available via WatchITGrow. T5-grain sorghum was not included in this study (refer to Sect. 2.1 for detail)

were input for the WOFOST model. Due to the extraordinary dry seasons in 2018 and 2019 modeling, the potential crop growth was strongly complicated. With the coordinates of the plots, the cloudless Sentinel-2 images were selected, providing *Weighted Difference Vegetation Index* (WDVI) data which were used to calculate the potato *Leaf Area Index* (LAI) (Fig. 17.3).

17.3   Technology Used and Yield Prediction

The DataBio technological components implemented in these pilots were developed and deployed by VITO, CREA (Consiglio per la Ricerca in Agricoltura e l'Analisi dell'Economia Agraria, Rome, Italy), and NB Advies. VITO provided the platform "WatchITGrow", while CREA and NB Advies deployed crop species tailored

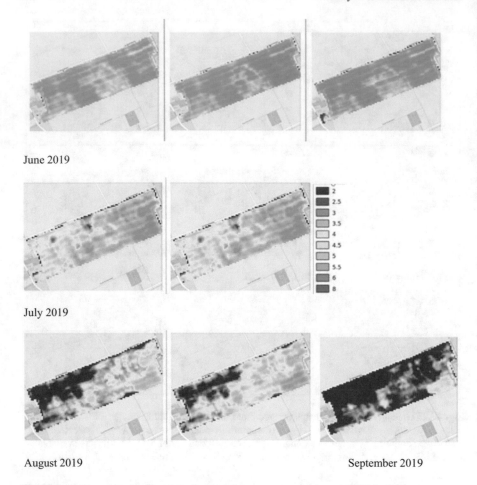

June 2019

July 2019

August 2019   September 2019

Fig. 17.3 Crop monitoring based on Sentinel-2 images expressing variability in LAI

machine learning technology, all of which were the backbone technology and end-to-end solutions of the pilot. The pilots were implemented in the form of advisory services under real-world commercial farms settings. The smart farming services were offered according to the specific cropping systems.

In biomass sorghum, services were centered around crop monitoring using proximal sensors to derive vegetation indices, and crop growth and yield modeling using fAPAR derived from satellite (Sentinel-2A and 2B) imagery and appropriate machine learning technologies.

The models used in this study were evaluated using symmetric mean absolute percentage error (SMAPE), mean absolute percentage error (MAPE), mean absolute error (MAE), and the coefficient of determination (R^2) as suggested in Habyarimana et al. [22]. The use of MAPE was justified as this metric allows the comparison of the values predicted from variables measured in different scales. On the other hand, the mean absolute error measures the magnitude but not the direction of the prediction errors; MAE is therefore an accurate representation of the average error and is considered as a better prediction metric in comparison with the root mean square error for dimensioned model assessments for the mean performance error. The symmetric mean average percentage error was implemented to account for the limitations observed in the mean absolute percentage error. SMAPE as well as MAPE average the absolute percentage errors, but in SMAPE, the errors are calculated using a denominator comprising the average of the predicted and observed values. The upper limit of the symmetric mean absolute percentage error is 200%, resulting in a 0–2 range that is suitable for evaluating the accuracy without the confounding effects of extreme values. In addition, the symmetric mean average percentage error corrects for the asymmetry in the computation of the percentage error. In this work, MAE was used to assess the reliability of the models during the cross-validated (CV) training (Fig. 17.4). A repeated CV was run for each model and produced resample vectors of mean absolute errors, each with 50 elements. We observed that the dispersion of the mean absolute errors at the training stage decreased in the order simple linear model > Bayesian generalized linear model > eXtreme Gradient boosting > Bayesian additive regression trees methods. Over the experimental duration evaluated, the simple linear model showed mostly higher prediction errors in the validation set; the coefficient of determination was also weakest in this model (Table 17.1). Overall, the Bayesian additive regression trees method displayed relatively high values of the coefficient of determination and the lowest prediction errors. The best explanatory variables were D.150 and D.165, i.e., the second half of May and the first half of June, respectively (Fig. 17.4). The days 240, 195, 210, and 120 of the year displayed minor effects,

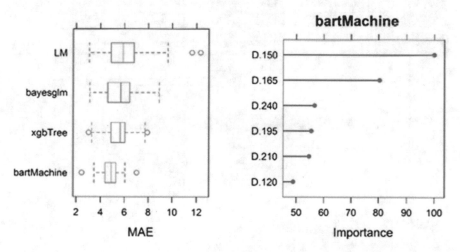

Fig. 17.4 From left to right: boxplot for models cross-validation MAE (t ha^{-1}) using fAPAR data. LM, bayesglm, xgbTree, bartMachine, respectively, simple linear model, Bayesian generalized linear model, eXtreme gradient boosting, and Bayesian additive regression trees. Relative importance of day of year (D) on sorghum biomass yields using bartMachine. Figure adapted from Habyarimana et al. [22]

Table 17.1 Model performance metrics

Model	SMAPE (%)	MAPE (%)	MAE (t ha^{-1})	R^2
LM	0.74	0.99	10.47	0.47
bartMachine	0.18	0.16	2.32	0.51
Bayesglm	0.74	0.98	10.34	0.48
xgbTree	0.44	0.36	4.07	0.62

SMAPE, MAPE, MAE, R^2, respectively, symmetrical mean absolute percentage error, mean absolute percentage error, mean absolute error, and coefficient of determination. LM, bartMachine, bayesglm, xgbTree, respectively, simple linear model, Bayesian additive regression trees (bartMachine method), Bayesian generalized linear model (bayesglm method), and eXtreme gradient boosting (xgbTree method)
Note Adapted from Habyarimana et al. [22]

while the days 135, 180, and 225 displayed no importance in terms of predicting ability [22].

In potato, the pilot's final result is a decision support system (DSS) for potato farmers that can provide data about the overall status of the crop and the potential yield based on EO, weather, and soil parameters. Figure 17.5 represents the concept of a simple (starch) potato DSS.

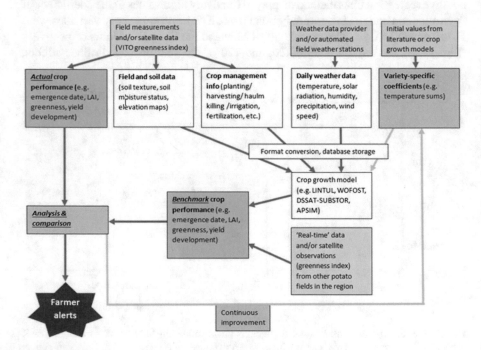

Fig. 17.5 Concept of the decision support system

The DSS involves the following data collection, processing, and visualization technology.

Data Collection: To provide benchmark data for potato crops, five types of data were collected: (1) *historical data* about crop performance in the past (i.e., emergency date, LAI, greenness, yield development, and actual yield and date of yield); (2) *historical data* about the field soil (soil texture, soil moisture status, and elevation maps); and (3) *actual data* about daily weather (temperature, solar radiation, humidity, precipitation; and wind speed); (4) reference values for indexes from literature; and (5) real-time EO data and IoT data (soil moisture status).

Data Processing: Data processing involved three steps: (1) calibration and calculation of a crop growth model, (2) real-time collection and processing of EO data, (3) benchmarking of the values, i.e., indexes resulting from the growth model and from the analysis of EO data. In the first step, the soil, crop, and weather data from field measurements, satellites, weather stations, literature, and other sources were collected, and after pre-processing, stored in a database and were used as input in a crop growth model. In order to benchmark crop performance, the WOFOST crop growth model (FAO) was introduced in the pilot and was calibrated using historical data (2017, 2018) and recent samples. Parallel to the calculation of the growth model, Sentinel-2 data were collected and calculated in real time, providing information about the most recent value of the indexes applied (LAI). The EO data processing involved the following steps: adjustment of the data with cloud mask and cloud-shadow mask, calculation of a-factor for Weighted Difference Vegetation Index (WDVI), calculation of WDVI from spectral data, and calculating LAI for potato fields based on WDVI-LAI correlation data. Finally, in the third step, the model then establishes the benchmark for crop performance: An estimate of the best possible performance under the given set of circumstances.

Data Visualization: The DSS is provided through an online platform, i.e., as data as a service for the farmers, in form of an early warning system that alerts farmers when their attention is needed. The online platform provides crop monitoring and benchmarking services that show the field variation. Sentinel-2 satellite images are very helpful for crop monitoring over a large area. But for use in a DSS, it is more useful to show just the field information and not the complete images.

17.3.1 Reflection on the Availability and Quality of Data

The Sentinel data proved very useful to extract the LAI information. However, during the growing season, there were quite extensive periods (15–20 days), in which no cloud-free images were available. Also, the cloud-shadow gave sometimes disturbing information. The historical yield data was collected and processed without the spatial location of the sample fields, which made them unusable for correlating it with the historical EO data. Privacy issues raised by the farmers prevented collecting this georeferencing information. The conclusion is that there is a lot of data available, but they are not always with a quality suitable for use. When the product is based on third-party service providers, a solid agreement about the availability is necessary. With more demands for service level agreements (SLA), the price of data-services may go up, making it less interesting to use for farmers. Reflecting on the big data technology (BDT) used in the sorghum pilots allows us to express a word of caution to scientists in the field. The IoT farm telemetry technology was used in year one for preliminary observation, but this technology revealed itself ill-adapted to biomass sorghum as the hardware, particularly the cables, were frequently damaged by rodents.

17.4   Business Value and Impact

The importance of sorghum as food, feed, and biofuel crop cannot be overemphasized. Biomass sorghum demonstrated higher yields with better energy balance relative to major crops of agroindustrial interest. As dedicated biomass sorghum crops are steadily increasing and precision farming is driving agricultural economies worldwide, harnessing satellite technology is well poised to bring about agricultural advantages, including cutting operational farming costs. The Sentinel-2-derived index describing the fraction of absorbed photosynthetically active radiation and the implementation of machine learning technology modeled in our sorghum pilots satisfactorily crop phenology and the aboveground biomass yields up to six months ahead of harvesting. In addition, we achieved promising key performance indicators as reflected in Table 17.2.

This study's outcomes can serve several purposes, including farmers being able to improve their sorghum biomass business operations through informed decision-making in planning field work, logistics, the supply chains, etc. Policy-makers and extension services will also benefit from the technologies implemented in this work

Table 17.2 KPIs of the biomass sorghum yield monitoring trials

KPI short name	KPI description	Goal description	Base value	Target value	Measured value	Unit of value	Comment
CREA-B1.3-KPI-01	Early in-season yield prediction error	Reduce prediction error	5	5	0.16	Percentage	MAPE (%, mean absolute percentage error)
CREA-B1.3-KPI-02	Early yields prediction	Increase the time (months) of prediction before harvest	0	2	6	Number of months before harvest	–

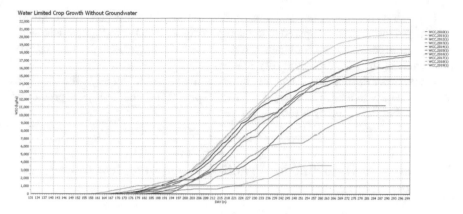

Fig. 17.6 Potential crop production

allowing early in-season information on potential biomass availability, which is critical to wider energy planning and avoiding energy-related crises.

In potato, the online platform shows the variability in Leaf Area Index (LAI). The LAI represents the area intercepting the solar radiation for crop growth. The online platform provided the farmers more insight by benchmarking their crop during the growth period with crops in the region, previous growing seasons, etc., and provided actionable information about the in-field variability and areas for inspection, and site-specific management, based on the relative performance of their field compared to the surrounding fields and the relative performance of their field compared to the potential. These new insights help farmers make better decisions for timely and more efficient, location-specific crop treatment. It was this benchmark information which was mostly appreciated by the farmers. The actual added value of the service is hard to tell because there is not really a baseline. The farmers were not used to an online crop monitoring system, so the pilot was much about raising awareness about the big data approach. The farmers appreciated much the field-specific information instead of a general satellite image, which needs to be interpreted by the farmer himself, the alerts when new data is available, avoiding the farmer's action to go and search for information, even when there is nothing new to find, and crop development benchmark. Farming is a business with a lot of variables, which not all can be controlled by the farmer. Therefore, a well-informed farmer has the advantage to be able to adapt to the circumstances. This benchmark enables farmers to spot problematic fields and areas in the field earlier and to react appropriately to save the crop and yield.

The crop growth model was used for potato yield prediction, which was calibrated with the yield data. The data for 2017 and 2018 was used to train the system and the data for 2019 was used to test the accuracy of the model. The potential crop growth was calculated only taking into account the solar radiation, assuming there were no limitations due to water or fertilizer shortages at any stage, whereas the water-limited crop growth was based on the actual rainfall in the growing season as the first limiting factor (Fig. 17.6).[1]

In general, the model has under-estimated the yield with water-limited growth and the potential yield compared to the samples for 2019. Due to limited data availability, the algorithm is not sufficiently trained yet for reliable yield predictions. The prediction of the potential yield (dry matter) based on the weather data of the last 10 years shows the relative differences between the years, but largely over-estimates the yield at harvest time. The crop growth model proves its benefit for yield prediction purposes, but the accuracy is too limited yet.

17.5 How to Guideline for Practice When and How to Use the Technology

Using satellite imageries and supervised machine learning technologies, it allowed us to model biomass sorghum phenology and carry out an early prediction of biomass yields up to six months before harvesting. This pilot combines expertise from Earth observation, ICT, artificial intelligence, and agricultural farming. The Earth observation data were mined to derive the biophysical parameter fAPAR, the agricultural farms provided the information that is critical for modeling farming outcomes, while the artificial intelligence expertise integrated the above information to model the solutions that would later be delivered to stakeholders in the form of advisory services. The equations produced in this pilot can easily be used in sorghum biomass farming businesses. As data science was done, the next big step should be putting the models into production, making them useful for any business. This is the beginning of our model operations life cycle including the following (but not necessarily limited to) key focus areas of machine learning engineering: the data pipeline (the data used to make the features used for model training such as fAPAR, phenology, biomass yields records), model training, model deployment, and model monitoring. At this level, the farmer knows how much he/she will produce early on in-season using only satellite imagery-derived fAPAR. In addition, the phenology stages can be monitored handily by the farmers using Web capable devices. In the real world, the farmer and other stakeholders will benefit from this technology as an advisory (Web) service either in-house or from third party, depending upon the expertise at the beneficiary level.

Like in any crop, potato farming is a business with many variables that not all can be controlled by the farmer. Therefore, a well-informed farmer has the advantage to be able to adapt to the circumstances. Therefore, there is a growing need for information generated over several cropping seasons and locations, which would allow for more reliable predictions. A farmer will be able to anticipate risk based on the big data analytics and subsequently change the management accordingly. Through big data sources and devices, the goals around profitability, efficiency, and cost management will be achievable. The availability of historical potato yield data with location information during the pilot was too limited to give reliable results. For training of the model much more field data is necessary to make the prediction

---

[1] WCC, WSO, respectively, WOFOST Control Center, dry weight of living storage organs

more reliable. Especially the yield data per field is essential field data. For the 2,500 farm members and about 44,000 ha (2017), with an average field size of 10 ha, this would mean that there would be 4,400 fields for collecting yield data every year. For farmers, the analysis provides them insight that would not have been available with only data about their own fields. In this respect, it is very important that farmers share their field data with each other or a trusted party. Privacy issues (and trade secrets) hinder the sharing of the data. A trusted party, like a cooperative, should provide farmers trust that their data will not be misused and thus facilitate the data sharing which will benefit them all.

17.6 Summary and Conclusions

These pilots were established as a solution to current limitations in crop monitoring in Europe Yield forecasting is based mainly on field surveys, sampling, censuses, and the use of coarser spatial (250–1000 m) resolution satellites (e.g., MODIS, SPOT-VEGETATION), all of which are undependable and/or costly. Our pilots were therefore designed to address these shortcomings. The main challenge in these pilots was being able to use high-resolution satellite images to predict sorghum biomass and potato yields early in the season, and with high precision to avoid stakeholders' aversion. The obtained results were encouraging. We were able to accurately predict aboveground sorghum biomass yields six months before harvesting with the best prediction times identified as days 150 and 165 of the year, i.e., late May and early June. These results show that crop monitoring can translate into global business without borders. They point on a remarkable opportunity for farmers and farming cooperatives for several business purposes. The models developed in this work can also help the extension services and other policy-makers in strategic planning, including assessing alternative means for energy supply and ways to avoid energy crisis. In the potato pilot, we gained insight about the possibility to apply the technologies provided by big data to smart farming services in order to gain a competitive advantage in terms of possible cost-effective services based on satellite imagery. Extensive field trials are expensive and will not predict yield in normal field conditions. The results from the DataBio project have been useful to speed up the process of improving the growth model on the basis of big data analysis. The approach contributed to better yield prediction based on the actual growing conditions with a limited number of samples or field trials. Once the model is validated through more empirical tests and observations, the processing industry will be able to enhance their sales process based on the yield prediction. Big data sources, like EO and sensor data, provide a continuous flow of data, which will certainly support the development of solutions that support the farmer in the decision process. New business opportunities can be found by implementing the yield prediction model that was tested in the pilot with AVEBE and other potato processing cooperatives, implementing a farmer decision support system, and elaborating on the potato growth model to create new services like variable rate application and irrigation planning.

References

1. Lobell, D. B., & Gourdji, S. M. (2012). The influence of climate change on global crop productivity. *Plant Physiology, 160,* 1686–1697. https://doi.org/10.1104/pp.112.208298
2. Yadav, S., & Mishra, A. (2020). Ectopic expression of C4 photosynthetic pathway genes improves carbon assimilation and alleviate stress tolerance for future climate change. *Physiology and Molecular Biology of Plants, 26,* 195–209. https://doi.org/10.1007/s12298-019-00751-8
3. Kralova, K., & Masarovicova, E. (2006). Plants for the future. *Ecological Chemistry and Engineering, 13,* 29.
4. Kussul, N. N., Sokolov, B., Zyelyk, Y. I., Zelentsov, V. A., Skakun, S. V., & Shelestov, A. Yu. (2010). Disaster risk assessment based on heterogeneous geospatial information.
5. Kussul, N., Shelestov, A., & Skakun, S. (2011). Flood monitoring from SAR data. In F. Kogan, A. Powell, & O. Fedorov (Eds.), *Use of satellite and in-situ data to improve sustainability* (pp. 19–29). Springer Netherlands.
6. Skakun, S., Kussul, N., Kussul, O., & Shelestov, A. (2014). Quantitative estimation of drought risk in Ukraine using satellite data. In *2014 IEEE Geoscience and Remote Sensing Symposium.* https://doi.org/10.1109/IGARSS.2014.6947642
7. Skakun, S., Kussul, N., Shelestov, A., & Kussul, O. (2016). The use of satellite data for agriculture drought risk quantification in Ukraine. *Geomatics, Natural Hazards and Risk, 7,* 901–917. https://doi.org/10.1080/19475705.2015.1016555
8. Habyarimana, E., Piccard, I., Catellani, M., De Franceschi, P., Dall'Agata, M. (2019). Towards predictive modeling of sorghum biomass yields using fraction of absorbed photosynthetically active radiation derived from Sentinel-2 satellite imagery and supervised machine learning techniques. *Agronomy, 9,* 203. https://doi.org/10.3390/agronomy9040203
9. Gallego, J., Kravchenko, A. N., Kussul, N. N., Skakun, S. V., Shelestov, A. Yu., & Grypych, Y. A. (2012). Efficiency assessment of different approaches to crop classification based on satellite and ground observations. *Journal of Automation and Information Sciences, 44.* https://doi.org/10.1615/JAutomatInfScien.v44.i5.70
10. Diouf, A. A., Brandt, M., Verger, A., Jarroudi, M., Djaby, B., Fensholt, R., Ndione, J., & Tychon, B. (2015). Fodder biomass monitoring in Sahelian Rangelands using phenological metrics from FAPAR time series. *Remote Sensing, 7,* 9122–9148. https://doi.org/10.3390/rs70709122
11. Duveiller, G., López-Lozano, R., & Baruth, B. (2013). Enhanced processing of 1-km spatial resolution fAPAR time series for sugarcane yield forecasting and monitoring. *Remote Sensing, 5,* 1091–1116. https://doi.org/10.3390/rs5031091
12. Johnson, D. M. (2016). A comprehensive assessment of the correlations between field crop yields and commonly used MODIS products. *International Journal of Applied Earth Observation and Geoinformation, 52,* 65–81. https://doi.org/10.1016/j.jag.2016.05.010
13. Kogan, F., Kussul, N., Adamenko, T., Skakun, S., Kravchenko, O., Kryvobok, O., Shelestov, A., Kolotii, A., Kussul, O., & Lavrenyuk, A. (2013). Winter wheat yield forecasting in Ukraine based on Earth observation, meteorological data and biophysical models. *International Journal of Applied Earth Observation and Geoinformation, 23,* 192–203. https://doi.org/10.1016/j.jag.2013.01.002
14. Kogan, F., Kussul, N. N., Adamenko, T. I., Skakun, S. V., Kravchenko, A. N., Krivobok, A. A., Shelestov, A. Yu., Kolotii, A. V., Kussul, O. M., & Lavrenyuk, A. N. (2013). Winter wheat yield forecasting: A comparative analysis of results of regression and biophysical models.
15. Kowalik, W., Dabrowska-Zielinska, K., Meroni, M., Raczka, T. U., de Wit, A. (2014). Yield estimation using SPOT-VEGETATION products: A case study of wheat in European countries. *International Journal of Applied Earth Observations and Geoinformation.* https://doi.org/10.1016/j.jag.2014.03.011
16. Davenport, J. R., Stevens, R. G., Perry, E. M., Lang, N. S. (2005). Leaf spectral reflectance for nondestructive measurement of plant nutrient status. *HortTechnology, 15,* 31–35. https://doi.org/10.21273/HORTTECH.15.1.0031

17. Peters, A. J., Ji, L., & Walter-Shea, E. (2003). Southeastern U.S. vegetation response to ENSO events (1989–1999). https://doi.org/10.1023/A:1026081615868

18. Sudbrink, D. L., Harris, F. A., Robbins, J., English, P. J., & Willers, J. L. (2003). Evaluation of remote sensing to identify variability in cotton plant growth and correlation with larval densities of beet armyworm and cabbage looper (Lepidoptera noctuidae).

19. Wang, X., Li, L., Yang, Z., Zheng, X., Yu, S., Xu, C., & Hu, Z. (2017). Predicting rice hybrid performance using univariate and multivariate GBLUP models based on North Carolina mating design II. *Heredity (Edinb), 118,* 302–310. https://doi.org/10.1038/hdy.2016.87

20. Bausch, W. C., & Duke, H. R. (1996). Remote sensing of plant nitrogen status in corn. *Transactions of the ASAE,* 1869–1875.

21. Osborne, S. L., Schepers, J. S., Francis, D. D., & Schlemmer, M. R. (2002). Detection of phosphorus and nitrogen deficiencies in corn using spectral radiance measurements.

22. Habyarimana, E., Piccard, I., Zinke-Wehlmann, C., De Franceschi, P., Catellani, M., & Dall'Agata, M. (2019). Early within-season yield prediction and disease detection using sentinel satellite imageries and machine learning technologies in biomass sorghum. In *Lecture notes in computer science 2019* (Vol. 11771, pp. 227–234). https://doi.org/10.1007/978-3-030-29852-4_19

23. Coping with drought: Stress and adaptive responses in potato and perspectives for improvement. ResearchGate. Retrieved July 27, 2020, from https://www.researchgate.net/publication/280908957_Coping_with_drought_Stress_and_adaptive_responses_in_potato_and_perspectives_for_improvement

24. World Food Studies Simulation Model (WOFOST). Retrieved July 27, 2020, from https://www.fao.org/land-water/land/land-governance/land-resources-planning-toolbox/category/details/en/c/1236431/

25. Bodemfysische Eenhedenkaart (BOFEK2012). In *WUR.* Retrieved July 27, 2020, from https://www.wur.nl/nl/show/Bodemfysische-Eenhedenkaart-BOFEK2012.htm

Chapter 18
Delineation of Management Zones Using Satellite Imageries

Karel Charvát, Vojtěch Lukas, Karel Charvát Jr., and Šárka Horáková

Abstract The chapter describes the development of a platform for mapping crop status and long-time trends by using EO data as a support tool for fertilizing and crop protection. The main focus of the pilot is to monitor cereal fields by high-resolution satellite imagery data (Landsat 8, Sentinel 2) and delineation of management zones within the fields for variable rate application of fertilizers. The first part of the paper is focused on analysis of strategies for recommendations derived from satellite data. The second part is focused on development of a software application with the goal to offer farmers a GIS portal. Here, users can monitor their fields from EO data, based on the specified period and select cloudless scenesfor further analysis. The tool supports collaborative communication between farmers and advisors.

18.1   Introduction, Motivation and Goals

Yield production zones are areas with the same yield level within the fields. Yield is the integrator of landscape and climatic variability and provides useful information for identifying management zones [1]. This work presents a basic delineation of management zones for site-specific crop management, which is usually based on yield maps over the past few years. Similar to the evaluation of yield variation from multiple yield data described by Blackmore et al. [2], the aim is to identify high yielding (above the mean) and low yielding areas expressed as the percentage of the mean value of the field. In addition, the inter-year spatial variance of yield data is important for agronomists to distinguish between areas with stable or unstable yields.

K. Charvát (✉) · V. Lukas · K. Charvát Jr. · Š. Horáková
Lesprojekt—služby s.r.o, Martinov 197, 277 13 Záryby, Czech Republic
e-mail: charvat@lesprojekt.cz

V. Lukas
Department of Agrosystems and Bioclimatology, Mendel University in Brno, Zemědělská 1665/1, 613 00 Brno, Czech Republic

K. Charvát · Š. Horáková
WirelessInfo, Cholinská 1048/19, 784 01 Litovel, Czech Republic

© The Author(s) 2021
C. Södergård et al. (eds.), *Big Data in Bioeconomy*,
https://doi.org/10.1007/978-3-030-71069-9_18

Complete series of yield maps for all fields are rare; thus, vegetation indices derived from remote sensing data are analysed to determine field variability of crops [3].

(1) **Diagnosis of the nitrogen status in crops by continuous monitoring of crop stands during vegetation**

This procedure is applied especially to crops with N-splitting fertilization and top-dressing during vegetation. It is based on the relationship between the crop biophysical properties and the spectral reflectance. The nutritional status is defined by the basic parameters of the crop stand, such as the nitrogen content [%] in the leaves of plants (or other parts of plants) and the amount of aboveground biomass [g/m^2]. Nutritional indicators, such as the N-uptake [g/m^2, kg/ha], are derived from this data. For this purpose, red-edge vegetation indices are most often used, which generally show a higher sensitivity to changes in chlorophyll content—NDRE, REIP, S2REP [4].

Evaluation of the relationship between N content and the amount of aboveground biomass during vegetation is analysed using the nitrogen nutrition index (NNI), which compares the current N content according to the critical N curve in various stages of plant development determined from the amount of aboveground mass [5]. The critical nitrogen absorption curve derived from the dilution curve developed by Justes et al. [6] is a common method in deciding whether the crops require additional N [7]. The value NNI = 1 indicates optimal nutritional level N, NNI < 1 insufficient nutritional level N and NNI > 1 excessive intake of N. Curves are defined for different crops. NNI is directly estimated from the empirical relationship with chlorophyll concentration within the canopy as measured by canopy reflectance. Leaf N concentration is estimated from the empirical relationship with chlorophyll concentration (Cab), and crop LAI is measured by remote sensing [8].

(2) **Variable rate applications according to yield potential zones**. In this case, fertilization is based on the requirements to cover the nutrient uptake for the expected yield. Yield levels are defined from yield production zones based on the analysis of a time series of yield maps or the trend of distribution of vegetation status from EO data (both 5–10 years) [9]. Production zones represent the percentage deviation from the average yield value on a given field, which is later determined in absolute values of the yield by multiplying with average expected yield values per each field.

This is a procedure suitable for crops with the recommendation of fertilization before the full coverage of crop stand, when it is not possible to use diagnostics of nutritional status based on continuous monitoring.

(3) **A combination of both mentioned principles**. In this case, coverage of the N-uptake by expected crop yield from productivity zones is corrected with splitted N-applications according to the actual diagnosis of crop stand by remote sensing. This approach includes the use of EO data or proximal sensing (N-sensors) with map overlay functionality [10] (Fig. 18.1).

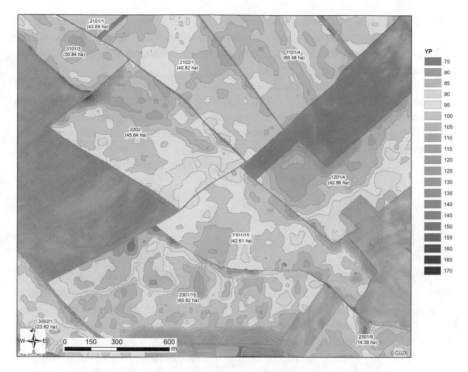

Fig. 18.1 Map of yield potential delineated from multi-temporal Landsat imagery

18.1.1 Nitrogen Plant Nutrition Strategies in Site-Specific Crop Management

The dose amount for individual management zones is determined based on two basic principles—increasing the N dosing in the zone with a higher yield (**yield-oriented**) or increasing the N rate in the below-average zones (**homogenization**).

The **yield-oriented strategy** is based on the principle of a higher requirement for nitrogen nutrient to cover a higher level of expected crop yield, which is spatially distributed by the yield productivity zones. The N rate is determined on the basis of a nitrogen balance modelling as part of the nutrient input. Areas with long-term lower crop yields are fertilized with lower N rates than places with expected higher yields. In graphical terms, this strategy is represented by a sloped curve whose inclination means the intensity of the N rate change. The curve has limit values at both ends—the minimum dose is for the plants in bad condition, which ensures at least a minimum supply of nutrients, and the maximum dose is for areas, where there could be a risk of lodging in the specific weather condition. The total amount of applied N can be specified during the growing season on the basis of continuous plant diagnosis, assessment of mineral N content in the soil or modelling of plant growth and the expected uptake of nitrogen by plants. This strategy follows the distribution of the yield potential zones within the field in a situation, where nitrogen is not considered as a yield-limiting factor. It is usually used for ear-types of cereal varieties, where the level of yield can be increased by supporting the formation of ears and ensuring an increased number of grains per ear.

The second strategy **homogenization** is based on the concept of agronomic and nutritional practice developed since the 1980s. The nitrogen is here considered a yield-limiting factor, and low-yielded areas are supported by higher doses of N. The dosing curve has a negative slope, includes capping at both ends, and its negative inclination can be specified by the user. This strategy is appropriate to increase the booting of cereals in weak places or to homogenize the qualitative parameters of the grain.

18.2   Pilot Set-Up

The pilot aimed at developing a platform for mapping of crop vigour status by using EO data (Landsat, Sentinel-2) as the support tool for variable rate application (VRA) of fertilizers and crop protection. This includes identification of crop status, mapping of spatial variability and delineation of management zones. Development of the platform was realized on a cooperative farm in Czech Republic; however, the basic datasets are already prepared for the whole Czech Republic. Therefore, the current pilot supports utilization of the solution on any farm in Czech Republic.

The pilot farm Rostenice a.s. with 8,300 ha of arable land represents a bigger enterprise established by aggregating several farms in the past 20 years. The main production is focused on the cereals (winter wheat, spring barley, grain maize), oilseed rape and silage maize for biogas power station. Crop cultivation is under standard practices, and partly conservation practices are treated on the sloped fields threatened by soil erosion. Over 1,600 ha has been mapped since 2006 by soil sampling (density 1 sample per 3 ha) as the input information for variable application of base fertilizers (P, K, Mg, Ca). Nowadays, the soil sampling covers the full area of the farm. Farm machinery is equipped by RTK guidance with 2–4 cm position accuracy. Until 2018, farm agronomists have not been using any VRA strategy of nitrogen fertilizers and crop protection because of lack of reliable solutions in Czech Republic (Fig. 18.2).

During the 2018 vegetation period, a field experiment was established for testing variable rate application of nitrogen fertilizer based on the yield potential maps computed from Landsat time-series imagery and digital elevation model (DEM). Testing was carried out on three fields with a total acreage of 133 ha. The main reason was to tailor nitrogen rates for spring barley according to the site-specific yield productivity and to avoid the crop lodging risk in the water accumulation areas. Plant nutrition of spring barley for malt production is more difficult than for other cereals because of limits for maximal N content in grain. Thus, balancing N rates to reach highest yield and simultaneously not exceeding N content in grain is crucial for successive production of spring barley.

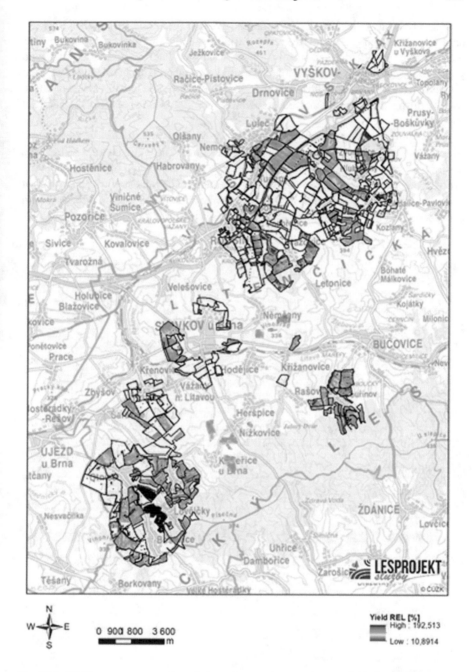

Fig. 18.2 Yield maps represented as relative values to the average crop yield of each field (harvest 2018)

For definition of yield productivity zones, a 8-year time series of Landsat imagery data was processed giving relative crop variability. The final map is represented as percentage of the yield to the mean value of each plot, later multiplied by expected yield [t ha^{-1}] as the numeric variable for each field and crop species. Values of yield potential were reclassified into three categories—high, middle and low-yielded areas, and the nitrogen rate was increased in the high expected yield areas (Figs. 18.3, 18.4 and 18.5).

Prescription maps for variable rate application of nitrogen fertilizers were prepared by reclassification and values editing tools in GIS. The valogen rate value was determined based on the agronomist experience and knowledge of the site-specific production conditions and crop variety requirements. The final step was an export of prepared maps into shapefile or ISO-XML format and upload into machinery board computers (mainly Trimble or Mueller Elektronik) (Figs. 18.6 and 18.7).

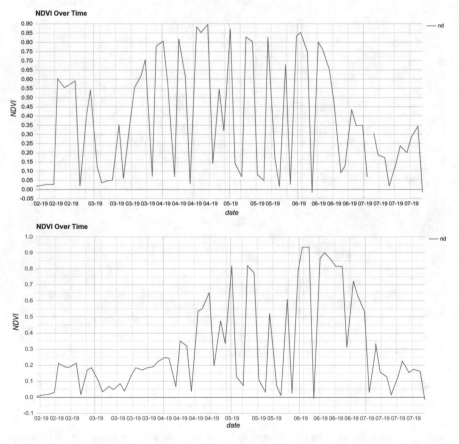

Fig. 18.3 Graphs of Sentinel-2 NDVI during the vegetation period 2019 for winter wheat (above) and spring barley (below) at locality Otnice (Rostenice farm). Low peaks indicate occurrence of clouds within the scene. *Source* Sentinel-2, Level L1C, Google Earth Engine

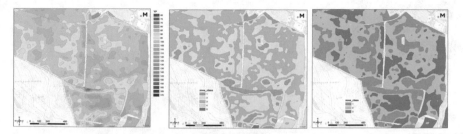

Fig. 18.4 Example of the output map products from yield potential zones classification from EO time-series analysis: classification into 5% classes (left), 5-zone map (middle) and 3-zone map (right). Blue/green areas indicate higher expected yield

Fig. 18.5 Map of yield potential zones (5-zone map) updated for 2019 season from 8-year time-series imagery; for southern (left) and northern (right) part of Rostenice farm

Fig. 18.6 Variable rate application of solid fertilizers by Twin Bin applicator on Terragator

Fig. 18.7 Variable rate
application of liquid N
fertilizers (DAM390) by
36 m Horsch Leeb PT330
sprayer

18.3 Technology Used

This work was supported by the development of a platform for automatic down-loading of Sentinel 2 data and automatic atmospheric correction. Through this platform, Lesprojekt is ready to offer commercial services around processing satellite data for any farm in Czech Republic. Another part in the platform development focused on transferring Czech LPIS into FOODIE ontology and on developing effective tools for querying data. Lespro did this together with PSNC, and the system is currently supporting open access to anonymous LPIS data through the FOODIE ontology and also secure access to farm data.

The main focus of the pilot discussed here is the monitoring of arable fields with high-resolution satellite imaging data (Landsat 8, Sentinel 2) and delineation of management zones within the fields for variable rate application of fertilizers. The main innovation is to offer a solution in the form of the Web GIS portal for farmers, where users can monitor their fields from EO data based on the specified time period, select cloudless scenes and use them for further analysis. This analysis includes unsupervised classification of a defined number of classes like identification of main zones, as well as generating prescription maps for variable rate application of fertilizers or crop protection products based on the mean doses defined by farmers in the Web GIS interface.

Spatial data about crop yields from the harvester were recorded in the period from June to September. Of the total 8300 ha acreage of the pilot farm, more than 3350 ha of arable land was covered by yield mapping in the cropping season 2018. We recorded crop yields specially for grain cereals (winter wheat, spring barley, winter barley), oilseed rape and for grain maize. Data were later processed for outlier analysis and by spatial interpolation techniques to obtain a final crop yield map in absolute [t ha^{-1}] and relative [%] measure.

To guarantee access for farmers and testing of yield potential, we calculated the yield potential for the 2017 season on a basic level for all Czech Republic, and

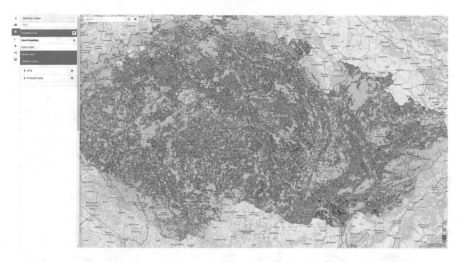

Fig. 18.8 Transformation and publication of Czech data as linked data with prototype system for visualizing

data are now available in open form on the Lesprojekt server for the whole Czech Republic. Farmers can freely test this basic data (Fig. 18.8).

Farm data

- Rostenice pilot farm data, including information about each field name with the associated cereal crop classifications arranged by year.
- Data about the field boundaries and crop map and yield potential of most of the fields in Rostenice pilot farm.
- Yield records from harvested crops on the fields in separate years.

Open data

- Czech LPIS data showing the actual field boundaries.
- Czech erosion zones (strongly/SEO and moderately/MEO erosion-endangered soil zones).
- Restricted area near to water bodies (example of 25 m buffer according to the nitrate directive) from Czech.
- The data about soil types from all over Czech (Fig. 18.9).

18.4   Exploitation of Results

The pilot's biggest success was the successful introduction of the variable application of nitrogen fertilizers based on satellite monitoring of the real plant operation on the farm fields. Although Rostěnice a.s. plays in its region a role of a pioneer in the use of precision farming technologies, they have long been hesitant about choosing the

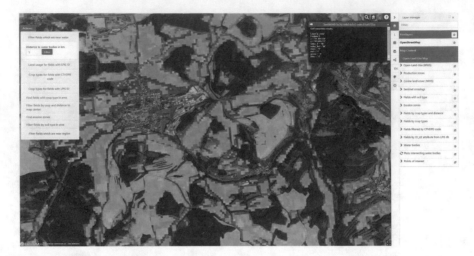

Fig. 18.9 Visualization of results

right technology for a variable N fertilizer application. After the initial scepticism towards the use of crop sensors in their operations, they finally decided to apply a variable application based on the delineation of the management zones from the yield potential maps and on the strategy of increasing the N dose in areas with higher expected yield. This strategy has proved to be a promising option for more arid farming conditions and when irrigation is difficult, because of low soil moisture. VRA testing started on the selected fields with spring barley (over 150 ha) in 2018. In this case, spring barley for beer production was chosen as the most sensitive crop for the N application, because it is difficult to achieve malting quality in these more arid conditions, where the sum of precipitation from March till July 2018 was at the level of 152 mm. Inadequate nitrogen nutrition of plants leads to significant yield reductions, while excessive N doses decrease grain malting quality. During the growing season 2019, a variable application of N fertilizers on an area of more than 3000 ha was launched. This included base N fertilizing before sowing spring barley and maize and first N application in top-dressing of winter cereals (winter wheat, winter barley). In addition, testing of variable application of crop growth regulators in spring barley by the combination of yield potential zoning from EO time-series analysis and actual crop status monitoring from Sentinel-2 imagery was also started.

References

1. Kleinjan, J., Clay, D., Carlson, C., & Clay, S. (2007). Developing productivity zones from multiple years of yield monitor data. https://doi.org/10.1201/9781420007718.ch4
2. Blackmore, S., Godwin, R. J., & Fountas, S. (2003). The analysis of spatial and temporal trends in yield map data over six years. *Biosystems Engineering, 84*(4), 455–466.

3. Charvát, K., Řezník, T., Lukas, V., Charvát, K., Horáková, Š., Kepka, M., & Šplíchal, M. (2016). Quo vadis precision farming. In *13th International Conference on Precision Agriculture*, St. Louis, MO, July 31–August 4, 2016.
4. Frampton, W. J., Dash, J., Watmough, G., & Milton, E. J. (2013). Evaluating the capabilities of Sentinel-2 for quantitative estimation of biophysical variables in vegetation. *ISPRS Journal of Photogrammetry and Remote Sensing, 82*, 83–92. ISSN 0924-2716.
5. Lemaire, G. (1997). *Diagnosis of the nitrogen status in crops*. Springer. ISBN 9783540622239.
6. Justes, E., Mary, B., Meynard, J.-M., Machet, J.-M., & Thelier-Huche, L. (1994). Determination of a critical nitrogen dilution curve for winter wheat crops. *Annals of Botany, 74*(4), 397–407. https://doi.org/10.1006/anbo.1994.1133
7. Delloye, C., Weiss, M., & Defourny, P. (2018). Retrieval of the canopy chlorophyll content from Sentinel-2 spectral bands to estimate nitrogen uptake in intensive winter wheat cropping systems. *Remote Sensing of Environment, 216*, 245–261. ISSN 0034-4257.
8. Lemaire, G., Jeuffroy, M. H., & Gastal, F. (2008). Diagnosis tool for plant and crop N status in vegetative stage. Theory and practices for crop N management. *European Journal of Agronomy, 28*(4), 614–624. ISSN 11610301.
9. Řezník, T., Pavelka, T., Herman, L., Lukas, V., Širůček, P., Leitgeb, Š., & Leitner, F. (2020). Prediction of yield productivity zones from Landsat 8 and Sentinel-2A/B and their evaluation using farm machinery measurements. *Remote Sensing, 12*(12), 1917. https://www.mdpi.com/2072-4292/12/12/1917
10. Mezera, J., Lukas, V., Elbl, J., Kintl, A., & Smutný, V. (2019). Evaluation of variable rate application of fertilizers by proximal crop sensing and yield mapping. Paper presented at the 19th International Multidisciplinary Scientific Geoconference, SGEM 2019.

Chapter 19
Farm Weather Insurance Assessment

Antonella Catucci, Alessia Tricomi, Laura De Vendictis, Savvas Rogotis, and Nikolaos Marianos

Abstract The pilot aimed to develop services supporting both the risk and the damage assessment in the agro-insurance domain. It is based on the use of remotely sensed data, integrated with meteorological data, and adopts machine learning and artificial intelligence tools. Netherlands and Greece have been selected as pilot areas . In the Netherlands, the pilot was focused on potato crops for the identification of areas with higher risk, based on the historical analysis of heavy rains. In addition, it covered automated detection of potato parcels with anomalous behaviours (damage assessment) from satellite data, meteorological parameters and soil characteristics. In Greece, the pilot worked with 7 annual crops of high economic interest to the national agricultural sector. The crops have been modelled exploiting the last 3-year NDVI measurements to identify their deviations from the normal crop health behaviour for an early identification of affected parcels in case of adverse events. The models were successfully tested on a flooding event that occurred in 2019 in the Komotini region. Even though the proposed methodologies should be tested over larger areas and compared against a larger validation dataset, the results already now demonstrate how to reduce the operating costs of damage assessors through a more precise and automatic risk assessment. Additionally, the identification of parameters that most affect the crop yield could transform the insurance industry through index-based solutions allowing to dramatically cut costs.

A. Catucci (✉) · A. Tricomi · L. De Vendictis
e-GEOS, Rome, Italy
e-mail: antonella.catucci@e-geos.it

S. Rogotis · N. Marianos
NEUROPUBLIC SA, Methonis 6 and Spiliotopoulou, 18545 Piraeus, Greece
e-mail: s_rogotis@neuropublic.gr

N. Marianos
e-mail: n_marianos@neuropublic.gr

© The Author(s) 2021
C. Södergård et al. (eds.), *Big Data in Bioeconomy*,
https://doi.org/10.1007/978-3-030-71069-9_19

19.1 Introduction, Motivation and Goals

Agricultural insurance protects against loss or damages to crops or livestock. It has a great potential to provide a value to farmers and their communities, both by protecting farmers when shocks occur and by encouraging greater investment in crops. This concept is particularly evident if considering current challenges related to climate change effects and increase of world population. However, in practice, insurance effectiveness has often been constrained by the difficulty of designing optimal products and by demand constraints. The objective of the pilot is the provision and assessment of services for the agriculture insurance market in selected areas based on the Copernicus satellite data series, also integrated with meteorological data, and other ground available data by using big data methods and AI tools.

Among the relevant needs of the insurances operating in agriculture, there are: the more consolidated procedures of damage assessment by means of earth observation techniques and the most promising evaluation of risk parameters down to parcel level.

For the **risk assessment phase**, the integrated usage of historical meteorological series and satellite derived indices, supported by proper modelling, allow to tune EO-based parameters in support to the risk estimation phase. The availability of this information allows a better estimation of potential risky areas and then a more accurate pricing and designing of insurance products. These advantages could positively impact the increase of insurance penetration. Moreover, the definition of key parameters related to the field lost by using machine learning-based approaches has the potential to support the design of innovative insurance products (such as parametric insurance) that are very promising for farmer protection.

For **damage assessment**, the operational adoption of remotely sensed data allows optimization and tuning of new insurance products based on objective parameters. This could imply a strong reduction of ground surveys, with positive impact on insurance costs and reduction of premium to be paid by the farmers.

19.2 Pilot Set-Up

The pilot included trial stages in two different areas of interest: the Netherlands and Greece.

In the Netherlands, the pilot has been realized considering potato crop that is particularly relevant for the national market. The pilot included the generation of different products to enable the detection of parcels with anomalous behaviours and the identification of the most influencing parameters of high impact on crop yield. Some examples of products are introduced here:

- **Weather-based risk map** that is intended to show occurrences of extreme weather events, heavy rains in particular, in order to identify areas with possible high damage frequency.

- **Intra-field analysis** that is aimed to detect the growth homogeneity and evidencing irregular areas within the parcel, providing an indicator of field anomalies.

Different partners have been involved in the pilot activities. Copernicus satellite data (both optical and SAR) and services have been provided by e-GEOS, the provision of machine learning technology by EXUS, meteorological data and services from MEEO. The involvement of end-users and the provision of local agronomic knowledge have been assured by NBAdvice.

In Greece, the pilot worked with annual crops (e.g. tomato, maize, cotton) of high economic interest to the Greek agricultural sector, in several regions of Northern Greece and in particular in Evros, Komotini and Thessaly. The pilot evaluated incidents like floods and heatwaves that fall under the definition of the climate-related systemic perils. The pilot effectively demonstrated how big data enabled technologies and services dedicated for the agriculture insurance market can eliminate the need for on-the-spot checks for damage assessment and promote rapid payouts. The role of field-level data has been revealed as their collection, and monitoring is important in order to determine if critical/disastrous conditions are present (heat waves, excessive rains and high winds). Field-level data can be seen as the "starting point" of the damage assessment methodology, followed within the Greek pilot case. Moreover, regional statistics deriving from this data can serve as a baseline for the agriclimate underwriting processes followed by the insurance companies who design new agricultural insurance products.

NP led the activities for the execution of the full lifecycle of this pilot case with technical support from FRAUNHOFER and CSEM. Moreover, a major Greek insurance company, INTERAMERICAN, was actively engaged in the pilot activities, bringing critical insights and its long-standing expertise into fine-tuning and shaping the technological tools to be offered to the agriculture insurance market.

The goal of this particular pilot case was to enable a better management of the damage assessment process (reduction of the required time) and to support other processes of the insurance companies.

19.3 Technology Used

19.3.1 Technology Pipeline

For the **trial stage in the Netherlands**, the pipeline has been composed of three main logical steps (Fig. 19.1):

Data Preparation: a set of data has been collected and properly pre-processed in order to get them ready for the processing phase. In particular, the following datasets have been considered: Sentinel-2 optical data, Sentinel-1 SAR data (soil moisture), Proba-V data, weather data including main parameters influencing crop growth (land

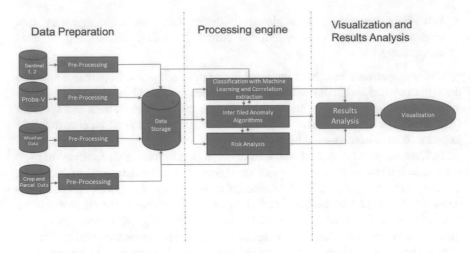

Fig. 19.1 Overview of the three main components of the pipeline for the trial in the Netherlands

surface temperature, 24-h precipitation accumulations, humidity, evapotranspiration) and crop data (crop type, parcel boundaries and location, soil type).

Processing Engine: the processing step includes different approaches implemented by means of proprietary algorithms that allows the extraction of relevant information that can be used by insurance companies and risk managers. In particular, the three main components are:

- **classification and correlation extraction** based on machine learning methods
- **inter-field anomaly detection and intra-field algorithms**
- **risk analysis tools.**

The processing engine is composed of different blocks that are part of the DataBio shared architecture.

Visualization: the visualization phase has been realised by components that are part of the DataBio architecture.

For the **trial stage in Greece**, a set of data collection, processing and visualization components has been used so as to technically support the pilot activities. More specifically the following technological components should be acknowledged:

In terms of **Data Collection**, a set of heterogeneous data is required in different spatial and temporal resolutions to provide services to the insurance companies. Moreover, historical data is critical for shaping insurance products and conducting effective assessments. Data abundancy holds the key for creating sound insurance products/tools. To collect all this data several data collection modules are used:

- In-situ telemetric stations provided by NP, so-called gaiatrons, that collect weather data,

- Modules for the collection, pre-processing of earth observation products, the extraction of higher-level products and assignment of vegetation indices at parcel level.

In terms of **Data Processing**:

- GAIABus DataSmart Machine Learning Subcomponent (NP): The specific component supports: EO data preparation and handling functionalities. It also supports multi-temporal object-based monitoring and modelling for damage assessment.
- GAIABus DataSmart Real-Time Streaming Subcomponent (NP): This component supports:

  - Real-time data stream monitoring for NP's gaiatrons installed in the pilot sites,
  - Real-time validation of data,
  - Real-time parsing and cross-checking.

- Neural Network Suite (CSEM): this component was used as a machine learning crop identification system for the detection of crop discrepancies that might derive from reported weather-related catastrophic events.
- Georocket, Geotoolbox and SmartVis3D (FRAUNHOFER): This component has a dual role: It is a back-end system for big data preparation, handling fast querying and spatial aggregations (data courtesy of NP), as well as a front-end application for interactive data visualization and analytics.

In terms of **Data Visualization**, the main component in this category is Neurocode (NP). Neurocode allowed the creation of the main pilot UIs in order to be used by the end-users (insurance companies). An additional DataBio component providing information visualization functionalities is Georocket (FRAUNHOFER).

19.3.2 Reflection on Technology Use

In the Netherlands, an **historical risk map** was generated based on SPOT-VGT/Proba-V 1 km fAPAR data from 2000 to 2017 (Fig. 19.2). The index was defined as the sum of fAPAR over the growing season. The risk map allows to detect zones with a higher damage frequency in the past. This technology seems to be effective to generate and to give an overview of the risk in a selected area. Nevertheless, more accurate datasets can be used to analyse more in depth the situation.

In addition, **weather-based risk maps** were produced to complement the historical risk map. The weather risk maps are intended to show the occurrence of extreme weather events in the past and are aimed to investigate if a reliable correlation between damages occurred to the crops and extreme weather events (heavy rains, in particular) occurs. The main goal was to define damage patterns and to zoom in on areas with a high damage frequency. At the end, eight different risk maps were calculated, one per threshold provided by end-users. Moreover, starting from the list of dates related

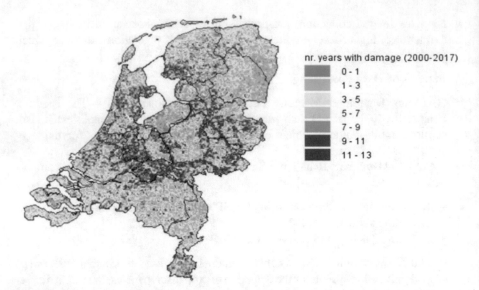

nr. years with damage (2000-2017)

- 0 - 1
- 1 - 3
- 3 - 5
- 5 - 7
- 7 - 9
- 9 - 11
- 11 - 13

Fig. 19.2 Map classifying the Netherlands territory in terms of number of years with damages

to damage claims and provided by the insurance companies for the years 2015–2018, the extraction of precipitation values (with the respective location coordinates) has been performed, in order to find further locations (in addition to those provided by the insurance company) where heavy rain events have occurred (see Fig. 19.3).

As concerning the detection of parcels with **anomalous behaviours** and identification of **influencing parameters**, the following approach was considered.

The dataset was split according to the different types of potato, and each group was clustered using satellite data, meteorological measurements and soil characteristics with a monthly aggregation.

Fig. 19.3 Map of precipitation extracted from KNMI dataset on date 30/08/2015. Yellow points: locations provided by the insurance company—blue points: further locations with 24-h precipitation values above the 50 mm threshold

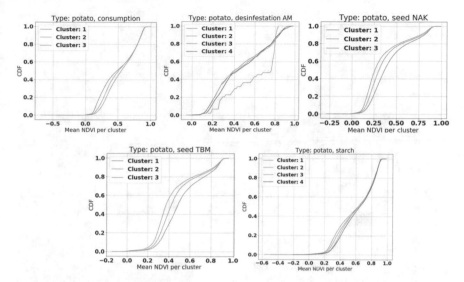

Fig. 19.4 Cumulative distribution function of mean normalized difference vegetation index (NDVI) grouped by cluster and type of potato

The clustering-based service has proved to be a very useful technique to identify parcels with anomalous behaviour and allows to consider in a single analysis all the variables that can affect the growth and the yield of a crop (Fig. 19.4). Unfortunately, it was not possible to validate the results due to lack of data from insurances but the approach seems to be very promising. Moreover, the performed activity reveals that temperature is a factor with high impact on NDVI of potatoes. See Fig. 19.5 where the first plot shows the average NDVI trends of parcels belonging to different clusters. The second one is related to the average temperature recorded over the area defined by the "blue" cluster, characterized by higher temperatures and lower NDVI values in the peak period, and over the area defined by the "red" cluster, characterized by lower temperatures and higher values of NDVI in the peak period.

Lastly, the **intra-field analysis** was performed over areas with a high presence of potatoes. The scope of the analysis has been to analyse each parcel to detect the growth homogeneity and evidencing irregular areas, providing an indicator of field anomalies. In order to resume the approach, a brief description of the intra-field analysis follows.

After creating an inner buffer in order to avoid border effects, the extraction of temporal profile at parcel level was performed. Some filters were applied in order to exclude parcels that were not cultivated or areas with high percentage of cloud coverage. Then, the observation that corresponded to the maximum growth stage of the crop was identified. At the end, each parcel was classified at pixel level according to statistical thresholds. See Fig. 19.6.

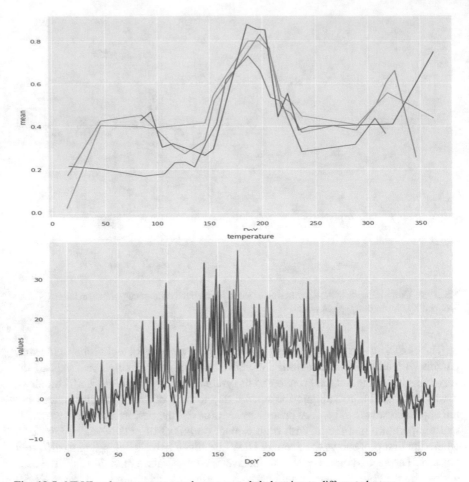

Fig. 19.5 NDVI and temperature trends over parcels belonging to different clusters

Intra-field service is extremely effective in detecting soil anomalies that do not allow crops to grow homogeneously within parcels. This service provides an indicator of soil goodness: texture and depth, for instance, have consequences on water consumption and on regular growth.

In Greece, crop type and area tailored crop models have been created for the whole Greek arable area making use of EO-derived NDVI measurements that have proven to be suitable for assessing plant health. In total, for each one of the 55 Sentinel-2 tiles that cover the whole Greek arable land, 7 major arable crops for the local agri-food sector were modelled and namely wheat, maize, maize silage, potato, tomato, cotton and rice ($55 \times 7 = 385$ models in total). The models were developed exploiting multi-year NDVI measurements from the available last three (3) cultivating periods and instead of using sample statistics (few objects of interest but many observations referring to them), population statistic methods (large number of objects of interest

Fig. 19.6 Areas with anomalous growth within a parcel (in red and orange)

but with few observations referring to them) were employed instead in order to identify NDVI anomalies. As sound insurance models are typically created using large multi-year historical records (~30 years), this approach is ideal for deriving robust estimates for setting anomaly thresholds (exploiting the space–time cube to have enough degrees of freedom). The goal is to detect deviations in NDVI measurements in respect to what is considered normal crop health behaviour for a specific time instance. Thereby, each crop model consists of 36 NDVI probability distributions that refer to all decades of the year. By adjusting these high and low thresholds (part of the strategy of the insurance company), it is evident that measurements found at the distribution extremes can be spotted and flagged as anomalies. Typically, insurance companies are looking for negative anomalies (below 15%) that provide strong indications of a disastrous incident (Fig. 19.7).

The figures (Table 19.1) graphically depict three different crop models created using the aforementioned procedure.

The effectiveness of the proposed monitoring methodology was tested against a flooding event (11/7/2019) in Komotini that affected cotton farmers in the region and led to significant crop losses (Fig. 19.8).

Initially, Gaiatron measurements confirmed that flooding conditions were present at the area as a result of increased volumes of rainfalls. This proves that the region might have been affected by the systemic risk and should be more thoroughly examined (Fig. 19.9).

This triggered an EO-based crop monitoring approach that captures the impact of the peril to crop's health. After only 2 weeks, the approach identified statistically significant differences compared to the respective crop model that indicates damages at field level. This validates the initial hypothesis that floods were responsible for severely affecting the region's crop health and consequently proves that the established methodology can be a powerful tool for early identification of

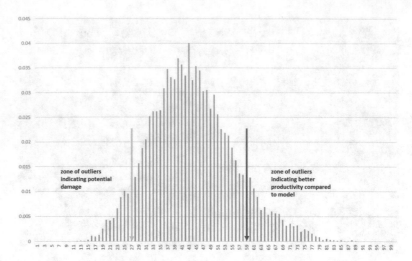

Fig. 19.7 Crop NDVI probability distribution referring to a single decade of the year (wheat-Larisa region-2nd decade of February). Anomalies can be found at the distribution extremes

Table 19.1 Crop models of cotton, maize and wheat

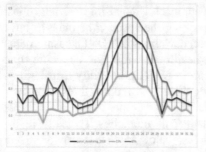	Cotton model in the Komotini region (T35TLF tile) by decade (horizontal axis). Light green threshold indicates lower 15% extremes while dark green threshold indicates upper 85% extremes of the probability distribution. Red line is presenting a single parcel status for the whole 2018 with its NDVI measurements staying within "normal" ranges for the critical cultivating periods
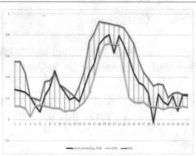	Maize model in Evros region (T35TMF tile) by decade (horizontal axis). Light green threshold indicates lower 15% extremes while dark green threshold indicates upper 85% extremes of the probability distribution. Red line is presenting a single parcel status for the whole 2018 with only one (1) NDVI measurement falling under the "normal" ranges for the critical cultivating periods (twenty-first decade)

(continued)

Table 19.1   (continued)

	Wheat model in Larisa region (T34SFJ tile) by decade (horizontal axis). Light green threshold indicates lower 15% extremes while dark green threshold indicates upper 85% extremes of the probability distribution. Red line is presenting a single parcel status for the whole 2018 with its NDVI measurements staying within "normal" ranges for the critical cultivating periods

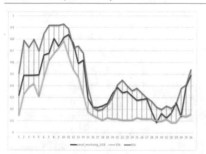

Fig. 19.8   Aftermath of the floods in Komotini region (11/7/2019)

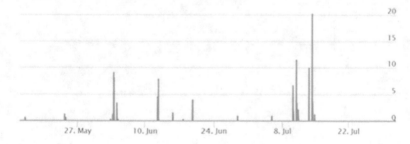

Fig. 19.9   Rainfall volume (mm) in the Komotini region

potentially affected/damaged parcels, crop types and areas. The findings have been presented both to the insurance company and the farmers in order to show how these technologies can bridge the gap among the farming and the insurance world (Fig. 19.10).

By mapping the outcome of the followed damage assessment procedures on top of a map, it is evident that high-level assumptions can be made. This involves the risk at which the insurance company is exposed to and prioritizing the work that needs to

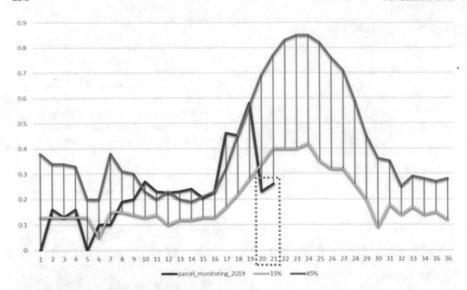

Fig. 19.10 Parcel monitoring at Komotini region (cotton) showing negative anomaly (deviation) for two consecutive decades just after the disastrous incident

be conducted by field damage evaluators (until now prioritization is not data-driven) that are advised to begin with parcels exhibiting higher damage estimates and steadily move to those with lower ones (Fig. 19.11).

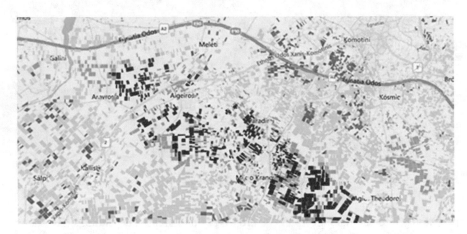

Fig. 19.11 High-level overview of the affected area, color coded with the output of the followed damage assessment procedures

19.4   Business Value and Impact

19.4.1   Business Impact of the Pilot

Business Impact of the Pilot—Netherlands

Results are promising in terms of general procedures and methods. These need to be tested over larger areas and compared with validation data provided by the final users (insurance). The data availability is a crucial challenge for this market considering the very restricted dissemination level of the information and the high competitive level. In fact, the insurance companies are not interested in supporting the development of products that can be available also for their competitors. To overcome these potential limitations, a set-up phase of the service in an operative environment is necessary in close cooperation with the insurance company involved. This collaboration has the potential to transform the tested methods into operative services filling the existing gap between prototype development and final product.

In order to analyse the benefit of the tested technology for the insurance industry (risk estimation also by means of machine learning), it is important to define the three levers of value in insurance market:

1.   Sell More
2.   Manage Risk Better
3.   Cost Less to Operate.

The activity performed in the pilot impacts essentially the point "Cost Less to Operate". One clear way to reduce operating costs in insurance is to add information and increase automation to complex decision-making processes, such as underwriting. To keep processing costs in check, many insurance carriers have a goal to increase the data available in support to a more precise and automatic risk evaluation in support of the underwriting. In fact, the use of decision management technologies like risk maps, machine learning and artificial intelligence can reduce the time spent to analyse each contract and focus team members on higher value activities. Moreover, the identification of parameters that most affect the crop yield performed in the pilot can support an innovative insurance typology called "parametric insurance". This particular insurance typology is revolutionizing the insurance industry allowing to dramatically cut operative costs removing the in-field direct controls.

The first step in building a parametric product is determining the correlation between the crop losses and a particular index representative of the climate event associated with the loss. The activity performed in the pilot by using a machine learning approach is to identify the most important parameter affecting the crop yield that can be the basis for a parametric or index-based insurance.

Quantifying the potential impact of the proposed solution for the insurance industry is a complex issue considering the work necessary to transform the methodology in an operative service. Just to provide some business projection, it can be considered that direct European agricultural insurance premiums in 2016 were 2.15 m€ (estimated by Munich RE) (Fig. 19.12).

Crop Insurance in the EU – Premium Volume m€[1]   Munich RE

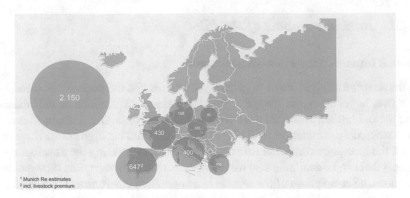

Fig. 19.12 Premium value distribution in Europe estimated by Munich RE

It can be considered that around 70% of this amount is spent by insurances to reimburse damages and the remaining 30% is used to pay internal costs and re-insurances. Considering this dimension and considering the row and very preliminary estimation obtained by the pilot, it is possible to assume that the cost that can be saved by using EO-based services in support of risk assessment is around 2% of the total cost used by the insurance to pay internal costs. Table 19.2 summarizes the potential available market for these services in Europe.

Business Impact of the Pilot—Greece

There is a constantly increasing need for agricultural insurance services, due to the adverse effects of climate change and the lack of sufficient compensation frameworks. From their side, insurance companies with offerings for the agricultural sector need to have precise and reliable systems that will facilitate the damage evaluation processes and will ensure swift and fair compensation to those who actually deserve it, thus

Table 19.2 Market projection in Europe

Market projections Market segment: insurance			
	Size by revenue	Market share	
Available market in Europe	2150 m€ 30% * 2% = 12.9 m€	100%	Potential cost reduction by using downstream services supporting the insurance industry is assumed the 2% of the total insurances income (30% of the insurance premium)

allowing follow-up/reactive measures to be undertaken and supporting food security in general.

In the two trial periods of DataBio, tailored agri-insurance tools and services have been developed with and for the agri-insurance companies that perform EO-based damage assessment at parcel level and target towards evolving to next-generation index-based insurance solutions. The pilot results clearly show that data-driven services can facilitate the work of the insurance companies, offering tools that were previously unavailable and were responsible for severe bottlenecks in their day-to-day activities including:

- long wait for official damage evaluation reports,
- dependence on the human factor,
- difficulties in prioritizing work after receiving several compensation claims.

19.4.2   Business Impact of the Technology on General Level

The remote sensing literature offers numerous examples proposing earth observation techniques to support insurance, for example in the assessment of damage from fire and hail [1, 2]. To date, however, few operational applications of remote sensing for insurance exist and are operative. Many scientific papers claiming potential applications of remote sensing [3–5], typically stress the technical possibilities, but without considering and proving its contribution in terms of "value" for the insurer. The discrepancy between the perceived potential and the actual uptake by the industry is probably the result of two main reasons:

- technological solutions not adequate and too expensive, in relation to the valued added
- over-optimistic assumptions by the remote sensing community, regarding the industry's readiness to adopt the information by remote sensing.

Despite this situation, EO can still play a central role in supporting the insurance market in agriculture trying to design services that can really bring value to the users. This is the case of supporting in field verification and parametric insurance products (innovative insurance products). The present pilot investigates these services demonstrating the potential and opening up the route for new collaboration with users.

19.5   How-to-Guideline for Practice When and How to Use the Technology

As said, the methodology needs to have a pre-operational set-up phase in close collaboration with the insurance company. In fact, the developed method can be applied to different areas and crops but only if an adequate training set of data related to occurred losses are available.

In Greece, the proposed solution is based on mature technologies and high-quality data, in order to ensure high accuracy and quality for the designed tools and services. EO-based methodologies were used in order to extract useful information from EO products for:

- damage assessment targeting towards a faster and more objective claims monitoring approach just after the disaster,
- the adverse selection problem. Through the use of high-quality data, it will be possible to identify the underlying risks associated with a given agricultural parcel, thus supporting the everyday work of an underwriter,
- large-scale insurance product/risk monitoring, that will allow the insurer to assess/monitor the risk at which the insurance company is exposed to from a higher level.

More and more insurance companies are interested in entering the agricultural market, which exhibits high value, due to its vulnerability to extreme weather phenomena. However, before they integrate such technology- and data-driven tools, they need to be persuaded that these tools will help them reduce operational costs by minimizing the human intervention and ensuring high quality of services. The involvement of one of the largest insurance companies in Greece in this pilot case (INTERAMERICAN) helps in bringing the proposed solution closer to the market, and with their precious feedback, it will be more easily available for commercial exploitation.

19.6   Summary and Conclusion

The objective for the pilot was to find useful services for the insurance to gain more insight about the risk and the impact of heavy rain events for crops. In the Netherlands, for instance, potato-crops are very sensitive to heavy rain, which may cause flooding of the field (due to lack of runoff) and saturation of the soil. This may cause the loss of the potato yield in just a few days. Areas of greater risk can be charged with higher costs for the farmer. The investigated correlation among precipitation and losses can support the identification of index for parametric insurance products. Moreover, instead of just raising the premium, the intention of the pilot was to be able to create awareness and incentives for farmers to prevent losses. Therefore, the services serve multiple purposes. Weather is an important factor in crop insurance, because it represents a critical aspect influencing yield. The analysis of the long-term precipitation, categorized in threshold values, for intense rain events, gave insight in the areas with higher risk. In the pilot, the relation between one single event and the potential yield loss has been analysed. For this purpose, an annotated set of data, where actual losses were determined, was necessary. Because of the privacy issues related to sharing the damage data, the location of damaged fields in the Netherlands could not be pinpointed precisely enough for correlation to the EO data. Without the details about historical events, this relationship could not be determined. In Greece, where a massive flood event occurred, impacts have been identified by analysing NDVI anomalies for the most common crop types. During the pilot activities, we realized that a service, based on the alert that a heavy rain event took place, would be useful for gaining insight about the impact on other locations. Additionally, in order to find the most limiting aspect in the crop development, we created a dataset based on the Sentinel-2 raster size to combine NDVI with SAR, precipitation (cumulative), temperature and soil type. The developed methodology, however, is valuable for further analysis, not limited to insurance topics and can be extended to other crops in support to risk assessment and also for design of new insurance products such as parametric insurance.

References

1. Young, F. R., Apan, A., & Chandler, O. (2004). Crop hail damage: Insurance loss assessment using remote sensing. In *Mapping and Resource Management: Proceedings of RSPSoc2004*.
2. Peters, A. J., Griffin, S. C., Viña, A., & Ji, L. (2000). Use of remotely sensed data for assessing crop hail damage. *PE&RS, Photogrammetric Engineering & Remote Sensing, 66*(11), 1349–1355.
3. Turvey, C. G., & McLaurin, M. K. (2012). Applicability of the normalized difference vegetation index (NDVI) in index-based crop insurance design. *Weather, Climate, and Society, 4*, 271–284. https://doi.org/10.1175/WCAS-D-11-00059.1
4. De Leeuw, J., Vrieling, A., Shee, A., Atzberger, C., Hadgu, K. M., Biradar, C. M., Keah, H., & Turvey, C. (2014). The potential and uptake of remote sensing in insurance: A review. *Remote Sensing, 6*(11), 10888–10912.
5. Savin, I. Yu., & Kozubenko, I. S. (2018). Possibilities of satellite data usage in agricultural insurances. *RUDN Journal of Agronomy and Animal Industries, 13*(4), 336–343.

Chapter 20
Copernicus Data and CAP Subsidies Control

Olimpia Copăcenaru, Adrian Stoica, Antonella Catucci, Laura De Vendictis, Alessia Tricomi, Savvas Rogotis, and Nikolaos Marianos

Abstract This chapter integrates the results of three pilots developed within the framework of the Horizon 2020 DataBio project. It aims to provide a broad picture of how products based on Earth Observation techniques can support the European Union's Common Agricultural Policy requirements, whose fulfillments are supervised by National and Local Paying Agencies operating in Romania, Italy and Greece. The concept involves the use of the same data sources, mainly multitemporal series of Copernicus Sentinel-2 imagery, but through three different Big Data processing chains, tailored to each paying agency's needs in terms of farm compliance assessment. Particularities of each workflow are presented together with examples of the results and their accuracy, calculated by validation against independent sources. Business value aspects for each use case are also discussed, emphasizing the way in which the automation of the CAP requests verification process through satellite technologies has increased the efficiency and reduced cost and time resources for the subsidy process. We end the chapter by highlighting the benefits of continuous satellite tracking as a substitute, but also complementary to the classical field control methods, and also the enormous potential of Earth Observation-based products for the agri-food market.

20.1 Introduction, Motivation, and Goals

In the framework of European Union (EU) common agricultural policy (CAP), farmers can have access to subsidies that are provided through paying agencies and authorized collection offices operating at national level or regional level [1]. For the

O. Copăcenaru (✉) · A. Stoica
TERRASIGNA SRL, 3 Logofatul Luca Stroici, 020581 Bucharest, Romania
e-mail: olimpia.copacenaru@terrasigna.com

A. Catucci · L. De Vendictis · A. Tricomi
e-GEOS, 965 Via Tiburtina, 00156 Rome, Italy

S. Rogotis · N. Marianos
NEUROPUBLIC SA, Methonis 6 and Spiliotopoulou, 18545 Piraeus, Greece

© The Author(s) 2021
C. Södergård et al. (eds.), *Big Data in Bioeconomy*,
https://doi.org/10.1007/978-3-030-71069-9_20

provision of the subsidies, paying agencies must operate several controls in order to verify the compliance of the cultivation with EU regulations. At present, the majority of the compliance controls are limited to a sample of the whole amount of farmers' declarations due to the increased costs of acquiring high and very high-resolution satellite imagery [2]. Moreover, they are often focused on a specific timeframe, not covering the whole lifecycle of the agricultural land plots during the year.

However, EU Regulation No. 746 of 18 May 2018 [3, 4] introduced the option for member states, starting from the 2018 campaign, to use an alternative methodology to that of field controls, using information from Copernicus Sentinel satellites, possibly supplemented by those of EGNOS/Galileo. Thus, paying agencies in several countries have set strategic targets to implement CAP subsidies control systems based on cost-efficient collection and processing of earth observation data [5] and efficiently converting them into added value operational services that can be embedded into the already existing workflows and integrated with the information already available in several institutional registers.

Therefore, the aim of the CAP support initiatives within the DataBio project was to provide products and services tuned in order to fulfill the requirements for the 2015–20 CAP [6], improve the CAP effectiveness, leading to a more accurate, and complete farm compliance evaluation provided to paying agencies operating in three EU countries: Greece, Italy, and Romania.

The technological core competency lies mainly in the implementation of specialized highly automated big data processing techniques, particularly based on multi-temporal series of Copernicus Sentinel-2 data, and directly addresses the CAP demands for agricultural crop-type identification, systematic observation, tracking, and assessment of eligibility conditions over the agricultural season.

The final products are tailored to the specific needs of the end-users and demonstrate the implementation of functionalities that can be used for supporting the subsidy process in verifying specific requests set by the EU CAP.

20.2 Pilot Set-Up

As the main goal of the approach was to provide services in support to the national and local paying agencies and the authorized collection offices for a more accurate and complete farm compliance evaluation, the pilot included trial stages in three different areas of interest.

- In **Romania**, TERRASIGNA ran CAP support monitoring service trials for a 10,000 km^2 area of interest (AOI) in the southeastern part of the country, thus aiming to provide crop-type maps for a large area, characterized by geographical variability and a broad number of crops, distributed over diverse locations and including small and narrow plots, making use of the Copernicus Sentinel-2 spatial and temporal resolution. Initially, the selection of the 10,000 km^2 AOI was done by performing a multi-criteria analysis based on three main elements: plots' size (a minimum degree of land fragmentation was desirable in order to properly test the methodology), crops diversity (the selected area included a large selection of agricultural crop types), and accessibility (any point or parcel within the area had to easily be accessed during field campaigns to collect observations needed for validation). For the 10,000 km^2 area of interest, intersecting three Sentinel-2 granules (35TLK, 35TMK, 35TNK), more than 150,000 plots of different sizes have been analyzed during each agricultural season. The analysis performed included parcels of over 0.3 ha, regardless of shape. Of course, the 10-m spatial resolution made the narrower parcels difficult to properly label. Starting from the 2018 agricultural season, TERRASIGNA has extended its CAP-related services and has monitored the declarations for the entire agricultural area of Romania, exceeding 9 million ha and corresponding to more than 6 million plots of various sizes and shapes, distributed across the 41 Sentinel-2 scenes, projected in 2 UTM zones, that intersect the territory of Romania (Fig. 20.1). The main end user was APIA—the Romanian National Paying Agency.

- In **Italy**, e-GEOS sets up a methodology that has been tested and applied for a 50,000-ha area of interest in the region of Veneto, Verona Province (Fig. 20.2), where the land parcel identification system (LPIS) 2016 data was available. The approach was based on the computation of markers, in relation to predefined scenarios in terms of crop types and reference periods for agricultural practices. It aimed to demonstrate and detect LPIS anomalies concerning crop types or crop families, with respect to the last update of the farmer's declaration integrated in the geospatial aid application (GSAA), and to re-classify the parcel itself. The end user, in this case, was AVEPA Paying Agency (Agenzia Veneta per I Pagamenti in Agricoltura), operating at regional level in one of the most important agricultural regions in Italy.

- In **Greece**, NEUROPUBLIC tested and evaluated a set of EO-based services designed to support specific needs of the CAP value chain stakeholders, for an area of interest covering 50,000 ha of annual crops with an important footprint in the Greek agricultural sector (rice, wheat, cotton, maize, etc.), located in the greater area of Thessaloniki (Fig. 20.3). The main stakeholders of the pilot activities were the farmers from the engaged agricultural cooperatives in the pilot area and GAIA EPICHEIREIN, that had a supporting role in the farmers' declaration process through its farmers service centers (FSCs). CSEM and FRAUNHOFER were also involved in the pilot, providing their long-standing expertise in the technological development activities. The pilot aimed at supplying EO-based products and services designed to support key business processes, including the farmer decision-making actions during the submission of aid application, and more specifically leading to an improved "greening" compliance in terms of crop diversification, which acts as a driver toward more sustainable ecosystems. Greening conditions dictate that farms with more than 10 ha of arable land should grow at least two crop types, while farms with more than 30 ha are required to maintain more than three crop types. The main crop type is not expected to cover more than 75% of the arable land. The ambition of this pilot case was to effectively deal with CAP demands for agricultural crop-type identification,

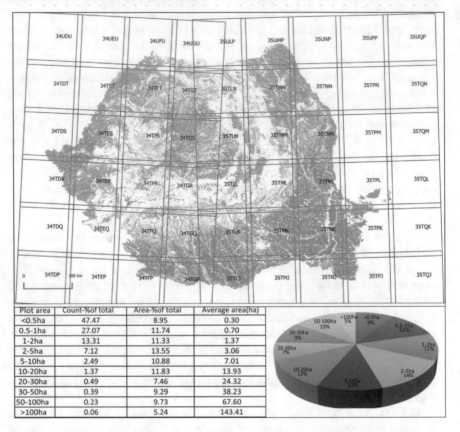

Fig. 20.1 Romania—total declared area and number of plots registered for CAP support (2019). Alphanumerics in the cells represent Sentinel-2 tiles. *Data source* Agency for Payments and Intervention in Agriculture (APIA), Romania

systematic observation, tracking, and assessment of eligibility conditions over a period of time.

20.3 Technology Used

20.3.1 Technology Pipeline

While the overall objective was similar, providing CAP-related services tailored to the specific needs of different stakeholders in charge of agricultural subsidies management, the three different approaches were based on technologies that have both similarities and important differences. Therefore, while the data collection and data

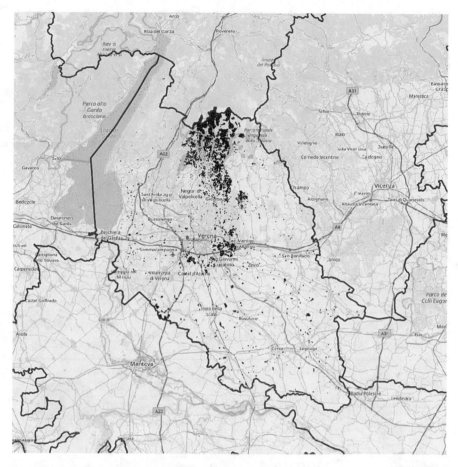

Fig. 20.2 Geographical distribution of the parcels analyzed within the trial stage in Italy (highlighted in black)

preparation phases follow very similar workflows, the data processing and analysis are based on separate technology pipelines (Fig. 20.4).

Technology Pipeline for the Trial Stage in Romania

For the trial stage in Romania, TERRASIGNA proposed an in-house developed fuzzy-based technique for crop detection and monitoring, based on combined free and open Sentinel-2 and Landsat-8 Earth Observation data image processing, data mining, and machine learning algorithms, all integrated in a toolbox for crop identification and monitoring [7].

The processing chain involves a series of well-defined steps:

- image preprocessing (numerical enhancements for Sentinel-2 and Landsat-8 scenes, ingestion of external data, clouds and shadows masking);

Fig. 20.3 Geographical distribution of the parcels that take part to the Greek pilot activities (highlighted with yellow color)

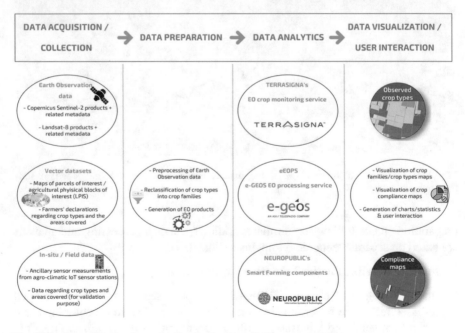

Fig. 20.4 Generic technology pipeline for the three CAP support trial stages

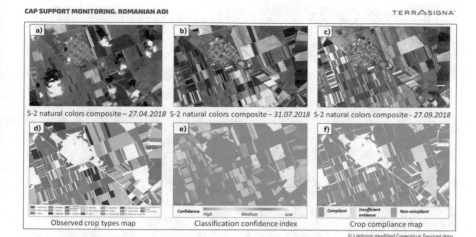

CAP SUPPORT MONITORING. ROMANIAN AOI   TERRASIGNA

Fig. 20.5 Romania—example of CAP support analysis results. **a–c** Sentinel-2 natural color mosaics (27.04.2018, 31.07.2018, 27.09.2018); **d** observed crop types map; **e** classification confidence index; **f** crop compliance map

- individual scene classification;
- the use of unsupervised machine learning techniques in order to obtain the crop probability maps at scene level;
- time series analysis, making the system capable of recognizing several types of crops, of the order of several tens and allowing the generation of overall crop probability maps and derived products.

The developed toolbox allows the automatic calculation of the following products (Figs. 20.5 and 20.6):

(1) Maps with the main types of crops, for a completed annual agricultural cycle;
(2) Intermediate maps with the main types of crops, during an ongoing annual agricultural cycle (which may serve as early alarms for non-observance of the declared crop type);
(3) Layers of additional information, showing the classification confidence index for the crop-type maps computed (values closer to 1 show higher trust levels for the assessed parcels);
(4) Maps with the mismatches between the type of crop declared by the farmer and the one observed by the application;
(5) Lists of parcels with problems, in order of the surfaces affected by inconsistencies, according to the data in product 4 above;
(6) RGB backgrounds with mediated aspect, uncontaminated by clouds and shadows, computed for a period of time, with national coverage. The computed synthetic images use the principle of weighted mediation, in a fuzzy logic, which guarantees a superior visual quality; they have a very natural look,

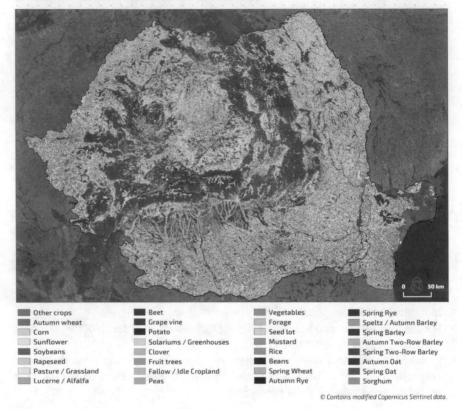

Fig. 20.6 Romania—observed crop-type map (2019) for the entire territory of the country, showing the 32 crop types that the algorithm is able to recognize, summing more than 97% of the total declared area in Romania

similar to a unique scene, however without the image being associated with a moment of time;

(7) RGB mosaics uncontaminated by clouds and shadows, computed for a period of time, with national coverage;

(8) NDVI maps uncontaminated by clouds and cloud shadows, computed for a period of time, with national coverage;

(9) Early discrimination maps between winter and summer crops.

Technology Pipeline for the Trial Stage in Italy

For the trial stage in Italy, a set of markers have been computed in relation to predefined scenarios in terms of crop families and reference periods during which agricultural practices have been defined. The methodology is working at parcel level, therefore computing several markers for each parcel depending on the specific crop family.

The full list of tuned markers includes plowing, vegetation presence\growing, harvesting and mowing. However, considering the typical phenological cycle and the agricultural practices for each crop class, not all the markers have been computed for all crop classes. For example, the markers considered for wheat (autumn–winter crop family) are plowing, vegetation presence/growing, harvesting and mowing, while for permanent grassland, only the presence/growing and mowing markers have been computed [7].

For the definition of markers, it should be considered that each of them should be defined according to the geographic location and specific algorithms and related parameters should be identified, therefore requiring a proper tuning by leveraging on time series analysis. This operation is supported by the analysis, for each crop family, of the spectral behavior along time, in order to identify from a mathematical point of view, markers related to specific activities.

For example, Fig. 20.7 shows the NDVI temporal trend of a corn parcel in the center of Italy, from which it is possible to identify, together with the support of false-color images, the relevant stages in the phenological cycle:

- Plowing: between January and April (false-color image A);
- Presence\growth: between April and August (false-color image B);
- Harvesting: between July and September (false-color image C);
- False-color image D shows the parcel after the harvesting.

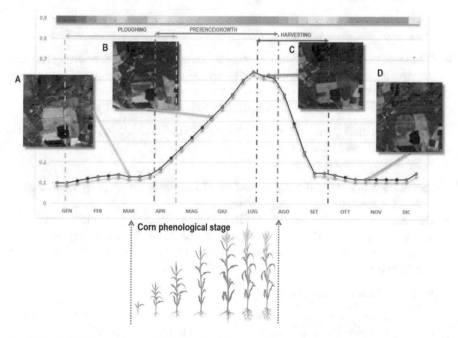

Fig. 20.7 NDVI temporal trend with identification of relevant stages in the phenological cycle

Once the markers are tuned according to type of interest, relevant periods, thresholds on NDVI values and geographic location for each crop type\families, they should be detected using a proper algorithm operating on time series. The results of the marker computation (positive\negative) can feed the internal workflow of the paying agencies, by:

- supporting the analysis and computation of parcel compliance versus administrative regulations of farmers' applications for subsidies;
- supporting the detection of LPIS anomalies (incorrect classification or update need) and then re-classification, testing the validity of markers of the other macro-classes.

Technology Pipeline for the Trial Stage in Greece

For the trial stage in Greece, a set of data collection, processing, and visualization components has been used to technically support the pilot activities [7]. More specifically, the following technological components should be acknowledged:

In terms of *data collection*:

- In-situ telemetric stations provided by NP, so-called gaiatrons, that collect ancillary weather data;
- Modules for the collection, preprocessing of Earth Observation products, the extraction of higher-level products, and assignment of vegetation indices at parcel level.

In terms of *data processing*:

- GAIABus DataSmart Machine Learning Subcomponent (NP), supporting EO data preparation and handling functionalities, multi-temporal object-based monitoring and modeling and crop-type identification;
- GAIABus DataSmart Real-time streaming Subcomponent (NP), supporting:
  - Real-time data stream monitoring for NP's gaiatrons installed in the pilot sites;
  - Real-time validation of data;
  - Real-time parsing and cross-checking.

- Neural Network Suite (CSEM), used as a machine learning crop identification system for the detection of crop discrepancies;
- Georocket, Geotoolbox and SmartVis3D (FRAUNHOFER), having a dual role: a back-end system for Big Data preparation, handling fast querying and spatial aggregations, as well as a front-end application for interactive data visualization and analytics.

In terms of *data visualization*:

- Neurocode (NP), the main component, allowing the creation of the main pilot UIs in order to be used by the end-users (FSCs of GAIA EPICHEIREIN);
- Georocket (FRAUNHOFER), an additional DataBio component providing information visualization functionalities.

20.3.2   Data Used in the Pilots

All the three trials (Romania, Italy, and Greece) aimed to demonstrate the advanced capabilities of Earth Observation data in monitoring agricultural areas [8].

Therefore, the input data consisted in:

- Sentinel-2 and Landsat-8 optical satellite data;
- The declarations of the farmers regarding cultivated crops and areas covered;
- The map of the parcels of interest or the map of the physical blocks of interest;
- List of crop codes used;
- List of crop classes to be followed (LCCF, i.e., very related groups of crops, which have similar aspect and phenological behavior);
- Ancillary sensor measurements from agro-climatic IoT sensor stations (used for the trial stage in Greece);
- A collection of a validation dataset, representative for the crop types/crop families distribution, derived from very high-resolution imagery (used for the trial stages in Romania and Italy).

20.3.3   Reflections on Technology Use

Reflections on Technology Use for the Trial Stage in Romania

The crop monitoring technology developed by TERRASIGNA is able to recognize a large number of crops families, of the order of tens. For Romania, it addressed the first most cultivated 32 crop families (according to the information provided by the National Paying Agency), which together cover more than 97% of the agricultural land [7]. The success rate in recognition was not equal between crops families, but an overall performance of 98.3% (Table 20.1) was obtained for the first most important 8 crops (winter wheat, sunflower, maize, green peas, winter barley, meadows and pastures, rapeseed, soybean). The countercheck data was obtained using a manual classification of a statistical sample in a test zone of the size of a Sentinel granule, supplemented with field-collected data regarding cultivated crop types and areas covered. The performance proved to be quite uniform reported to the size of the plots and remained high even for parcels smaller than 1 ha (Table 20.1).

At the moment, taking into account the agricultural specificity of Romania, defined by excessive land fragmentation, as a result of the existing legislation, the developed technology is using only optical data, consisting in both Copernicus Sentinel-2 and Landsat-8 imagery. According to the Romanian National Paying Agency, out of the total of 6 million plots for which payments have been granted, 2.7 million plots have an area smaller than 0.5 ha (44% of the total number), while 1.8 million plots consist of an area between 0.5 and 1 ha. Therefore, the small narrow plots are not suitable for SAR analysis for crop-type identification, taking into account the noise level, despite the good spatial resolution of Sentinel-1 images. Moreover, as stated before, in terms of overall accuracy (OA), the classification result using only Sentinel-2

Table 20.1 Romania—results of the validation based on independent data consisting of very high-resolution imagery and field-collected data

	<1 ha	1–1.5 ha	1.5–2.5 ha	2.5–5 ha	5–10 ha	10–20 ha	>20 ha	Overall performance
Winter wheat	*99.1%* N = 130 ha	*98.6%* N = 226 ha	*97.5%* N = 619 ha	*98.2%* N = 1,919 ha	*98.5%* N = 3,073 ha	*98.3%* N = 4,494 ha	*99.4%* N = 22,208 ha	**98.7%** Omissions: 0.73%
Maize	*99%* N = 22 ha	*94.4%* N = 30 ha	*90.7%* N = 81 ha	*88.1%* N = 216 ha	*90.1%* N = 396 ha	*93.1%* N = 679 ha	*99.1%* N = 4,877 ha	**93.7%** Omissions: 4.93%
Sunflower	*97.8%* N = 25 ha	*97.8%* N = 41 ha	*99.6%* N = 109 ha	*96.7%* N = 320 ha	*99.3%* N = 664 ha	*99.1%* N = 1,008 ha	*99.5%* N = 3,663 ha	**98.8%** Omissions: 3.99%
Soybean	*100%* N = 4 ha	*92.1%* N = 18 ha	*90.1%* N = 41 ha	*99.6%* N = 186 ha	*99.9%* N = 558 ha	*100%* N = 800 ha	*100%* N = 2,370 ha	**99.3%** Omissions: 2.12%
Rapeseed	*99.7%* N = 77 ha	*99.6%* N = 93 ha	*98.9%* N = 268 ha	*99.6%* N = 811 ha	*98.6%* N = 1,107 ha	*99.4%* N = 1,633 ha	*99.8%* N = 8,346 ha	**99.5%** Omissions: 1.16%
Pastures	*98.1%* N = 111 ha	*97.3%* N = 165 ha	*99.2%* N = 393 ha	*99.2%* N = 1,199 ha	*99.1%* N = 2,161 ha	*99.5%* N = 3,306 ha	*99.5%* N = 7,302 ha	**99.3%** Omissions: 2.64%
Peas	n.a.	*96.5%* N = 2.4 ha	*100%* N = 4 ha	*100%* N = 23 ha	*100%* N = 75 ha	*86.9%* N = 93 ha	*100%* N = 348 ha	**97.4%** Omissions: 1.44%
Winter barley	*100%* N = 10 ha	*100%* N = 3.5 ha	*91.5%* N = 24 ha	*95.3%* N = 58 ha	*100%* N = 1,477 ha	*100%* N = 442 ha	*100%* N = 1,682 ha	**98.7%** Omissions: 2.79%
All crops	**97.8%** N = 336 ha 434 plots	**97.4%** N = 543 ha 436 plots	**96.6%** N = 1446 ha 722 plots	**97.5%** N = 4454 ha 1,216 plots	**98%** N = 8075 ha 1,109 plots	**98.2%** N = 12,261 ha 865 plots	**98.2%** N = 50,106 ha 1,060 plots	**98.3%**

The italic values represent the percentage of accuracy for the specific size classes analyzed for each crop category

The bold values represent the overall percentage of accuracy for each crop category and also the overall percentage of accuracy for each size class

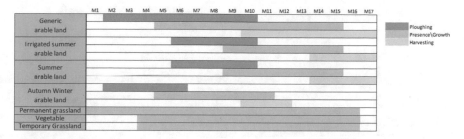

Fig. 20.8 Example of predefined scenarios regarding agricultural practices for the crop categories analysed

imagery reached 0.98. Thus, a major increase in overall accuracy using SAR data was not foreseen.

Reflections on Technology Use for the Trial Stage in Italy

The crop monitoring technology developed by e-GEOS for the trial stage in Italy was based on NDVI profile trends [7], which allowed the computation of a set of markers related to agricultural practices that should take place (e.g., plowing, vegetation presence/growth, and harvesting), in relation to predefined scenarios (Fig. 20.8), in terms of:

- selected macro-crop type;
- reference periods;
- NDVI thresholds.

At the beginning of the trial activities, the LPIS crop types have been aggregated in macro-classes (23 families) and the predefined scenarios have been tuned for the seven classes suitable for the automatic detection of anomalies and reclassification, based on the Sentinel-2 time series.

Analyzing their distribution and considering that the largest part (about 67%) of the agricultural crop families in the AOI belongs to 2 main groups, permanent grassland and arable land, only the crop families of these 2 groups have been considered in order to test the algorithm of anomalies detection and re-classification at macro-class level.

The markers computed in relation to predefined scenarios have been implemented in a decision model to verify their correct classification. The model has been run for each parcel of the macro-classes considered as suitable for the automatic detection of anomalies. Examples of parcels for which the original macro-class has been confirmed or detected as anomalous through the automatic analysis based on the related markers are displayed in Fig. 20.9.

Parcels detected as anomalous have been automatically re-classified testing the validity of the markers of the other macro-classes, thus updating the LPIS in terms of macro-classes (Fig. 20.10).

As expected in the arable land area, due to the usual crop rotation practice, the largest part of parcels changed their agricultural use between 2016 and 2018

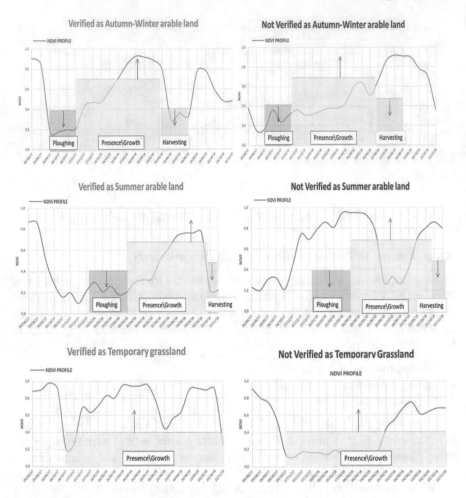

Fig. 20.9 Example of marker analysis based on predefined scenarios

(Fig. 20.11). In most cases, it is simply a change from winter–autumn to summer or temporary grassland and vice versa (Fig. 20.12).

The results are confirmed by the pie charts (Figs. 20.13 and 20.14) that describe, for different crop families (autumn–winter arable land, summer arable land and irrigated summer arable land) the percentage of parcels for which the crop family has been confirmed (in green) and the percentages of anomalous parcels, re-classified as other crop families.

Irrigated summer arable land parcels (e.g., rice paddies) are mostly confirmed (few anomalies) probably because these types of crop field, supported by irrigation systems, are not subject to crop rotations (Fig. 20.15).

In terms of permanent grassland areas, as expected, the percentage of anomalies is meaningful lower, considering the fact that usually the agricultural use of these

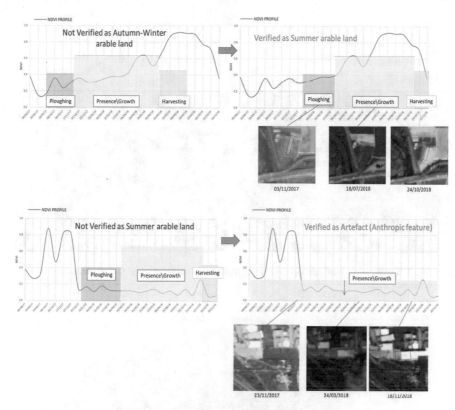

Fig. 20.10 Examples of non-compliant (left) and re-classified (right) parcels

Fig. 20.11 LPIS arable land parcels classified as verified (green), anomalous (red) and not analysed (gray)

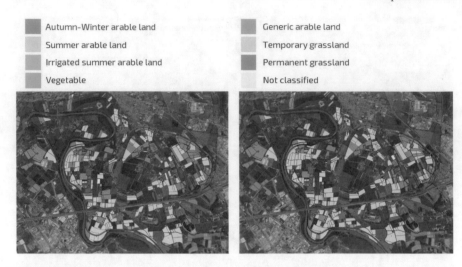

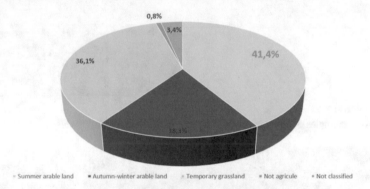

Fig. 20.12 LPIS arable land parcel classes in 2016 (left) versus 2018 (right), after re-classification of anomalous parcels

Summer arable land · Autumn-winter arable land · Temporary grassland · Not agricule · Not classified

Fig. 20.13 2016 LPIS summer arable land parcels updated to 2018

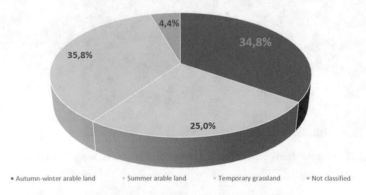

· Autumn-winter arable land · Summer arable land · Temporary grassland · Not classified

Fig. 20.14 2016 LPIS winter–autumn arable land parcels updated to 2018

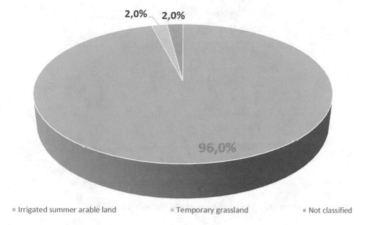

2,0%   2,0%

96,0%

▪ Irrigated summer arable land   ▪ Temporary grassland   ▪ Not classified

Fig. 20.15   2016 LPIS irrigated summer arable land parcels updated to 2018

parcels is stable for several years (a grassland field is defined as permanent if it is not plowed for 5 years, at least) (Figs. 20.16 and 20.17).

The accuracy of the methodology proposed for the LPIS anomalies detection and reclassification has been assessed through a validation activity based on data extracted from very high-resolution imagery. About 1000 parcels have been considered for

Fig. 20.16   LPIS permanent grassland parcels classified as verified (green), anomalous (red) and not analysed (gray)

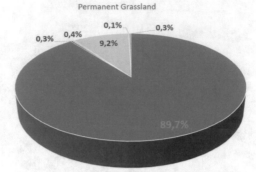

Fig. 20.17 2016 LPIS permanent grassland parcels updated to 2018

Table 20.2 Results of the validation based on reference data extracted from very high-resolution imagery

Crop family	Parcel number	Accuracy (%)
Autumn–winter arable land	26	84.6
Summer arable land	55	96.4
Permanent grassland	973	96.5
Temporary grassland	73	38.2

the accuracy assessment (Table 20.2). The resulting validation dataset consisted of four main crop families: autumn winter arable land, summer arable land, permanent grassland, and temporary grassland, reflecting the crop families' distribution over the entire area. Other crop families, considered statistically insignificant in terms of number of parcels, have not been taken into account in the accuracy assessment.

The results reveal very high accuracy for permanent grassland and summer arable land (more than 95%), high for winter arable land (85%). However, the computed accuracy for the temporary grassland crop family with respect to the farmers' declarations is just around 40%. The remaining 60% mis-classified parcels are distributed, according to farmers' declarations, mainly as permanent grassland (33%) and they require an additional refinement of marker rules in order to improve the accuracy.

The performances will be further tested in wider areas in order to evaluate the potential to be used in operative scenarios.

Reflections on Technology Use for the Trial Stage in Greece

In Greece, "greening" compliance was assessed for the 2019 cultivation year and the respective aid applications [7]. The farmers that could benefit from the methodology were the ones holding parcels larger than 10 ha, eligible for checks for greening requirements related to crop diversification. The crop types that have been modeled by the GAIABus DataSmart Machine Learning Subcomponent were seven (7) in total and more specifically: wheat, cotton, maize, tobacco, rapeseed, rice, and sunflower.

Table 20.3 Normalized crop classification confusion matrix (horizontal axis corresponds to the true label, whereas the vertical one to the predicted label)

	Maize	Cotton	Rapessed	Sunflower	Tobacco	Rice	Wheat
Maize	0.932	0.004	0.022	0.021	0.004	0.003	0.005
Cotton	0.009	0.954	0.006	0.019	0.079	0.002	0.008
Rapeseed	0.000	0.000	0.713	0.000	0.000	0.000	0.000
Sunflower	0.023	0.007	0.019	0.824	0.061	0.000	0.017
Tobacco	0.000	0.001	0.000	0.005	0.712	0.000	0.001
Rice	0.002	0.001	0.000	0.000	0.000	0.004	0.000
Wheat	0.032	0.032	0.239	0.131	0.144	0.000	0.968

If seen as a multiclass classification problem, the performance of the trained crop models to the 2019 testing data are offered at the confusion matrix, in Table 20.3.

Using the trained models as the backbone of the CAP support methodology, the assessment of "greening" compliance was conducted over 2019s aid applications. A traffic light system was employed to inform the farmers that there could have been a problem within their declarations:

(a)   if the confidence level of the classification result was >85% and the declared crop type of the farmer was confirmed by the classification, traffic light should be green;
(b)   if the confidence level of the classification result was <85% and the declared crop type of the farmer was confirmed by the classification, traffic light should be yellow;
(c)   if the declared crop type of the farmer was not confirmed by the classification, traffic light should be red.

According to this approach, the farmer is more protected in order to receive the payment as robust and reliable feedback is provided to him/her. The farmer is notified for issues (especially when the main crop seems to cover more than 75% of the cultivated land—mandatory condition for ensuring crop diversification) that put at risk his/her eligibility for greening compliance, thus contributing to raising awareness and allowing follow-up activities to be taken. An example regarding greening eligibility assessment is shown in Table 20.4.

Moreover, in order to support the Greek pilot activities, an integrated analytics platform has been finalized and deployed (Fig. 20.11). The use of machine learning

Table 20.4 Greening eligibility assessment using a traffic light system (table and map projection)

Parcel	Crop group		DataBio Assessment		Traffic Light	Area (ha)	Map projection
ID	Declared	Detected	Status	Result			
001	Wheat	Wheat	Assessed	Compliant		2.08	
002	Wheat	Wheat	Assessed	Compliant		1.67	
003	Maize	Wheat	Assessed	Not compliant		1.1	
004	Maize	Maize	Assessed	Insufficient evidence		1.46	
005	Wheat	Wheat	Assessed	Insufficient evidence		1.25	
006	Cotton	Cotton	Assessed	Compliant		0.82	
007	Cotton	Wheat	Assessed	Not compliant		0.73	
008	Wheat	Wheat	Assessed	Compliant		1.88	
					Total	10.99	

services provided a proof of concept for its use in CAP support scenarios. FRAUN-HOFER was responsible for the development of the UI, integrating pixel heat maps from the different classifiers and information visualization capabilities. A CSEM developed system for the management of machine learning models was used to facilitate the simple and retraceable management of models. RESTful services, combined with security features in the form of JSON Web Tokens (JWT) and encryption with Hypertext Transfer Protocol Secure (HTTPS), were implemented and integrated into the service. The service has also been containerized to allow simple deployment. This service enables the communication with FRAUNHOFER's component GeoRocket and UI for on-demand crop-type classification, in both pixel and parcel levels (Fig. 20.18).

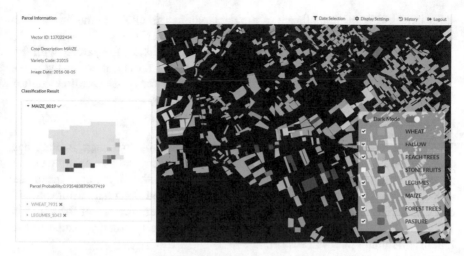

Fig. 20.18 User interface created by FRAUNHOFER for the Greek CAP support trial. The user interface integrates CSEM's classification results into pixel heat maps

20.4   Business Value and Impact

20.4.1   Business Impact of the Pilot

All the three CAP support trials developed within the DataBio project were tailored according to the needs of specific end-users (national and regional paying agencies), and, therefore, the business impact of the pilots is closely related [9]. The added value of the three pilots effectively consists in the increase of efficiency that the payment authorities and other end-users experience in using satellite monitoring and big data technologies.

Business Impact for the Trial Stage in Romania

The possibilities for exploitation of the project's result for TERRASIGNA focused on proving a concept and attracting a long-term collaboration with the National Agency for Payments and Investments in Agriculture (APIA), holding responsibility in Romania of the implementation of CAP mechanisms for direct payments. A cooperation agreement was signed with the agency, in order to offer and test the results of the pilot—crop compliance maps in support of APIA's activity of monitoring the subsidies payments [9].

The CAP support pilot, through its EO crop monitoring component, offered the stakeholder the possibility to check, in a more efficient way, the compliance between the declarations made by the farmers in request of the subsidy payments and the real crop in the fields. While currently a minimum of 5% from the applications is cross-checked either by field sampling or by remote sensing, the developed methodology allowed checking the compliance of the declarations submitted by the farmers for all agricultural parcels with individual areas exceeding 0.3 ha.

Moreover, as the service automatically detects and signals the parcels with the highest probability not to grow the crop declared by the farmer, it makes the regulator's decision more efficient in selecting the parcels for field control or control through very high-resolution imagery.

Business Impact for the Trial Stage in Italy

The marker-based approach proposed by e-GEOS for the trial stage in Italy has demonstrated its applicability not only for CAP monitoring, but also opening up the street for future innovation in the market. e-GEOS is active in the agro-insurance and CAP market segments, with a network of actual and potential customers and users, including the paying agencies operating in different regions of the country. Therefore, the CAP-related developed products have been strongly related to the Italian agricultural policy needs [9].

Business Impact for the Trial Stage in Greece

GAIA EPICHEIREIN, through its associated network of farmer service centers (FSCs), provides collection and advisory services to the Greek Farmers concerning the submission of the aid application for direct payments, including eligibility pre-check mechanisms for error reduction and proof provision. The total number of holdings in Greece for 2016 was 686.818. GAIA subsidy services are mainly oriented to aging small-sized farmers, which own 80% of the holdings in Greece. Over the last two annual periods, GAIA EPICHEIREIN provided collection services and cross-compliance checks to 76% of the holdings. Even if GAIA EPICHEIREIN has a market share of 76%, the ongoing CAP changes and trends, the differentiations in the internal market and the new business plans for smart farming (driven by the evolution in sensor and space technology) indicate that GAIA EPICHEIREIN needs to evolve its services in order to keep its competitive advantage and sustain its market share [9].

For the Greek scenario, the offered DataBio solutions allow the farmer (beneficiary) to deal effectively with the greening requirements. More specifically, DataBio solutions will be a valuable tool within the suite of digital CAP support services offered by GAIA EPICHEREIN's and its FSCs that support the crop declaration process. During the process and usually after the declaration period closes and error-checking tools are applied, the FSC would be able to check the farmer's claim for the greening requirements, examine the results, and inform the farmer for follow-up activities that better serve his/her interests.

Apart from the exploitation value for the partners involved, the pilot introduced concrete benefits for the farmers and the agri-food sector as well. The results of the pilot effectively showed that EO-based crop identification services, tailored for monitoring greening compliance, offered a layer of protection against errors in the declaration process which could lead to a significant financial impact for the farmer. Additionally, and from a higher level, agricultural monitoring approaches could contribute to more efficient funding absorption, thus securing investments and progress in the agri-food sector.

20.4.2   Business Impact of the Technology on General Level

The added value of multi-temporal copernicus sentinel data and applied new technologies (automated detection and determination using machine learning) in the context of CAP support can be explained through two different, but complementary aspects:

Copernicus Sentinel Data Stream

For the first time in the history of Earth Observation, almost every single region in Europe benefits from repetitive (5 days in average) observations with solid spatial and spectral resolution. Generally speaking, this stack of information enables early usage of EO data in the agricultural season, which in turn allows the extraction of preliminary conclusions that can be used within control with remote sensing (CwRS) decision trees (e.g., detection of winter crops, post-winter water ponding).

The 10 m spatial resolution enables the survey of the smaller plots, that in many European countries (including Romania), represent a significant number of CAP applications.

The spectral resolution provides all the necessary information (visible, NIR, SWIR) for observing the crops phenology and for distinguishing additional features (e.g., water, burned area, built-up).

The "turning data into information" policy is fully exploited, by transforming the wealth of satellite and in-situ data into valued-added services based on processing and analyzing the data, monitoring changes and making the datasets comparable, integrating them with other sources and, finally, validating the results.

Finally, the no-cost policy of the EC and the unprecedented volume of data on a full, free and open basis foster new business opportunities and job creation and provide the necessary sustainability to invest in developing copernicus data-based workflows.

Application of New Technologies

The usage of time series of Sentinel-2 satellite images for crop detection can increase the results precision, as the 5-days revisit time almost triples the number of surveys compared to the Landsat feed (16-days revisit time). An automatic nation-wide technology will warrant constant quality of the results over large areas and time periods. On the short time scale, this will allow avoiding human subjectivity.

The usage of the new technology is significantly decreasing the time, money, and human power required currently by the control with remote sensing (CwRS) campaigns. Instead of one year of administrative actions (from very high-resolution data selection to the real photo interpretation), the decision maker will have the option to solely focus on areas already marked as red lights or on areas known as prone to risk.

The usage of the early results will also enable the use of technology as a deterrent tool; e.g., the farmer's declaration lists a winter crop, and no winter crop was observed at the end of March.

20.5 How-to-Guideline for Practice When and How to Use the Technology

The three CAP support approaches use earth observation data time series, thus providing wide and repetitive homogeneous coverage, translated into an unprecedented amount of information. The technologies benefitting from these data volumes represent a solid solution for a continuous monitoring of CAP compliance. The EU Copernicus Sentinel-2 satellites hold an enhanced revisiting time, delivering regular coverage over large areas and allowing a uniform observation of the agricultural plots. The superior spectral resolution allows the identification of the phenological growth stages and the distinction between various crop types or classes.

However, the pilots also had to overcome some major drawbacks, mainly related to data fusion, georeferencing errors (deeply affecting the quality of the crop mapping results for narrow or small plots), cloud and shadow masking, or semantic confusions between crop classes.

The highly automated proposed approaches allow the implementation of big data analytics using various crop indicators, resulting in reliable, cost and time saving procedures, and allowing a more complete and efficient management of EU subsidies, strongly enhancing their procedure for combating non-compliant behaviors.

The developed techniques have undergone continuous development and improvements, are replicable at any scale level and can be implemented for any other area of interest. Any further developments of the CAP monitoring technologies will be able to provide products tuned in order to fulfill the requirements of the present and future EU common agricultural policy. This application of big data processing technologies based on copernicus sentinel data will also significantly improve the way which farmers are doing online aid applications and, for the paying agencies, will help to keep the LPIS up-to-date and to move to the new checks by monitoring.

Moreover, the Copernicus free and open data policy, together with the long-term availability certainty, are important factors that highly help the developed solutions enter the European market and trigger collaborations between government agencies (regional or national paying agencies) and private sector companies.

The DataBio European Lighthouse project offered new business opportunities and aimed to directly improve a series of CAP support activities for providing supporting tools and services, in line with the commands of the EU's new agricultural monitoring approach. The effort is expected to continue in the next years for all the three companies, setting strategic targets such as integration of information available in several institutional registers, active use of technologically most relevant and cost-efficient remote sensing services and proactive cooperation with rural communities and farmers.

20.6   Summary and Conclusion

Common agricultural policies and activities from national and regional paying agencies can radically benefit from the use of continuous satellite monitoring instead of random and limited controls.

The DataBio European Lighthouse project, with its three different CAP Support approaches, in Romania, Italy, and Greece, demonstrates the potentiality for final users to exploit Copernicus data in the agriculture domain, a key economic sector for most of the European countries. The proposed methodologies have undergone continuous development and improvements over the last years, offering a wide range of opportunities in order to enhance the implementation of the CAP. The continuous agricultural monitoring services, based on the processing and analysis of Copernicus satellite imagery time series, are not just CAP compliance tools, but can also offer a great range of supplementary information for both public authorities and farmers and can support the set-up of more environmentally friendly and efficient agricultural practices.

This market is one of the more promising in terms of exploiting the full potential of earth observation deployment and represents a successful example of how policies and strategies drive advancement in big data processing technologies, encourage innovation within the public sector and trigger long-term private–public partnerships.

References

1. DG Agri—European Commission. (2007). *Europeans, agriculture and the common agricultural policy (CAP)*. Retrieved August 16, 2020, from https://ec.europa.eu/commfrontoffice/publicopi nion/archives/ebs/ebs_410_en.pdf
2. Ivanov, D. (2017). Future development of common agriculture policy after 2020. *Trakia Journal of Science, 15*(Suppl. 1), 158–162.
3. Commission Implementing Regulation (EU) 2018/746 of 18 May 2018 amending Implementing Regulation (EU) No 809/2014 as regards modification of single applications and payment claims and checks C/2018/2976 (OJ L 125, 22.5.2018, pp. 1–7).
4. Commission Implementing Regulation (EU) No 809/2014 of 17 July 2014 laying down rules for the application of Regulation (EU) No 1306/2013 of the European Parliament and of the Council with regard to the integrated administration and control system, rural development measures and cross compliance (OJ L 227, 31.7.2014, p. 69).
5. Bartolini, F., Vergamini, D., Longhitano, D., & Povellato, A. (2020). Do differential payments for agri-environment schemes affect the environmental benefits? A case study in the North-Eastern Italy. Land Use Policy. https://doi.org/10.1016/j.landusepol.2020.104862
6. Erjavec, K., & Erjavec, E. (2015). Greening the CAP—Just a fashionable justification? A discourse analysis of the 2014–2020 CAP reform documents. *Food Policy, 51*, 53–62.

7. Estrada, J., et al. (2020). DataBio deliverable 1.3—Agriculture pilot final report. Retrieved 3.11.2020 from https://www.databio.eu/wp-content/uploads/2017/05/DataBio_D1.3-Agriculture-Pilot-Final-Report_v1.2_2020-01-21_TRAGSA.pdf
8. Klien, E., et al. (2018). DataBio deliverable 5.3—EO services and tools. Retrieved 3.11.2020 from https://www.databio.eu/wp-content/uploads/2017/05/DataBio-D5.3-EO-Ser vices-and-Tools_v1.0_2018-06-15_Fraunhofer.pdf
9. Stanoevska-Slabeva, K., et al. (2020). DataBio deliverable D7.2—Business plan v2. Retrieved 3.11.2020 from https://www.databio.eu/wp-content/uploads/2017/05/DataBio_D7.2.Business-Plan-v2_v1.2_2020-03-19_UStG.pdf

Chapter 21
Future Vision, Summary and Outlook

Ephrem Habyarimana

Abstract The DataBio's agriculture pilots were carried out through a multi-actor whole-farm management approach using information technology, satellite positioning and remote sensing data as well as Internet of Things technology. The goal was to optimize the returns on inputs while reducing environmental impacts and streamlining the CAP monitoring. Novel knowledge was delivered for a more sustainable agriculture in line with the FAO call to achieve global food security and eliminate malnutrition for the more than nine billion people by 2050. The findings from the pilots shed light on the potential of digital agriculture to solve Europe's concern of the declining workforce in the farming industry as the implemented technologies would help run farms with less workforce and manual labor. The pilot applications of big data technologies included autonomous machinery, mapping of yield, variable rate of applying agricultural inputs, input optimization, crop performance and in-season yields prediction as well as the genomic prediction and selection method allowing to cut cost and duration of cultivar development. The pilots showed their potential to transform agriculture, and the improved predictive analytics is expected to play a fundamental role in the production environment. As AI models are retrained with more data, the decision support systems become more accurate and serve the farmer better, leading to faster adoption. Adoption is further stimulated by cooperation between farmers to share investment costs and technological platforms allowing farmers to benchmark among themselves and across cropping season.

21.1 Summary of the Agriculture Pilots Outcomes

The agriculture pilots Chaps. (14–20) discuss the applications of big data technology in the agricultural arable farming, horticulture, and EU Common Agricultural Policy (CAP) support as well as in insurance assessment. The main focus of the pilot "Smart farming for sustainable agricultural production" (Chap. 15) was to offer

E. Habyarimana (✉)
CREA Research Center for Cereal and Industrial Crops, via di Corticella 133, 40128 Bologna, Italy
e-mail: ephrem.habyarimana@crea.gov.it

© The Author(s) 2021
C. Södergård et al. (eds.), *Big Data in Bioeconomy*,
https://doi.org/10.1007/978-3-030-71069-9_21

smart farming advisory services for the cultivation of olives, peaches, grapes, and cotton, based on a unique combination of Earth observation (EO), big data analytics, and Internet of Things (IoT). The two-year trials' results showed a significant reduction of the number of crop protection sprays, nitrogen fertilizer applications, and irrigation water; all of which resulted in decreased production costs and increased yields. It is expected that the results achieved will be further improved as more data are produced to better train the models.

In the Genomics pilot (Chap. 16), genomic prediction and selection (GS) modeling was implemented to accurately estimate the genetic merit upon which superior cultivars can be selected, leading to simplified breeding schemes and shorter breeding cycles, all of which results in increased yields and genetic gains per unit time and cost. The GS technology showed meaningful and attractive predictive performance: the evaluated genomic selection models performed comparably across traits and were found suitable to sustain sorghum breeding for several traits including the production of antioxidants. In comparison with conventional phenotypic breeding, the genomic predictive and selection modeling allows cutting costs five times and cutting four times the time of cultivar development. These findings can lead to potential business applications such as genetic improvement of sorghum for several traits including grain antioxidants for health-promoting and specialty foods, and the use of the next-generation genotyping platforms (NGS) validated in this pilot for sequencing and genotyping services in other plant species and animal husbandry.

In Chap. 17 "Yield Prediction in Sorghum (*Sorghum bicolor (L.) Moench*) and Cultivated Potato (*Solanum tuberosum* L.)," the main objective was in-season yield prediction using satellite imageries and machine learning techniques. These pilots were established as a solution to current limitations in crop monitoring in Europe: yield forecasting approaches based mainly on field surveys, sampling, censuses, and on the use of coarser spatial (250–1000 m) resolution satellites (e.g., MODIS, SPOT-VEGETATION), all of which are unreliable and/or costly. In sorghum, it was possible to accurately predict above-ground sorghum biomass yields six months before harvesting with the best prediction times identified as days 150 and 165, i.e., late May and early June, respectively. The results from this study represent a remarkable opportunity for farmers and farming cooperatives to use this information for several business-related purposes. The models developed in this work will also help the extension services and other policymakers for strategic planning purposes, including assessing alternative means for energy supply. The potato pilot showed that smart farming services based on satellite images offer to the farmers a clear competitive advantage through better cost-effectiveness. The results from DataBio have been useful to improve the potato growth model on the basis of big data analysis. The approach contributed to better yield prediction based on the actual growing conditions with a limited number of samples or field trials. New business opportunities can be found by implementing the yield prediction model that was tested in the pilot, implementing a farmer decision support system and by further developing the potato growth model to create new services like irrigation planning and a variable rate application of fertilizers.

Chapter 18 discusses the delineation of management zones using satellite imageries based on areas with the same yield level within the fields. The method provides useful information for identifying management zones. This strategy is based on two basic principles—increasing the nitrogen (N) dosing in the zone with a higher yield (yield-oriented) or increasing the N rate in the below-average zones (homogenization). In the yield-oriented method, the N rate is determined on the basis of a nitrogen balance modeling, identifying areas with long-term lower crop yields to be fertilized with lower N rates than places with expected higher yields. In homogenization, low-yielded areas are supported by higher N doses. Homogenization is carried out when nitrogen is a yield-limiting factor and when it is appropriate to increase the booting of cereals in weak places or to homogenize the qualitative parameters of the grain.

Chapter 19 discusses farm weather insurance assessment to protect against loss or damage to crops or livestock, and to provide a value to farmers and their communities. This assessment has the potential to encourage greater agricultural investments. Copernicus satellite data series, big data technologies, and AI were used for this purpose in order to meet the most pressing needs of the insurance companies operating in agriculture: damage assessment and risk parameters estimation down to parcel level. Risk and damage assessment maps and indices were built, and this resulted in promising parametric insurance for farmer protection, and in strong reduction of ground surveys, with positive impact on insurance costs and reduction of the premium to be paid by the farmers.

Chapter 20 deals with Copernicus data and control of common agricultural policies (CAP) subsidies. The aim was to provide services to help the authorities to fulfill the requirements for the 2015–20 CAP and improve the CAP effectiveness. This should lead to a more accurate and complete farm compliance evaluation provided to paying agencies. The piloting took place in three EU countries: Greece, Italy, and Romania. Multi-temporal series of Copernicus Sentinel-2 data were deployed in this pilot to address the CAP demands for agricultural crop type identification, systematic observation, tracking, and assessment of eligibility conditions over the whole agricultural season. The results from this pilot showed that the CAP, and activities from national and regional paying agencies can benefit from the use of continuous satellite monitoring instead of random and limited controls. Stakeholders were offered the possibility to check, in a more efficient and accurate way, the compliance between the declarations made by the farmers in request of the subsidy payments and the real crop in the fields. While conventionally a minimum of 5% of the applications are cross-checked either by field sampling or by remote sensing, the methodology developed in this pilot allowed checking the compliance of the declarations submitted by the farmers for all agricultural parcels above the 0.3 ha threshold.

21.2 Evaluation of the Implemented Technologies and Future Vision

The extensive DataBio's agricultural trials were designed and conducted at the demonstration level in real production environments, i.e., in commercial fields. The outcomes from these studies were encouraging. Since several environments and business models were trialed, the conclusions and recommendations from these works are meaningful for farming business purpose on a broad scale. The agriculture pilots were run mostly as advisory services across Europe and in different areas of precision agriculture or smart farming. Whole-farm management was implemented using information technology, satellite positioning data, remote sensing and proximal data gathering, and Internet of Things technology. The overall goal was to optimize the returns on inputs while reducing environmental impacts, on the one hand, and streamlining the CAP monitoring, on the other.

Several technology adoption options were studied. The findings shed light on the potential of digital (or smart) agriculture to solve one of the major concerns in Europe, i.e., the declining workforce in the farming industry. Indeed, the high-throughput crop monitoring and risk/damage assessment, automated and intelligent agricultural input applications, in-season crop performance and yield prediction, and the (IoT) Internet connectivity solutions help to run farms with a lot less workforce and manual labor. These methods can also open up business at a global level. All in all, the main drivers of big data technologies in agriculture as implemented in the DataBio project are: (1) autonomous machinery, (2) mapping of yield and variable rate of agricultural inputs application, (3) input optimization (irrigation water, nitrogen, crop protection compounds, variable inputs application rate maps), and (4) crop performance and in-season yields prediction. Genomic prediction and selection (GS) (5) is a new and highly promising plants and husbandry breeding method which gets much attention by the main stakeholders—public and private research and development entities. The favorable GS attributes are expected to have wide-ranging implications in plant breeding as the cost and duration of cultivar development are reduced and farmers can grow a better variety faster. This helps to make more income.

21.3 Outlook on Further Work in Smart Agriculture

According to FAO, achieving global food security and eliminating malnutrition are among the most challenging issues humanity is facing. By 2050, a societal challenge will be to almost double food production from existing land areas in order to feed more than nine billion people [1, 2], while facing yield-reducing consequences of climate change and the spread of a wide range of pests and diseases [3]. Therefore, agricultural development must combine fundamental research and advanced technologies to produce more healthy food with less input. The DataBio project tackled that important challenge through a multidisciplinary approach that delivered, within three years, new knowledge to help stakeholders toward a more sustainable agriculture with reduced ecological footprint [4]. In the coming years, farmers will have to face a series of challenges such as climate change adversities (mainly drought and heat stress, and nitrogen scarcity), shrinking agricultural land areas, and depletion of finite natural resources, e.g., irrigation water. All these challenges show that the need to enhance farm yield is real and critical.

New information technologies and AI breakthroughs will impact farming in Europe and worldwide, helping reduce hunger and improve food quality. The results from the agricultural pilots confirmed the benefits gained from applying these technologies in the European farming industry. However, it is less clear in which form the technologies will be adopted and at what speed. Data is central here; as more data is gathered and AI models retrained with this data, the decision support systems become more accurate and serve the farmer better, which leads to faster adoption. It must also be noted that agriculture is less technological than other major industrial sectors [5], meaning that new technologies will meet resistance from some farmers. It is also clear that cooperation between farmers is needed to share investment costs.

The technologies implemented in the DataBio's agricultural pilots have shown their potential to transform agriculture in several aspects. Of these aspects, predictive analytics is expected to play a fundamental role in transferring big data technology into the production environment. Indeed, according to the Department for Environment Food and Rural Affairs [6], the two most common reasons for adopting precision farming techniques such as those developed in DataBio were the improved accuracy in farming operations and the reduced input costs. Likewise, crop performance monitoring and yield predictions will play a key role when they are accurately supporting the decisions of the farmer and other parties at interest. Therefore, refining the predictive analytics, especially with more historic data, will help both the farmer and the technology provider to stay on the market.

References

1. Spiertz, J. H. J. (2010). Nitrogen, sustainable agriculture and food security. A review. *Agronomy for Sustainable Development, 30,* 43–55. https://doi.org/10.1051/agro:2008064.
2. Ray, D. K., Mueller, N. D., West, P. C., & Foley, J. A. (2013). Yield trends are insufficient to double global crop production by 2050. *PLoS One, 8.*https://doi.org/10.1371/journal.pone.0066428.
3. Pandey, P., Irulappan, V., Bagavathiannan, M. V., & Senthil, M. (2017). Impact of combined abiotic and biotic stresses on plant growth and avenues for crop improvement by exploiting physio-morphological traits. *Frontiers in Plant Science, 8,* 537. https://doi.org/10.3389/fpls.2017.00537.
4. Walter, A., Finger, R., Huber, R., & Buchmann, N. (2017). Opinion: Smart farming is key to developing sustainable agriculture. *PNAS, 114,* 6148–6150. https://doi.org/10.1073/pnas.1707462114.
5. Schellberg, J., Hill, M. J., Gerhards, R., et al. (2008). Precision agriculture on grassland: Applications, perspectives and constraints. *European Journal of Agronomy, 29,* 59–71. https://doi.org/10.1016/j.eja.2008.05.005.

6. Department for Environment Food and Rural Affairs (DEFRA). (2013). Farm practices survey. In: GOV.UK. https://www.gov.uk/government/collections/farm-practices-survey. Accessed on 25 Sep 2020.

Part VI
Applications in Forestry

Chapter 22
Introduction—State of the Art of Technology and Market Potential for Big Data in Forestry

Jukka Miettinen and Renne Tergujeff

Abstract Forest monitoring is undergoing rapid changes due to the growing data volumes, developing data processing technologies and increasing monitoring requirements. The DataBio forestry pilots set out to demonstrate how big data approaches can support the forestry sector to get full benefit of the evolving technologies and to meet the increasing monitoring requirements. In this introductory chapter, we describe underlying technical and market forces driving the forestry sector toward big data approaches, and give short overviews on the forestry pilots to be presented in the following chapters.

22.1 Evolving Technologies and Growing Data Volumes

The forestry sector has been one of the forerunners in processing and analysis of large datasets. Particularly, remote sensing-based forest monitoring has utilized large datasets in the form of digital imagery since the 1970s when the first Landsat satellite was launched [1]. Satellite-based inventory approaches have been integrated into large area forest inventory programs since the 1990s [2, 3]. But in many ways, the launch of the Google Earth Engine (https://earthengine.google.com/), a cloud-based platform for planetary-scale geospatial analysis, in 2010, and the first global multi-year tree cover clearance analysis produced on the platform by Hansen et al. in 2013 [4], can be seen as the start of the big data era in forest monitoring. The platform and the innovative tree cover clearance product very much showed the direction for future big data development in the forestry sector.

Since then, data volumes have continued to grow rapidly, and the availability of different types of datasets has improved, increasing potential use cases for forestry big data. While in 2014, there were only around 200 active Earth observation (EO) satellites in orbit, in 2019, there were nearly 700 of them [5]. Simultaneously with the increasing number of EO satellites, also the temporal and spatial resolutions have improved rapidly. The 10–30 m spatial resolution Landsat 8 and Sentinel-1

J. Miettinen (✉) · R. Tergujeff
VTT Technical Research Centre of Finland, 02044 VTT, Espoo, Finland
e-mail: jukka.miettinen@vtt.fi

© The Author(s) 2021
C. Södergård et al. (eds.), *Big Data in Bioeconomy*,
https://doi.org/10.1007/978-3-030-71069-9_22

and Sentinel-2 satellites are replacing coarse (250–1000 m) spatial resolution data in many large area forest monitoring applications, e.g., for burnt area [6], forest distur- bance [7], and health [8] monitoring. The Copernicus Sentinel program alone (with its six satellites in orbit) produces over 12 TB of data per day [9]. On the commer- cial front, companies like Planet Labs (https://www.planet.com/) are able to scan the entire globe every day in 3–5 m spatial resolution, providing high-frequency data for forest monitoring in unprecedented spatial detail. Other companies concentrate on less than 1 m very high spatial resolution imagery, which can be used as reference data in big data forest monitoring approaches.

The increase in EO data volumes is combined with the escalation of drone surveil- lance (including hyperspectral cameras, etc.), and the continuing national moni- toring campaigns with airborne optical and LiDAR sensors [10]. Furthermore, field measurements are increasingly taken with electric devices, increasing the speed and volume of data collected. Field measurement campaigns are supplemented by contin- uous data collection from machinery used in forest operations (e.g., location and measurement data from the cutting-heads of harvester machines [11]). And most recently, the launch of crowdsourcing-based data collection approaches allows fast and effective collection of large field observation datasets.

The most effective way for the forestry sector to utilize the great volumes of data produced by modern technology is through centralized storage and processing plat- forms. Since the early days of Google Earth Engine, numerous other online platforms have been set up. Nowadays, many online platforms operate in clusters that provide the resources to implement big data forest applications in an effective and innova- tive manner. Platforms like the Copernicus Data Access and Information Services (DIAS; https://www.copernicus.eu/en/access-data/dias) offer direct access to EO big data and processing capabilities, while other, often domain-specific platforms, such as the Forestry Thematic Exploitation Platform (Forestry TEP; https://f-tep.com/), additionally provide tools and services designed particularly for utilization of big data, e.g., in forestry. The platforms form a hierarchical offering, from data stor- ages and processing platforms to nuanced application platforms and interactive user interfaces. This network of platforms allows creation of delivery pipelines that can maximize the benefits of big data in forestry, by providing users with timely datasets and analysis results that meet their specific information requirements.

22.2 Expanding Market

Forests are in focus nowadays perhaps more than ever before. Both political and market interest in bioeconomy, growing recognition of the importance of forests in climate change mitigation, and increasing requirements on forest management (e.g., in the field of sustainability) demand timely and affordable information on forest resources. Forestry stakeholders, like government entities, non-governmental organizations, private companies, and forest owners, are bound by a wide range of international and national strategies and legislation. For example, in Europe, forestry stakeholders are not only affected by the European Forest Strategy [12], but also, e.g., by the Biodiversity Strategy [13] and Bioeconomy Strategy [14]. While the Forest Strategy provides a policy framework that coordinates and ensures coherence of forest-related policies, the Biodiversity Strategy aims to protect ecosystems (including forests) and biodiversity, and the Bioeconomy Strategy aims to serve as an umbrella for long-term sustainable development. These European wide strategies are reflected in national-level legislation in member states, requiring stakeholders to report and monitor increasing number of forestry indicators, ranging from harvest levels and reforestation status to biological diversity, carbon balance, forest health, and many more.

In most European countries, traditional methods for forest management are based on static management plans, created at the planting stage and reviewed after long periods (typically in five to ten years intervals). This type of management process does not meet the needs of modern requirements of manifold up-to-date information described above. Furthermore, in recent years, the management plans have become declarations of intentions, including objectives for multifunctional forests (non-wood products and services). However, the management system lacks effective monitoring methods that allow forest owners, managers, and regulators to validate the progress in achieving the target objectives set out in the management plan.

Forest owners, forestry operators, and companies using wood as raw material are also affected by various voluntary certification schemes like the Forest Stewardship Council (FSC; https://fsc.org/en) and the Program for the Endorsement of Forest Certification (PEFC; https://www.pefc.org/). They both aim to ensure sustainable forest management using a set of criteria ranging from sustainable wood production to biodiversity, forest health, and carbon balance. Independent auditors need access to a wide variety of forest variable information and change statistics to be able to verify that the certification standards have been followed correctly. Overall, the rising interest in forests and the widening range of forest management aspects included in strategies, legislation, and certification schemes are rapidly growing the market for timely forest information. The modern technology outlined in the previous sections can be used to establish operational monitoring systems providing transparent products helping to meet the increased monitoring and reporting requirements.

Big data can benefit both the provider and the customer side of the forest monitoring market. On the provider side, one of the main stakeholder groups in Europe consists of the Earth observation (EO) data, product, and service providers. According to the European Association of Remote Sensing Companies (EARSC), the EO service sector employed over 7000 people in 500 companies with over 900 million € revenue already in 2014 [15], with a strong growing trend. Forest monitoring is among the main focus areas of the European private EO sector. Although EO data cannot be used to monitor all of the variables required in modern forest management (e.g., plant and animal biodiversity in fine detail), it does provide the means to monitor key variables like the structural forest characteristics and forest health data, as demonstrated by the DataBio forestry pilots presented in the following chapters. In addition to the EO, forestry big data market benefits, e.g., consultants and forestry experts, IT specialists, and data analysis specialists.

The customer side of the market is likewise varied and expanding. The public sector has their monitoring requirements defined by national forest legislations, and non-governmental organizations want to monitor development on forest resources to support their activities. Companies directly involved in forestry activities need to have timely information on the forest resources, not only to support their own forest management, but also to provide data for certification purposes. Even companies that are not directly involved in forest management activities increasingly choose to get involved in the forest monitoring market due to the increasing legislation, certification, or consumer pressure. For example, food manufacturers, energy companies, and sellers of wood-based products (e.g., furniture) may have compulsory obligations or voluntary interest in forest monitoring. This trend is expected to grow in the future, as environmental issues are becoming an increasingly important part of business practices.

Overall, information on forest resources is nowadays needed frequently and in high spatial detail to meet the requirements of various reporting and regulative monitoring schemes. Moreover, the information is expected to be available in short notice and in easily reachable online platforms to allow direct integration of the data into the stakeholders' databases and operational analyses. These are hard demands, but forestry big data has a great potential to meet these demands, with appropriate network of online storage, processing, and distribution platforms. This is what the DataBio forestry pilots aimed to demonstrate.

22.3 DataBio Forestry Pilots

The objective of the DataBio forestry pilots was to demonstrate how Big Data could boost the Forestry sector. The pilots, carried out in four countries (Belgium, Czech Republic, Finland, and Spain), were built around practical forestry cases. They validated the use of Big Data technologies and assessed how the expectations of user communities can be met. Overall, the pilots sought to demonstrate how big data approaches could be used to:

1. *Improve presentation and delivery of crowdsourced forest data* and introduce new functionalities on data distribution and analysis platforms. Crowdsourced data is among the newest types of data used in forestry. The best practices for data utilization are still very much in development. At its best, crowdsourced data provides an effective way to gather information, e.g., on forest damages after storm events. However, its usability may be affected by issues like bias or unreliability. Furthermore, new technical solutions are needed for both data collection as well as distribution of crowdsourced data. One of the DataBio pilots (Chap. 23) concentrated on crowdsourced data collection and utilization.

2. *Optimize the use of tree resources.* Detailed characterization of trees and forest characteristics is used to determine the optimal use of trees for a given output (e.g., pulp, paper, textile, and biofuels) in order to guarantee that supply

meets demand. To enable reliable optimization of forest management activities, information on forest structural variables (e.g., species, height, and stem number) need to be kept up-to-date. Outdated forest information is one of the major hindrances in forest management around the world. With the increased temporal and spatial resolution, forestry big data allow improved timeliness of accurate forest variable data provision, and thereby improved optimization of tree resources. Provision of up-to-date forest characteristics utilizing online platforms was looked into in one of the DataBio pilots presented in Chap. 24.

3. *Improve identification of forest health and damages* caused by biotic (such as pests and diseases) or abiotic (such as snow, wind, dryness, rains, and fires) agents using remote sensors. Biotic forest damages are expected to become increasingly common in the near future due to rising temperatures [16]. Similarly, the frequency of extreme weather events is expected to rise due to the climate change, increasing the risk for abiotic damages. Big data processing and analysis allows implementation of time series approaches that allow forest health and damage monitoring for large areas in high spatial and temporal detail. Two pilots dealing with forest health monitoring are presented in Chaps. 25 and 26.

An overarching idea in all of the DataBio forestry pilots was to develop integrated tools to support management planning that is based on online platform infrastructures. Several of the pilots were linked to the Wuudis platform (https://www.wuudis.com/), which was used as the central piece to develop and demonstrate usability of inter-platform approaches. The Wuudis Service developed by Wuudis Solutions Oy is a commercial service on the market for forest owners, timber buyers, and forestry service companies. It enables the management of forestry activities (e.g., thinning and harvesting) and forest resources (e.g., forest estate evaluation) through a single tool. It can be used to obtain real-time information about the forest and its timber resource, track executed silvicultural and harvest activities, and plan the needed forest management activities.

The Wuudis platform (Fig. 22.1) was linked with other platforms in the pilots to highlight the possibilities of inter-platform connections in big data processing pipelines. Most notably, satellite image processing and analysis capabilities of the EO Regions! (https://www.eoregions.com/) platform and the Forestry Thematic Exploitation Platform (Forestry TEP; https://f-tep.com/) were used to feed user specific forest variable information into the Wuudis system. The EO Regions! platform provides services, information, and products specially created for service providers in Wallonia and Europe, while the Forestry TEP platform enables commercial, research, and public sector users in the forestry sector to efficiently access satellite data-based processing services and tools for generating value-added forest information products. Via the Forestry TEP platform, the users can also create and share their own services, tools, and generated products.

Similarly, the Metsään.fi service (https://www.metsaan.fi/) was linked with Wuudis Service to enable the exchange of data in both directions, to expand the data resources and functionality of both services. Wuudis users benefit from the

Fig. 22.1 Forestry estate borders and data transferred into Wuudis from the Metsään.fi service

open source data available in Metsään.fi, while users of Metsään.fi benefit from the additional functionalities available in Wuudis. The Metsään.fi service is provided by the governmental body Finnish Forest Center to make forest resource information available for citizens free of charge. The platform serves forest owners and Forestry service providers. Through the portal, forest owners in Finland can conduct business related to their forests at home from their own desktops. Metsään.fi connects forest estate owners with related third parties, including providers of Forestry services. This makes it easy to manage forestry work and to be in touch with forestry professionals.

In the following chapters of the book, four DataBio forestry pilots are presented. The presentations outline the pilot structure and highlight the main technical results. They also analyze the technological and market aspects of the usability and potential of big data in forestry. The chapters include:

Chapter 23—*Finnish Forest Data-Based Metsään.fi-services:* The best ways to utilize crowdsourced data in forestry are still very much in development. The pilot aimed to trial crowdsourced forest data presentation and new functionalities related to it. The launch of a new open forest data service, as well as related crowd-sourcing services, was included in this pilot. Two areas for crowdsourcing solutions were implemented: (1) showing quality control data for young stand improvement and early tending for seedling stand, and (2) storm damage data. Other possible crowdsourced data, such as other forest damage than storm damage data, were also evaluated.

Chapter 24—*Forest Variable Estimation and Change Monitoring Solutions Based on Remote Sensing Big Data:* Lack of up-to-date information on forest structural characteristics commonly prevents optimal forest management in large parts of the world. The pilot aimed to demonstrate the feasibility of online platform-based forest inventory approaches. The pilot focused on developing the forest inventory system on the Wuudis platform, which is based on remote sensing data and field surveys. The pilot was started in Finland and Belgium, but later expanded into Spain. The goal was to evaluate the usability of the technologies and processing methods of the project partners in different conditions varying from the Northern Boreal forests in Finland, through temperate forests in Belgium to the Galician forests in the Atlantic coastline in Spain. The pilot demonstrated inter-platform capabilities for comprehensive and near real-time quantitative assessment of forest cover over the interest areas.

Chapter 25—*Monitoring Forest Health: Big Data Applied to Diseases and Plagues Control:* Forest health monitoring is increasingly important due to the changing climate, and Big Data has the potential to offer means for effective large-scale forest health monitoring. The pilot set up the first version of a methodology and mathematical model based on remote sensing images (Sentinel-2 + Unmanned Aerial Vehicle) for the monitoring of health status of forests in the Iberian Peninsula. The work focused on the monitoring of *Quercus* forests affected by *Phytophthora cinnamomi* Rands and on the damage in eucalyptus plantations affected by *Gonipterus scutellatus*. After the definition of the big data algorithms and image processing techniques development, an EO-based system for monitoring the health of big forest areas was proposed, in order to enable public administrations to optimize their forest management resources.

Chapter 26—*Forest Damage Monitoring for the Bark Beetle:* Bark beetle outbreaks cause widespread ecological and economic damage in central Europe on a yearly basis, and are predicted to become even more severe in the near future. The pilot aimed to develop a new methodology for forest health assessment based on Copernicus satellite data (Sentinel-2). An approach was designed for assessment of forest health of the entire area of Czech Republic and other temperate forest regions in Europe, while reducing costs for field surveys. The method supports government officials by enabling effective identification of forest owners eligible for subsidies/tax relief. In addition, forest owners benefit from publicly available map server, where all forest health status maps are made available to allow pro-active management of forest properties.

After the individual pilot descriptions, a summary *Chap. 27—Conclusions and Outlook—Summary of Big Data in Forestry* will draw together the main findings of the DataBio project on the usability and potential.

Acknowledgements The authors would like to thank Seppo Huurinainen from Wuudis Solutions Oy, for leading the forestry pilots for most of the duration of the DataBio project, and everybody who contributed to the work in the forestry pilots during the project.

References

1. Lauer, D. T., Morain, S. A., & Salomonson, V. V. (1997). The landsat program: Its origins, evolution, and impacts. *Photogrammetric Engineering and Remote Sensing, 63,* 831–838.
2. Tomppo, E. (1996). Multi-source national forest inventory of Finland. In: Päivinen, R., Vanclay, J., Miina, S. (Eds.), New thrusts in forest inventory. Proceedings of the Subject Group S4.02-00,

Forest Resource Inventory and Monitoring, and Subject Group S4.12-00, Remote Sensing Technology, Vol I. IUFRO XX World Congress, Tampere, Finland, August 6–12, 1995. European Forest Institute, Joensuu, Finland.

3. Food and Agriculture Organization. (1996). Forest resources assessment 1990: Survey of tropical forest cover and study of change processes. FAO Forestry Paper 130. Food and Agriculture Organization (FAO), Rome.
4. Hansen, M. C., Potapov, P. V., Moore, R., et al. (2013). High-resolution global maps of 21st-century forest cover change. *Science, 342,* 850–853.
5. The GeoSpatial. (2019). How many earth observation satellites are there in space right now? https://www.thegeospatial.in/earth-observation-satellites-in-space. Accessed on 26 Sept 2020.
6. Chuvieco, E., Mouillot, F., van der Werf, G. R., et al. (2019). Historical background and current developments for mapping burned area from satellite earth observation. *Remote Sensing of Environment, 225,* 45–46.
7. Cohen, W. B., Yang, Z., Healey, S. P., et al. (2018). A LandTrendr multispectral ensemble for forest disturbance detection. *Remote Sensing Environment, 205,* 131–140.
8. Hirschmugl, M., Gallaun, H., Dees, M., et al. (2017). Methods for mapping forest disturbance and degradation from optical earth observation data: A review. *Current Forestry Reports, 3,* 32–45.
9. Copernicus. (2018). Develop your Copernicus-based business with the Copernicus incubation programme. https://www.copernicus.eu/en/develop-your-copernicus-based-business-cop ernicus-incubation-programme. Accessed on 26 Sept 2020.
10. National Land Survey of Finland. (2020). Laser scanning data 2008–2019: https://www.maa nmittauslaitos.fi/en/maps-and-spatial-data/expert-users/product-descriptions/laser-scanning-data. Accessed on 26 Sept 2020.
11. Maltamo, M., Hauglin, M., Næsset, E., et al. (2019). Estimating stand level stem diameter distribution utilizing harvester data and airborne laser scanning. Silva Fennica. https://doi.org/10.14214/sf.10075.
12. European Commission. (2013). A new EU forest strategy: For forests and the forest-based sector. https://eur-lex.europa.eu/resource.html?uri=cellar:21b27c38-21fb-11e3-8d1c-01aa75ed71a1.0022.01/DOC_1&format=PDF. Accessed on 26 Sept 2020.
13. European Commission. (2011). The EU biodiversity strategy to 2020. https://ec.europa.eu/env ironment/nature/info/pubs/docs/brochures/2020%20Biod%20brochure%20final%20lowres.pdf. Accessed on 26 Sept 2020.
14. European Commission. (2018). A sustainable bioeconomy for Europe: Strengthening the connection between economy, society and the environment—Updated bioeconomy strategy. https://ec.europa.eu/research/bioeconomy/pdf/ec_bioeconomy_strategy_2018.pdf. Accessed on 26 Sept 2020.
15. European Association of Remote Sensing Companies. (2016). Creating a European marketplace for earth observation services. https://earsc.org/wp-content/uploads/2020/08/EARSC-Marketplace-Alliance-PressRelease-V1.pdf. Accessed on 26 Sept 2020.
16. Seidl, R., Schelhaas, M.-J., Rammer, W., et al. (2014). Increasing forest disturbances in Europe and their impact on carbon storage. *Nature Climate Change, 4,* 806–810.

Chapter 23
Finnish Forest Data-Based Metsään.fi-services

Virpi Stenman

Abstract This chapter introduces the Finnish forest data ecosystem and its role in DataBio project pilots. In these DataBio pilots, the main objective is to improve the use of the Finnish forest resource data. The Finnish forest data provides a foundation for the forest big data-based online and e-services. The technical solution elements for the introduced DataBio pilots are based on standardized XML data sets, X-Road data transfer protocols, open forest data application programming interfaces (APIs) and crowdsourcing applications. The Metsään.fi-services including the open forest data APIs and Wuudis-mobile application are the key components for the customer's user interface. In the end of the chapter, the pilot-specific business benefits and key performance indicators are decribed showing clear positive impacts of the pilots. At the end of this chapter, visions for the future of public online services are discussed.

23.1 Introduction

Private forests are in a key position as raw material sources for traditional and new forest-based bioeconomy. In addition to wood material, the forests produce non-timber forest products (e.g. berries and mushrooms), opportunities for recreation and other ecosystem services.

In Finland, private forests cover roughly 60% of forest land, but about 80% of the domestic wood used by the forest industry. Today, the value of the forest industry production is 2.1 billion euros, which is a fifth of the entire industry production value in Finland. The forest industry export in 2017 was worth about 12 billion euros, which covers a fifth of the entire export of goods. Therefore, the forest sector is important for Finland's national economy [1].

The Finnish Forest Centre (FFC) is a public organization and operates under the steering of the Ministry of Agriculture and Forestry of Finland (https://www.metsakeskus.fi/en/finnish-forest-centre-focusing-people-and-forest). Gathering the forest resource data from privately owned forests in Finland is one of the FFC's

V. Stenman (✉)
Suomen Metsäkeskus, Aleksanterinkatu 18 A, 15140 Lahti, Finland
e-mail: Virpi.Stenman@metsakeskus.fi

© The Author(s) 2021
C. Södergård et al. (eds.), *Big Data in Bioeconomy*,
https://doi.org/10.1007/978-3-030-71069-9_23

statutory tasks, and today around 1.5 million ha of private forest inventories are annually updated. The inventory cycle for all of the private forests in Finland takes around 10 years and covers 14 million ha of privately owned forestland.

Gathering and maintenance of remote sensing and airborne laser scanning-based forest resource data started in the beginning of 2010 by the FFC. At present, the forest resource data covers almost 90% of the surface area of productive forest land in private forests. The forest resource data is utilized by forest owners and forestry actors. The forest resource data is constantly updated and maintained with the subsidy applications, forest use declaration notifications as well as with the update requests provided by the forest owners via Metsään.fi-service. Furthermore, the stand growth is added to all forest stand compartments in the forest resources database annually, and the forest management or felling proposals are simulated for the compartments accordingly.

The monetary benefits of this forest resource data ecosystem have been estimated by the Natural Resources Institute Finland (https://www.luke.fi/en/) as well as by Metsäteho Oy (https://www.metsateho.fi/briefly-in-english/), and they are annually over 26 million euros [2]. The potential monetary benefits are annually around 110–120 million euros. Furthermore, the forest resource data provides additional and indirect benefits for the forest service providers and via the investments around 1.95 billion euros.

23.2   Background and Objectives

The objectives of the Finnish forest data ecosystem are to ensure the high-quality and comprehensive forest inventory, which is standardized, up-to-date and easy to use. Furthermore, the forest data is an enabler for the FFC to produce the public services as well as data products based on the forestry sector demand.

The Metsään.fi-service is based on forest resources data that has been collected by remote sensing since 2011. Forest data can be utilized in, for example, the regional planning of forests and commercial forestry, to support the assessment of wood use possibilities and generally for developing forest businesses.

The Metsään.fi-service included in the Metsään.fi-website is a free e-service for forest owners and corporate actors (companies, associations and service providers) in the forest sector. The service aims to support active decision-making among forest owners by offering forest resource data and maps on forest properties, by making contacts with the authorities easier through online services and to act as a platform for offering forest services, among other things (Fig. 23.1). In addition to the Metsään.fi-service, the website includes open forest data services that offer the users national forest resource data that is not linked with personal information.

The Metsään.fi-service was launched in November 2012 as a version that was subject to charge, and was changed to a service free of charge for forest owners in 2015. By the end of 2018, about 110,000 forest owners had logged into the service. The forest owners that use the service own forest properties that are larger than

Fig. 23.1   Example of Metsään.fi map layer consisting of multiple data sets

average. The Metsään.fi-service's usage activity was increased in particular by forest owners experiencing that the presented recommendations for forest management matched their own objectives.

A central challenge in developing the website is to integrate several different sources of information into one entity that offers forest owners and actors all forest and nature data simultaneously. From the perspective of both forest owners and actors, the up-to-dateness of forest resource data and improvement of quality is one of the most important development objectives.

It is inherent for a service that is maintained with public funds that it is seen to be necessary and that it is being used. By the end of 2018, already over 100,000 forest owners had logged into the service. This is about a third of forest properties measuring over 2 ha. The forest owners and other industry actors see the service useful in many ways, but there are also areas that need improvement. It is important for future use and usefulness of the service to improve it and its content continuously.

23.3   Services

The Metsään.fi-website was also further developed through the DataBio project, where the objective was to improve the use of forest resource data and Metsään.fi-service [3]. The pilot focused on Metsään.fi databases and e-service integration to the national service architecture of Finland (based on X-road approach) where important features were, for example, data and user security, single-login and easy user role-based authentication and data access permissions (https://esuomi.fi/?lang=en). Furthermore, the launch of open forest data service, as well as related crowdsourcing services, was included in this pilot. These new types of data gathering methods were also expected to increase the availability of FFC's forest resource data.

The two recognized areas for crowdsourcing solutions were as follows: showing quality control data for young stand improvement and early tending for seedling stand, and storm damage data. Other possible crowdsourced data, such as other forest damage than storm damage data, was also evaluated during the project. Another pilotable topic was the open-data interface to environmental and other public data in Metsään.fi databases. This topic was highly dependent on the development of the Finnish forest legislation.

In these pilots, the requirements were specified for refining and showing the crowdsourced forest data to Metsään.fi users [4]. The implementation of the new functionalities and data presenting was carried out in collaboration with Metsään.fi's development team and other FFC's projects.

23.4   Technology Pipeline

The technology pipeline was specifically tailored for this pilot; however, the Suomi.fi-based data transfer service enables the data transfer in a standardized way between the FFC and other partners [5]. Also standardized forest data can be utilized for other purposes and on different scenarios. Suomi.fi-service is also applied for the user identification and authentication by Metsään.fi-service and many other public organizations in Finland.

The technology pipeline-related components consisted of Metsään.fi-service, open forest data service and Wuudis solution for mobile data gathering (Fig. 23.2).

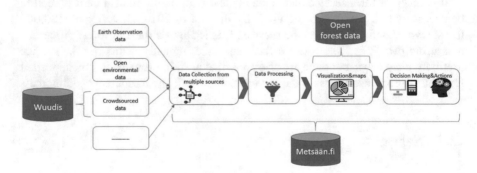

Fig. 23.2 Example of pilot data processing pipeline on a high abstraction level

Table 23.1 List of pilot specific components

Component name	Purpose for the pilot	Available at
Wuudis solution	Data sharing platform between authorities and end users	https://www.wuudis.com/
Metsään.fi-service	To improve the current component by joining the National service architecture for digital services (Suomi.fi) and to implement Suomi.fi e-identification and e-authorization for Metsään.fi users	https://www.metsaan.fi/

Table 23.2 List of pilot specific data assets

Data type	Dataset	Dataset location
Oracle database model in XML standard	Forest resource data	Finnish Forest Centre
Open environmental data in XML and OGC Geopackage standards	Open forest data	Finnish Forest Centre
Finnish Forest Centre CRM database (Legacy system)	Customer and forest estate data	Finnish Forest Centre
Mobile application dataset in XML format	Storm and forest damages observation and possible risk areas	Finnish Forest Centre/Wuudis solutions Oy

23.5   Components and Data Sets

Technical components listed in Table 23.1 and data assets listed in Table 23.2 were utilized in the pilot [6].

23.6   Results

The pilot deliverables consisted of integration of the Metsään.fi-service with the national service architecture of Finland (based on X-road approach). This phase consisted of important features such as data and user security, single-login and easy user-role-based authentication and data access permissions. Open forest data service was launched in March 2018, and related crowdsourcing services, including Wuudis based Laatumetsä mobile application for the forest damages as well as quality control monitoring, were published in the end of 2018.

In the beginning of 2019, the required XML standard schema version was released, and after that, the X-road approach was applied also for the crowdsourcing solutions regarding the forest damages reported by the Laatumetsä mobile application (Fig. 23.3). This activity was successfully implemented and finalized in September

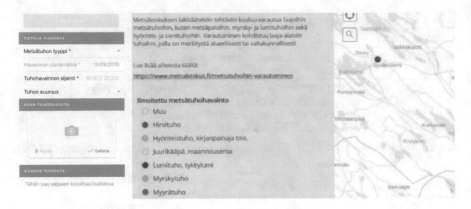

Fig. 23.3 Example of Laatumetsä mobile solution with related map service

2019, and it was mainly a technical solution improvement activity and therefore not visible for the end users.

In the beginning of the project, a top-level evaluation criteria for the pilot were agreed, and these were preliminary based on the Finnish Forest Act at the time being. However, the Finnish Forest Act was revised in March 2018, and the project evaluation criteria were updated accordingly. Additionally, more detailed key performance indicators were chosen to evaluate the results more precisely on the pilot level [7]. The updated top-level evaluation criteria with achieved results were as follows:

- In the beginning of the project 2017, the amount of FFC's forest resource data was around 200 GB. The amount was expected to increase by approximately 100 GB per year during the project, amounting to around 500 GB by the end of 2019. The result in the end of October 2019 was 574 GB.
- The coverage of forest resource data in Metsään.fi-service was in the beginning of 2017 around 11 million ha. The amount was expected to increase by 800,000 ha per year, amounting to around 13.4 million ha by the end of 2019. The result at the end of October 2019 was 12.5 million ha. The target was not completely achieved due to the fact that the data was getting outdated for the areas where the laser scanning was done over 10 years ago.
- The amount of data available for downloading for forestry operators' own information systems was at the beginning of the DataBio project around 1.5 million ha. The amount was expected to increase by 1 million ha per year, amounting to around 4.5 million ha by the end of 2019. The result at the end of October 2019 was 8.2 million ha.
- The amount of forest owners as Metsään.fi end users was at the beginning of the DataBio project around 70,000. The amount was expected to increase as follows: 85,000 in the end of 2017, 100,000 in the end of 2018 and 110,00 in the end of 2019. The result at the end of October 2019 was 119,046 forest owners.
- The amount of forestry service providers, i.e. so-called actors using the Metsään.fi-service, was in the beginning of the project around 380 pcs. The amount was

expected to increase as follows: 550 in the end of 2017, 650 in the end of 2018 and 750 in the end of 2019. The result at the end of October 2019 was 794 users.

Based on the above top-level evaluation, criteria and achieved results can be stated that the pilot targets were well achieved and exceeded. The results of the pilot were very promising, and they clearly indicate that by standardized solutions, i.e. with standardized data and data transfers as well as application programming interfaces, it is possible to build a completely new type of ecosystem, which is utilizing multiple data sources. In this type of ecosystem, the data sources can be scalable from closed data sets to open data as well as the data can be further enriched with crowdsourcing solutions, where citizens are acting as observers. This type of ecosystem consisting of the pilot specific pipelines is fully scalable and exploitable for the European forestry sector or even globally. By applying the same data standards, also the forestry sector businesses could be expanding their business opportunities across country borders.

23.7   Perspective

Related to the launch of the open-data interface to environmental and other public data in Metsään.fi databases, the main finding was that simple solutions do work; however, it is good to plan and reserve enough resources, not only for the development activities but also for the maintenance, end user support as well as training.

Regarding the shared multiuser data environment and Metsään.fi-services, certain purpose limitation factors were hindering to apply similar authorization processes for all of the end users. The backend service provider Suomi.fi could not provide the needed option for the user role specific authorization profiles. This type of factors could have been perhaps identified and mitigated during the pilot's risk management planning phase.

The findings related to the crowdsourcing solutions was that due to the available technologies, it is easy to implement and launch new types of data gathering solutions. However, the difficulty is in motivating the citizens to produce the information with new types of tools especially when the information is not necessarily fully integrated with the processes of the public authorities.

23.8   Benefits and Business Impact

The pilot-specific business impact and benefits were further analysed during the pilot with technical key performance indicators (KPIs), which were identified in the beginning of the pilot (Table 23.3). Most of the indicators are indicating very positive business impacts based on the pilot findings.

Table 23.3 Finnish Forest Data pilot KPI table

KPI short name	KPI description	Goal description	Base value	Target value	Measured value	Unit of value
Net Promoter Score (NPS)	Increased Metsään.fi user satisfaction regarding the e-services flexibility and quality	Measured with the Net Promoter Score (NPS) index on scale -100-0-100	0	>0	48	NPS
Data quality on range 1–5	Improvement in data quality measured via the end user survey and on scale 1–5	Measured on scale 1–5	3	>3	3.65	Scale
Operative cost savings	Based on the fact that utilization of the e-services and especially e-applications will save 75% costs compared to the traditional way of working with paper applications	The baseline value for this indicator is the year 2017 value, when 26% of all the applications were processed as e-applications	26%	>26%	35%	%
Revenue/employee	Employee productivity is expected to increase from the year 2017	Baseline of 59%, which is the amount of the contacted (phone, meeting) forest owners or service providers who joined the Metsään.fi-service	59%	>59%	66%	%
Sustainability	The amount and coverage of the data related to nature objects. This is measured as million hectares	Expected to increase as it is easier to capture high biodiversity profile candidates, for instance valuable habitats via the online services when the related data sets are available online	0	>0	25.73 Mha	Mha

(continued)

Table 23.3   (continued)

KPI short name	KPI description	Goal description	Base value	Target value	Measured value	Unit of value
Data amount for open forest data service	The total amount of open forest data available via the Metsään.fi-service implemented during DataBio project	Expected to increase	0	>0	439.3	GB
Data amount for open forest data service	The total amount downloaded data via the Metsään.fi, open forest data service implemented during the DataBio project	Expected to increase	0	>0	16,295	GB
Quantity of visits in open forest data service	The total quantity of visits and data loadings of open forest	Expected to increase	0	>0	10,928,529	pcs

23.9   Future Vision

The Metsään.fi-service is equated to several other authoritative online services that have been developed in Finland over recent years: Suomi.fi, vero.fi, kanta.fi, among others. The supply of online services is meant to increase the opportunities of citizens, companies and communities to use public services, regardless of time and place. E-services are usually the easiest and fastest way to contact authorities regarding, for example, forest use declarations and cost-sharing applications. When the use of online services increases, the public service production becomes more efficient and common tax money is saved. The starting point is that the public administration's online services are functional, safe and easy to use. The customer-centred planning, renewal of service processes, the interoperability of services and the data security and protection are central when building online services.

The main topics in developing Metsään.fi-webpages and services in future include

- one entity that offers forest owners and actors all forest and nature data simultaneously,
- the service remains free of charge for forest owners and actors with possible supplementary services subject to charge,
- marketing the service especially to new forest owners,
- easy to use, clarity should not decrease and an improved mobile application should be offered to the users,

- informing the users of the purpose, method and limits of the forest resource data offered by the Metsään.fi-website should be emphasized further than before, so that the expectations for the material become more realistic,
- material related to nature and leisure values and more diverse forest treatment options will have their own user base in the future,
- the control of global warming and the support of the biodiversity of nature will likely receive more attention in the future: Metsään.fi-service acts as an important platform for relaying information, and it makes it more effective to focus counselling towards forest owners.

23.10 More Information

Please find more information about Metsään.fi-services on the report prepared jointly with DataBio-project: Finland's model in utilizing forest data—Metsään.fi-website's background, implementation and future prospects (https://www.metsak eskus.fi/sites/default/files/ptt-report-261-finlands-model-in-utilising-forest-data. pdf). Furthermore, information regarding the pilot can be found from the DataBio pilot documentation (https://www.databio.eu/wp-content/uploads/2017/05/Dat aBio_D2.3-Forestry-Pilot-Final-Report_v1.1_2020-03-04_VTT.pdf). Technical solutions applied in this pilot have been defined as part of the DataBio technical documentation available at DataBioHub website (https://www.databiohub.eu/).

Literature

1. Valonen, M., et al. (2019). Finland's model in utilising forest data—Metsään.fi-website's background, implementation and future prospects. https://www.metsakeskus.fi/sites/default/files/ptt-report-261-finlands-model-in-utilising-forest-data.pdf. Accessed on 4 Dec 2020.
2. Räsänen, T., Hämäläinen, J., Rajala, M. ja Ritala, R. Metsävaratiedon hyödyntäminen puuhuollossa. Forest Big Data -hankkeen osaraportti. Metsätehon raportti 245, 29.12.2017. In Finnish. Available from: https://www.metsateho.fi/wpcontent/uploads/Raportti_245_Metsavaratiedon_hyodyntaminen_puuhuollossa.pdf.
3. Tergujeff, R., et al. (2017). DataBio Deliverable D2.1—Forestry Pilot Definition. https://www.databio.eu/wp-content/uploads/2017/05/DataBio_D2.1-Forestry-Pilot-Definition_v1.0_2017-06-30_VTT.pdf. Accessed on 4 Dec 2020.
4. Bozza, A., Russo, R., et al. (2017). DataBio Deliverable D6.6—Data Driven Bioeconomy Pilots v2. https://www.databio.eu/wp-content/uploads/2017/05/DataBio_D6.6-Data-driven-Bio economy-Pilots-v2_-v1.0_2019-05-31_CiaoT1.pdf. Accessed on 4 Dec 2020.
5. Södergård, C., et al. (2018). DataBio Deliverable D4.1—Platform and Interfaces. https://www.databio.eu/wp-content/uploads/2017/05/DataBio_D4.1-Platform-and-Interfaces_v1.0_2018-05-31_VTT.pdf. Accessed on 4 Dec 2020.
6. Habyarimana, E., et al. (2017). DataBio Deliverable D6.2—Data Management Plan. https://www.databio.eu/wp-content/uploads/2017/05/DataBio_D6.2-Data-Management-Plan_v1.0_2 017-06-30_CREA.pdf. Accessed on 4 Dec 2020.

7. Stanoevska-Slabeva, K., et al. (2020). DataBio Deliverable D7.2—Business Plan v2. https://www.databio.eu/wp-content/uploads/2017/05/DataBio_D7.2.Business-Plan-v2_v1. 2_2020-03-19_UStG.pdf. Accessed on 4 Dec 2020.

Chapter 24
Forest Variable Estimation and Change Monitoring Solutions Based on Remote Sensing Big Data

Jukka Miettinen, Stéphanie Bonnet, Allan A. Nielsen, Seppo Huurinainen, and Renne Tergujeff

Abstract In this pilot, we demonstrate the usability of online platforms to provide forest inventory systems for exploiting the benefits of big data. The pilot highlights the technical transferability of online platform based forest inventory services. All of the services tested in the piloting sites were technically implemented successfully. However, in new geographical areas, strong user involvement in service definition and field data provision will be needed to provide reliable and meaningful results for the users. Overall, the pilot demonstrated well the benefits of technology use in forest monitoring through a range of forest inventory applications utilizing online big data processing approaches and inter-platform connections.

24.1 Introduction, Motivation, and Goals

Remote sensing data from traditional aerial platforms, unmanned aerial vehicles (UAV), and satellite sensors presents an optimal way to timely collect information on forest cover and characteristics over large and small interest areas. The amount of available remote sensing data has greatly escalated during the past decade. This escalation is caused by a growing number of sensors, more frequent observations, and an increasing spatial and spectral resolution of the sensors. The remote sensing data boom enables implementation of more frequent and detailed remote sensing-based forest monitoring approaches than previously possible.

J. Miettinen (✉) · R. Tergujeff
VTT Technical Research Centre of Finland, 02044 VTT Espoo, Finland
e-mail: jukka.miettinen@vtt.fi

S. Bonnet
Spacebel, Liege Science Park, 4031 Angleur, Belgium

A. A. Nielsen
Technical University of Denmark, 2800 Kgs. Lyngby, Denmark

S. Huurinainen
Wuudis Solutions Oy, Kiillekuja 1, 50130 Mikkeli, Finland

© The Author(s) 2021
C. Södergård et al. (eds.), *Big Data in Bioeconomy*,
https://doi.org/10.1007/978-3-030-71069-9_24

At the same time, new big data processing approaches need to be developed to fully exploit the potential provided by the increasing data volumes. Particularly, the availability of the Copernicus Sentinel-2 multispectral optical data and its applicable free data policy present a great opportunity for developing low-cost commercial applications for environmental monitoring. Online platforms, such as the Forestry Thematic Exploitation Platform (Forestry TEP; https://f-tep.com/) and the EO Regions! (https://www.eoregions.com/), enable creation of services for efficient processing of satellite data to value-added information.

The goal of this pilot was to develop a forest inventory system on the Wuudis Service (https://www.wuudis.com/) based on remote sensing data and field surveys. Selected DataBio project partners integrated their existing market-ready or almost market-ready technologies into the Wuudis Service, and the resulting solutions were piloted with the Wuudis users, forestry sector partners, associated partners, and other stakeholders.

24.2   Pilot Set-Up

The consortium for this pilot consisted of: (1) Wuudis Solutions, Finland, (2) VTT Technical Research Centre of Finland (VTT), (3) Spacebel, Belgium, and (4) Technical University of Denmark (DTU). In addition, Forest Management Institute (FMI) from the Czech Republic coordinated their own pilot activities with this pilot. All activities were linked to the Wuudis platform, and inter-platform connections were developed between Wuudis and two other platforms coordinated by consortium members: Forestry TEP coordinated by VTT and EO Regions! coordinated by Spacebel.

Three different test sites were used in the pilot: (1) the Hippala forest estate in Southern Finland, (2) Walloon Region, Southern Belgium, and (3) the forest property 'Barbanza, Enxa, Xian, Dordo, Costa de Abaixo e O Sobrado', located at the municipality of Porto do Son, A Coruña province in Galicia, in Northwestern Spain. The Galician site is owned by the rural community 'Comunidade de Montes Veciñais en Man Común (CMVMC) de Baroña' and managed by the Asociación Sectorial Forestal Galega (ASEFOGA), a forest owners' association based in Santiago de Compostela.

In the pilot sites, VTT, Spacebel, and DTU conducted demonstrations and further development of their forest monitoring applications and services. In parallel, FMI was developing a new methodology for forest health assessment, which allows assessment of forest health in the entire area of Czech Republic. The results of the FMI work (described in detail in 'Chapter 26 Forest damage monitoring for the bark beetle') can be linked to the online platforms used in the pilot through the Open Geospatial Consortium (OGC), Web Map Service (WMS), and Web Map Tile Service (WMTS) interfaces.

24.3   Technology Used

24.3.1   Technology Pipeline

In this pilot, technology pipelines were established to facilitate smooth utilization of remote sensing data for forest inventory purposes in an online environment. The pipelines combined data sources, processing software components, and inter-platform communication into continuous processing chains that enable fast data processing and smooth delivery of the results. Figure 24.1 presents an example of the forest inventory pipeline that was created to allow efficient forest structural variable estimation with VTT proprietary software utilizing the *Probability* [1] method and to connect this process with the Wuudis platform. The pipeline utilizes the Forestry TEP platform for data sourcing and processing, feeding information to the Wuudis platform. The VTT software Envimon and Probability are used in data analytics. The four main components of the DataBio generic pipeline (i.e., data acquisition, data preparation, data analytics, and data visualization and user interaction) are marked in red text in Fig. 24.1.

The pipeline presented in Fig. 24.1 generates layers of forest structural variable estimates, by combining information derived from the 10 m resolution Sentinel-2 data with field sample plots. Sample plot data collected by the Finnish Forest Center (FFC) is used as a reference in the estimation model training.

For easy integration of satellite maps and the analyzed (highlighted) theme maps, standard OGC WMS or WMTS interfaces were used as a starting point. The Wuudis

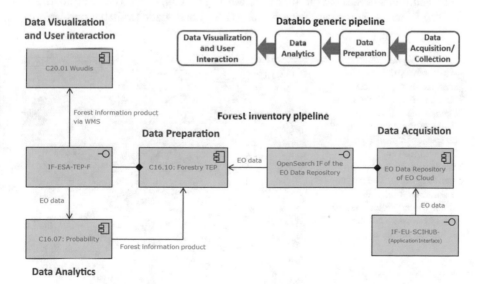

Fig. 24.1 Example of the forest inventory pipeline established in the pilot, with reference to the generic DataBio pipeline concept

Service is using OpenLayers as the mapping client library. In the first stage of the project, the forest variable estimates produced by VTT were presented as image raster data (GeoTIFF format) with 10 m pixel resolution, with one image band per variable and each pixel containing the estimated variable value. The output was made available for integration in the Wuudis end user system via WMS interface from the Forestry TEP.

In the second stage of the pilot, the system was further developed to enhance the connection between Forestry TEP and the Wuudis platform. Delivery of the VTT forest variable estimates produced with the *Probability* [1] method was enabled in the Extensible Markup Language (XML)-based Finnish Forest Information Standard [2] format. This approach allows to use forest management plan geometries as a baseline; remote sensing based, pixel-format information is expanded to these geometries and stored back in an updated forest management plan. In this enhanced system, the data is provided from the Forestry TEP in a ready-to-use format (for the end users), which could be used in Wuudis or any other online platform with no further calculations needed. The Forestry TEP service also allows retrieving the forest variable estimates in a standard Geographic JavaScript Object Notation (GeoJSON) format.

In the Hippala pilot area, Finland, the estimated forest variables include: stem number; stem volumes for pine, spruce, broadleaved, and total; diameter; basal area; and height. Figure 24.2 illustrates the species-wise volumes estimated for the Hippala forest estate.

The technology pipeline presented above relates to a larger context of connection and optimal utilization of various types of online platforms. Figure 24.3 presents Forestry TEP as an online platform that enables efficient exploitation of the Copernicus Sentinel satellite data in forest monitoring and analysis. The satellite data is sourced from the European Space Agency (ESA) and made available on platform,

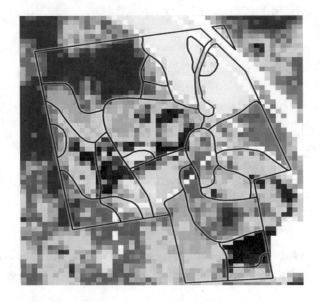

Fig. 24.2 Visualization of species-wise volumes generated using the Forestry TEP platform at Hippala. Shown is the estimated stem volume of the dominating tree species in each 10 m by 10 m area (red = broadleaved, blue = pine, green = spruce). The darker the color, the higher the volume (range around 0–300 m³/ha). Forest stands are outlined by red lines

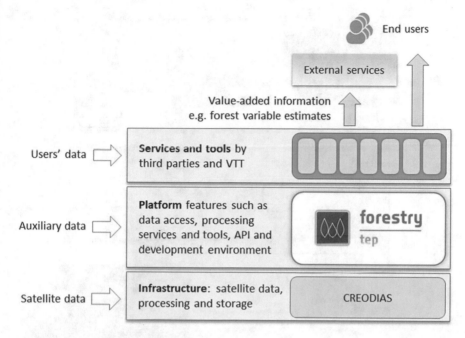

Fig. 24.3 Forestry TEP is an online platform for efficient exploitation of Copernicus Sentinel and other satellite data in forest monitoring and analysis. Along with the data, the platform offers processing services and tools and allows to develop and share new services

via the underlying infrastructure and data services of CREODIAS, one of the five Copernicus Data and Information Access Services (DIAS) platforms. Forestry TEP provides processing services and tools and serves also as a platform for new services, such as the Envimon and Probability tools of VTT that were used in this pilot. Subsequently, the results produced in Forestry TEP can serve as input for various external service platforms (like Wuudis or EO Regions!). In the expanding landscape of different types of platforms with increasing data volumes, efficient inter-platform pipelines are essential in enabling exploitation of the full potential of EO big data for forest inventory purposes.

In the DataBio project, inter-platform connections between EO Regions! and Wuudis were also developed, enabling numerous possibilities to feed the Wuudis Service in geographical and dendrometric content. EO Regions! is a commercial showcase of a satellite image processing system (e.g., Sentinel-2) allowing automated processing. The products can be downloaded by the customer (after online ordering) or directly connected to another platform dedicated to a specific theme (e.g., forest management). The connections between EO Regions! and Wuudis (Fig. 24.4) allow several scenarios for combined use of the two platforms. Users can, e.g., (1) work independently on either platform to import their data, or (2) use mobile applications to encode dendrometric data, or (3) order forestry services from either platform. In all

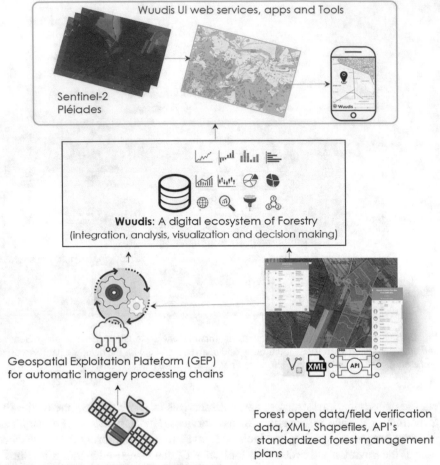

Fig. 24.4 EO Regions! platform provides access to various remote sensing services based on earth observation data, in particular the Copernicus data, allowing automated processing and connection with other platforms

of these cases, the users will benefit from the increased offering and functionalities provided by the connection between EO Regions! and Wuudis.

In addition, a study on the usability of Senop hyperspectral camera for boron deficiency mapping was performed at a test site in Finland. Finally, several demonstrations of the transferability of the technical capabilities were performed in a test site in Galicia, Spain, where teams from DTU, FMI, and VTT applied their methods in coordination with Wuudis platform. Figure 24.5 shows the Spanish study site stand boundaries in the Wuudis platform. The user interface of Wuudis platform provides stand-wise information that can be used for forest management planning and monitoring decision making. In addition to basic information (like property codes, area,

Fig. 24.5 Pilot site in Galicia populated with forest estate data in Wuudis Service. The user can browse through information such as ID, area, stem count, volume, and tree value for each forest stand, and visualize supporting material such as field photos

etc.), forest variable information such as development level, stem count, and volume can be provided, as well as derived information including, e.g., the value of trees. In addition, the system allows inclusion of remote sensing imagery and photos, as well as other supporting material such as field measurement results.

Figure 24.6 illustrates a demonstration of tree height estimation in the Galician pilot area by the *Probability* [1] method, visualized on the Forestry TEP platform. The estimation was conducted using Mar 29, 2019, Sentinel-2 satellite imagery and field measurements by Wuudis Solutions staff. There are some higher than expected values for open areas and shrublands, since these areas were not represented in the limited field reference data, but overall the forest areas clearly stand out with a range of tree height values around 8–18 m.

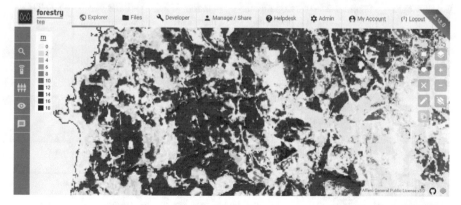

Fig. 24.6 Tree height (in meters) estimation in the Galician conditions, visualized in the Forestry TEP platform (legend pasted on the image)

24.3.2 Data Used in the Pilot

The pilot utilized several different types of remotely sensed datasets as well as field data (Table 24.1). Remotely sensed datasets included Sentinel-1 and Sentinel-2 satellite data and airborne hyperspectral remote sensing data. In the Finnish test site, sample plot data by the Finnish Forest Center was used as reference in the estimation model training. In the Galicia test site, Wuudis Solutions staff conducted field work, collecting forest variable information from ten forest stands. The measured information included six different forest structure variables: (1) species, (2) age, (3) basal area, (4) stem count, (5) mean diameter, and (6) mean height. The field data was recorded in the Wuudis platform, together with photographs. In addition, all available information from the forest estate stands were recorded into the Wuudis system.

The Sentinel satellite data was found to be very useful for operational forest monitoring applications in online platforms. The systematic acquisition scheme and high temporal frequency (i.e., short revisit time) provide large amounts of data suitable for high temporal resolution service provision. The high number of spectral bands (10) usable for forest monitoring purposes in the Sentinel-2 satellites, combined with the

Table 24.1 Data assets utilized in the pilot. GB stands for gigabyte, TB for terabyte, and PB for petabyte

Data type	Dataset	Dataset original source	Dataset location	Volume (GB)	Velocity (GB/year)
Satellite data	Sentinel-1	Copernicus program	Online repositories such as DIAS or the Copernicus Open Access Hub	1–8 GB per scene	~1.5 PB
Satellite data	Sentinel-2	Copernicus program	Online repositories such as DIAS or the Copernicus Open Access Hub	~1 GB per scene	~800 TB
Airborne data	Hyperspectral remote sensing data	Senop hyperspectral camera	Wuudis platform	*n.a	n.a
Field data	Forest plot data for Finland	Finnish Forest Center	Metsään.fi [3]	n.a	n.a
Field data	Forest plot data for Galicia	Wuudis staff	Wuudis platform	n.a	n.a

*n.a. not available

10–20 m spatial resolution, enables development of high-quality forest monitoring applications. Furthermore, Sentinel-1 and Sentinel-2 data is stored in centralized platforms, such as the Copernicus Data and Information Access Service (DIAS) platforms or the Copernicus Open Access Hub, and can be accessed directly with processing platforms like Forestry TEP.

In the pilot, two different types of field datasets were used. The national coverage sample plot data by the Finnish Forest Center that was used as reference in Finland was confirmed to be very suitable for the online applications demonstrated in this pilot. However, such field datasets are not available in all countries. The amount, quality, and timeliness of field data often play a crucial role in EO big data-based forest inventory applications, and therefore, operational collection of such data is very important, for example, the pilot in Galicia, Spain, depended on collection of on-site field data for the testing and demonstration of the products and services. However, due to limited resources, only a small amount of data could be collected, which considerably limited the scope of demonstration that could be conducted.

24.3.3 Reflection on Technology Use

Overall, the pilot demonstrated well the benefits of technology use in forest monitoring through a range of forest inventory applications utilizing EO big data and online big data processing approaches. These applications and services were further developed to improve user experience. One of the key development aspects in the pilot was the inter-platform operability. The services were integrated with the Wuudis platform, demonstrating the possibilities and benefits of inter-platform interactions. The resulting solutions were piloted with Wuudis users, forestry sector partners, associated partners, and other stakeholders.

The experiences from the pilot confirm the value of big data in forest monitoring and encourage further development of big data approaches for forest monitoring purposes. The massive increase in remote sensing data volumes over the past decade has enabled remote sensing-based forest monitoring in unprecedented levels of frequency and detail. Big variety of data sources is available, each with their own characteristics in, e.g., spatial resolution, update frequency, level of detail and accuracy for the thematic task at hand, and cost. This allows picking the most suitable data for the need or to combine various approaches for the best overall effect. The freely available satellite data from the Copernicus Sentinel program is a key opportunity for many tasks, especially when aiming to cover large areas. Standardized processing pipelines in the online environment, such as the ones developed in this pilot, are crucial in taking full advantage of the high volumes of data in an operational and effective manner.

24.4   Business Value and Impact

The entire pilot focused on development and integration of marketable forest inventory services into the Wuudis platform and other related platforms. Overall, the pilot results were successful in demonstrating the usability of a range of forest inventory applications on the platform. The pilot demonstrated the functionality of inter-platform connections and service provision, which enables wider exploitation of the services developed in and outside of this pilot. The services are applied on the respective platforms and exploitation of the services is growing.

Table 24.2 presents the key performance indicators (KPIs) measured during the pilot. The Wuudis tree-wise monitoring MVP (minimum viable product) service was launched in June 2018 and sold to leading forest management associations (forest management associations of Pohjois-Karjala, Savotta, and Päijänne) and forest industries in Finland. Over 5000 ha were monitored by the Wuudis network of service providers.

In addition to the measurable KPIs, the pilot aimed at testing and demonstrating new services for forest damage monitoring. Several services were successfully tested and demonstrated in Belgium, Finland, and Spain, utilizing several online platforms and inter-platform connections. This will increase the service offering in all the involved platforms (Wuudis, Forestry TEP, and EO Regions!) and enable higher revenue in the future.

The pilot is a good example on how research results are used in business development. The pilot brought together new commercial partners for added-value services on top of Wuudis platform. Business agreement between Wuudis Solutions and Spacebel regarding the distribution of the Wuudis Services to the forest users of the EO Regions! platform and the commercialization of Spacebel's earth observation forest products in the Wuudis platform were set up during the project. Negotiations on operational-level inter-platform connections between Wuudis and Forestry TEP

Table 24.2 Pilot KPIs

KPI description	Goal description	Base value	Target value	Measured value	Unit of value
Usability of tree-wise monitoring service MVP	Goal is to sell the tree-wise monitoring service MVP to forest management associations in Finland	0	No target value	3	Number of customers
Surface processed with MVP	The goal is to increase the area processed using the MVP service	0	4000	5000+	ha

as well as with the Finnish state forest enterprise Metsähallitus are also ongoing at the time of writing.

Because of the pilot, Wuudis Solutions is now able to better understand the needs of the Spanish market. Wuudis Service was tested in a real business environment, and the results were encouraging. Wuudis Solutions is expanding its customer base in Spain through establishment of a subsidiary, Wuudis Solutions S.L. in November 2019 and partnering with local airborne data service providers like Agresta. Wuudis Solutions has already secured new implementations and R&D projects in Spain/Galicia (e.g., TEMPO, ICEX, Galician Wood Cluster).

24.5 How-to-Guideline for Practice When and How to Use the Technology

For forest monitoring stakeholders, be they private forest owners, forestry companies, or public entities, the best avenue for big data utilization for forest monitoring purposes is through online platforms. As described in Sect. 24.3 above, there are several levels of online platforms enabling utilization of EO data for forest monitoring purposes. These include, for example, the DIAS platforms providing data access to forestry-related Big Data and several platforms providing forestry-related applications and services. These platforms include, for example, the Forestry TEP, EO Regions!, and Wuudis platforms used in this pilot.

The application platforms provide direct access to satellite data and auxiliary datasets, and ready-made applications for the utilization of the data for forest monitoring purposes. In addition, e.g., Forestry TEP offers an application development interface, where users can develop their own applications utilizing the Big Data available on the platform. Inter-platform connections bring further benefit to the users through wider service offering.

On a general level, more effort is needed to increase the interest toward platform services in the forestry community and to ensure smooth user experience. In many parts of Europe, the forestry sector has a long history with strong traditions in forest management practices. It may take some time to change the perspective of the forestry stakeholders to fully approve big data-based approaches. In order to increase the interest from the user side, the service providers now need to (1) further develop methods to fully exploit EO big data for forest monitoring, (2) convince the forestry stakeholders about the concrete benefits of online services in efficient utilization of big data, and (3) further improve cooperation between service providers to ensure smooth and effortless user experience and increased interest.

The importance of local promotional activities and locally tuned services cannot be overemphasized. This can be achieved through strong involvement of local level actors (such as regional forest administrations or local forest associations), which enables direct connection to local datasets and stakeholders. This, in turn, allows fine-tuning of the provided services according to local practices and requirements.

24.6 Summary and Conclusion

Overall, the pilot demonstrated well the benefits of technology use in forest monitoring through a range of forest inventory applications utilizing big data and online big data processing approaches. In addition, the pilot highlighted (1) the technical transferability of online platform-based forest inventory services and (2) importance of local involvement in fine-tuning services to meet local needs. All of the services tested in the pilot areas were technically implemented successfully. However, stronger user involvement in service definition and field data provision would be needed to provide more reliable and meaningful results for the users.

The pilot was very successful in further developing capabilities to perform comprehensive and near real-time quantitative assessment of forest cover over the project pilot areas. This type of near real-time forest monitoring allows monitoring of forest damages, deforestation, and forest degradation.

The pilot was also generally successful in creating the inter-platform connections. However, the challenges of integration of services between platforms and service providers became clear during the pilot. Best practices for inter-platform cooperation between service providers (both technical and financial) need to be further developed to enable smooth and effortless user experience, to gain the maximum benefit from the range of service providers working together.

Acknowledgements The authors would like to thank everybody who contributed to the work presented in this chapter. Particularly, we would like to acknowledge the inputs of Roope Näsi from the Finnish Geospatial Research Institute, National Land Survey of Finland as well as Jesús Martínez Ben and Sudip Kumar Pal from Wuudis Solutions Oy.

We also want to acknowledge the following organizations for their provision of open source EO and field datasets: the European Space Agency and the Finnish Forest Center.

References

1. Häme, T., Stenberg, P., Andersson, K., et al. (2001). AVHRR-based forest proportion map of the Pan-European area. *Remote Sensing of Environment, 77,* 76–91.
2. Finnish Forest Centre. (2014). Finnish forest information standard. https://www.metsatietostand ardit.fi/en/. Accessed on 26 Sep 2020.
3. Finnish Forest Centre. (2017). General information about open forest Data. https://www.met saan.fi/en/general-information-about-open-forest-data. Accessed on 26 Sep 2020.

Chapter 25
Monitoring Forest Health: Big Data Applied to Diseases and Plagues Control

Adrian Navarro, María Jose Checa, Francisco Lario, Laura Luquero, Asunción Roldán, and Jesús Estrada

Abstract In this chapter, we present the technological background needed for understanding the problem addressed by this DataBio pilot. Spain has to face plagues and diseases affecting forest species, like Quercus ilex, Quercus suber or Eucaliptus sp. Consequently, Spanish Public Administrations need updated information about the health status of forests. This chapter explains the methodology created based on remote sensing images (satellite + aerial + Remotely Piloted Aircraft Systems (RPAS)) and field data for monitoring the mentioned forest status. The work focused on acquiring data for establishing the relationships between RPAS generated data and field data, and on the creation of a correlation model to obtain a prospection and prediction algorithm based on spectral data for early detection and monitoring of decaying trees. Those data were used to establish the links between EO image-derived indexes and biophysical parameters from field data allowing a health status monitoring for big areas based on EO information. This solution is providing Public Administrations with valuable information to help decision making.

A. Navarro · M. J. Checa · F. Lario · L. Luquero · A. Roldán · J. Estrada (✉)
TRAGSA Group, Conde de Peñalver 84, 28006 Madrid, Spain
e-mail: jmev@tragsa.es

A. Navarro
e-mail: anavar30@tragsa.es

M. J. Checa
e-mail: mjca@tragsa.es

F. Lario
e-mail: flario@tragsa.es

L. Luquero
e-mail: lluquero@tragsa.es

A. Roldán
e-mail: aroldan@tragsa.es

© The Author(s) 2021
C. Södergård et al. (eds.), *Big Data in Bioeconomy*,
https://doi.org/10.1007/978-3-030-71069-9_25

25.1 Introduction, Motivation, and Goals

Spain has to face worrying situations derived from plagues and diseases that are affecting forest species, like *Quercus ilex*, *Quercus suber* or *Eucaliptus sp*, in the Iberian Peninsula, causing high economic losses.

Spanish Public Administrations and forest owners need updated information about the health status of forests to perform a sustainable and suitable forest management. The optimal combination of different Earth observation (EO) data and field data allows the creation of new products for forest monitoring and effective tools for decision making with a good balance between results obtained and cost of use.

Therefore, the goal of this DataBio Pilot is the creation of a methodology based on remote sensing images (satellite + aerial + Remotely Piloted Aircraft Systems (RPAS)) and field data for monitoring the health status of forests in large areas in two different scenarios. This work focused on monitoring the health of Quercus sp. forests affected by the fungus *Phytophthora cinnamomi Rands* and the damage in *Eucalyptus* plantations affected by the coleoptera *Gonipterus scutellatus Gyllenhal*.

Phytophthora cinnamomi severely affects several tree species, like *Quercus ilex* and *Q. suber*, in different areas in Spain (Extremadura, Andalucia, Castilla y León, Castilla La Mancha, Madrid) causing a great ecological and economic problem. Detection is currently performed on the field through direct observations or through data sampling and analysis in the laboratory.

However, big data sets as very high resolution (VHR) EO data (Orthophotos and RPAS images with visible and near infrared bands) can be used to identify dead trees and locate possible affected areas of *Quercus* forests and to analyze their evolution. Detailed RPAS-generated visible and multispectral images as well as field data were collected from selected sampling plots and analyzed. Those data were used to establish the links between EO image-derived indexes and biophysical parameters from field data allowing a more general health status monitoring for big areas based on EO information.

A similar approach has been used for monitoring the damage in *Eucalyptus* plantations caused by the *Gonipterus scutellatus*. In this case, the main motivation for this pilot was to develop an efficient mapping and assessment tool for monitoring and assessing the damages in order to adapt management procedures and minimize economic losses.

Gonipterus scutellatus defoliates *Eucalyptus* plantations severely. *Eucalyptus* is one of the main commercial species in the North of Spain (Galicia, Asturias, and Cantabria), where *Gonipterus* produces huge economic losses by impeding the development and growth of trees. Authorities (Xunta de Galicia), industrial companies from the paper sector (Empresa Nacional de Celulosas—ENCE)) and forest owners need an economic, systematic, and objective tool for affected areas identification and damage assessment, in order to adapt management and minimize economic losses.

EO images-derived vegetation indices can be used for a systematic monitoring of the health status in the selected study areas; anomalies will show areas where

Gonipterus can be defoliating, which will be checked on the field, either visually or using RPAS. These two information sources will be combined to define an optimal methodology for data acquisition and analysis. The rate of defoliation must be analyzed and linked to EO data.

In both cases, those EO-based solutions are providing Public Administrations with valuable information to help decision making. The EO-based system provides forest health monitoring of big forest areas including mapping and assessment tools so Public Administrations can optimize forest management resources.

25.2 Pilot Setup

The work focused on acquiring data for establishing the relationships between RPAS generated data and field data, and on the creation of a correlation model to obtain a prospection and prediction algorithm based on spectral data for early detection and monitoring of decaying trees affected by *Phytophthora*. The general methodology applied in the pilot is very briefly summarized in Fig. 25.1

In the case of *Quercus*, the aim is to monitor the state of the rees in the areas of open forest, "dehesas", in order to follow the evolution of the "seca" disease. Therefore, WHR images are required to identify the trees individually (spatial resolution <= 50 cm).

In the case of *Eucaliptus*, the pilot aim is the development of a conceptual model for estimating the defoliation degree at tree-level according to the user's requirements defined by the main paper manufacturer in Spain. A simplified model at "tree level" to assess defoliation and assign treatment priorities was obtained by establishing the correlation between EO (RGB, multispectral, thermal) and field data. This model is adjusted to the criteria established by the customer. According to these criteria, the treatment against *Gonipterus* is only applied to trees defoliated at a degree of 10–60% (Fig. 25.2).

General Methodology

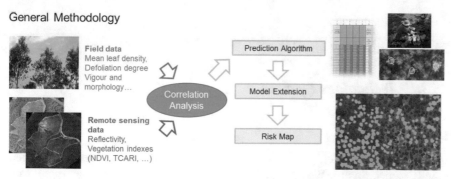

Fig. 25.1 General methodology for the classification of vigour/decay status from field and multispectral data from RPAS flights

Fig. 25.2 Maps which show the "degree of defoliation" (upper figure), and the "treatment priorities" (lower figure) at "tree level"

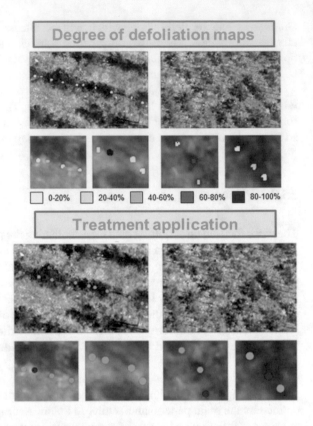

Results obtained so far allow to state that it is possible to assess defoliation and assign treatment priorities by using RPAS data. Nevertheless, some problems appear due to the low density of *Eucalyptus* crowns. In addition, the variety of the land cover makes it impossible to detect trees automatically, which is necessary for extending the model to plantation level and obtaining the risk maps.

25.3 Technology Used

25.3.1 Technology Pipeline

Data collection:

In the case of oaks, a field campaign was launched in July 2018 in Haza de la Concepción (Cáceres, Spain). We analyzed 380 ha of sparse forest ("dehesa" type), from which 100 ha were selected for data acquisition. Vegetation consisted of *Quercus ilex* and *Quercus suber* showing different degrees of affection by *Phytophthora*. 81

Quercus ilex trees were sampled in 9 plots. Measured parameters were: mean leaf density (measured with a specialized camera as LICOR 2200), mean leaf surface and biomass (green and dry), pigment concentration from leaves (chlorophyll and carotenes), crown and trunk morphology, health status inventory and damage assessment, analysis of soil and roots for determining the presence/absence of *Phytophthora cinnamomi*.

The RPAS data collected were obtained using a eBee+[1] platform with SODA[2] RGB camera Sequoia multispectral camera (Green, Red, RedEdge, and NIR bands) over the study site.

Regarding *Eucaplitus*, there were several field campaigns such as:

- July 2017 > Socastro (Pontevedra, Spain). Timber company ENCE manages 14 ha. Plantation of 6–7 years old *Eucaliptus globulus* with crown mean size 7–10 m. Dense understory vegetation dominated by *Ulex europaeus* and *Rosa sp*. Ninety-six *Eucaliptus* trees were sampled in 8 plots, 12 trees per plot in the different existing strata. Measured parameters in each tree were: (i) % defoliation of the crown's upper third (according to ENCE's protocol); (ii) defoliation, trunk, and crown morphology (according to PLURIFOR project's protocol); (iii) mean leaf density (measured with LICOR 2200).
- April, 2018 > Loureza (Pontevedra, Spain). Here, ENCE manages 120 ha. Commercial plantation of 6–8 years old *Eucaliptus globulus*. The plantation showed very different degrees of affection by *Gonipterus*. In this case, 210 trees were sampled (10 trees per 21 plantation lines). Measured parameters in each tree were the same as in Socastro.

Data processing:

We calculated several spectral indexes related to vegetation activity and pigments from multispectral RPAS data: normalized difference vegetation index (NDVI), green normalized difference vegetation index (GNDVI), normalized red-green difference index (NGRDI), soil-adjusted vegetation index (SAVI), optimized soil-adjusted vegetation index (OSAVI), anthocyanin reflectance index (ARI1-ARI2), and transformed chlorophyll absorption reflectance index (TCARI).

The general data flow with VHR EO data was:

1. Field campaign: acquisition of RPAS data and tree samples.
2. Image processing: orthorectification, orthomosaics generation, radiometric calibration, etc. (The software used was Pix4D and specific remote sensing programs: ERDAS Image and PCI Geomatics).
3. Calculation of vegetation index.
4. Generation of a binary tree/non-tree mask (object-based image analysis, OBIA algorithm implemented in eCognition) from the multispectral information of the RPAS/orthophotos images.

---

[1] https://www.sensefly.com/drone/ebee-mapping-drone/.

[2] https://www.sensefly.com/camera/sensefly-soda-photogrammetry-camera/.

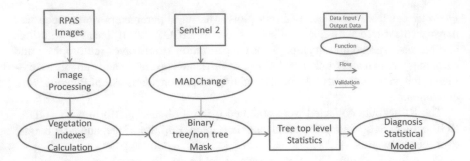

Fig. 25.3 Monitoring health pipeline

5. Extraction of statistics (minimum and maximum value, mean, mode and standard deviation) at treetop level from EO data and correlation with field data (biophysical parameters).
6. Construction of the statistical model of diagnosis and extension to the study area. The goal of these models is to optimize the monitoring of health status and to guide field visit.

The general pipeline is shown in Fig. 25.3.

MADchange[3] is a DataBio component [1] that detects change over time in multi- or hyper-spectral data as acquired from space or airborne scanners and it has been used as a validating system for the results obtained in specific areas of the *Eucalyptus* areas of study.

The initial correlation model was obtained from the first campaign (100 ha) to obtain a prediction algorithm for the early detection of decaying trees affected by *Phytophthora* based on spectral data. The extrapolation of the algorithm to the whole area (Haza "dehesa", 380 ha) was developed in a second stage (Fig. 25.4).

In the case of *Eucaliptus*, processing of RPAS data was complex, due to the low leaves density of *Eucaliptus* canopies, which makes it hard to distinguish them from the land cover. This makes the automatic extraction of crowns very difficult (Fig. 25.5).

Due to the previously mentioned tree density problem, this pilot has been working with images provided by airborne cameras and published by the Spanish National Geographic Institute as Spanish National Plan PNOA[4] orthophotos.

The analysis of historic RGB and NIR images to analyze the evolution of the disease impact at the study site was considered of great interest and priority in Spain. Therefore, efforts were focused on developing a methodology for the automatic/semi-automatic detection of surviving trees. The methodology should be affordable and capable of detecting dead/surviving trees on a multitemporal and regional scale.

---

[3] https://www.databiohub.eu/registry/#service-view/MADchange/0.0.1.

[4] Plan Nacional de Ortofotografía Aérea—Aerial orthophotos National Plan.

Fig. 25.4 RGB (left), multispectral (center), and NDVI (right) mosaics of the study area derived from RPAS data

Fig. 25.5 RGB mosaic of the study area derived from RPAS data (left); detail of *Eucaliptus* canopies (right), where their low leaves density can be seen

This work was performed in collaboration with the Spanish Ministry of Agriculture and Environment MAPA (Área de Recursos Genéticos Forestales) in the framework of the "Phytophthora Working Group" coordinated by MAPA. All Spanish Autonomous Communities affected by Phythopthora participate in this Working Group, as well as research centers and universities.

The general methodology designed by DataBio and proposed to MAPA for the automatic/semi-automatic detection of surviving trees using is shown in Fig. 25.6.

The data processing steps applied were:

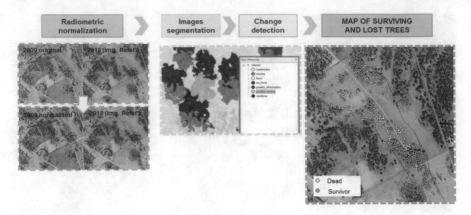

Fig. 25.6 General methodology proposed for the assessment and monitoring of *Phythophthora* in dehesas at a regional scale

- A radiometric normalization of the historic series of images with different acquisition dates to allow multi-temporal analysis. This is a highly resource-consuming process; however, the "Image Enhancer Framework" mentioned in chapter 212 was used.

- An object-based image analysis (OBIA) algorithm for automatic/semi-automatic detection of surviving trees using aerial images was developed (see Fig. XX). This algorithm was employed for generating a mask based on the segmentation and classification of tree crowns from each image set (2009 and 2018). The OBIA algorithm uses image segmentation techniques, grouping pixels into homogeneous areas named segments or objects. This process takes into account spectral, textural, neighborhood, and shape parameters in the identification of tree canopies from multispectral RPAS images and orthophotos. In a second step, objects are classified from the vegetation indices in order to obtain a trees/no trees mask ("tree crowns mask"). This binary mask is used to define the objects of interest and extend the diagnostic model (Fig. 25.7).

- A change detection analysis among the two historic sets was performed by comparing the "tree crowns mask" from different dates (Fig. 25.8), thus allowing identification of surviving and dead trees. Finally, a shapefile was obtained containing the location of surviving/dead *Quercus ilex* trees in ten study areas.

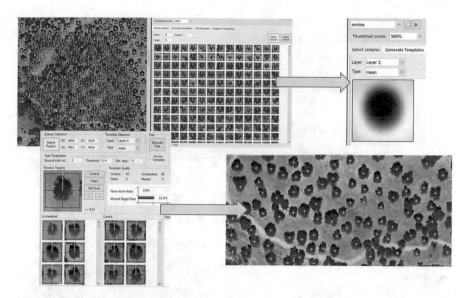

Fig. 25.7 Training of the OBIA algorithm employed for the semi-automatic detection of individual trees

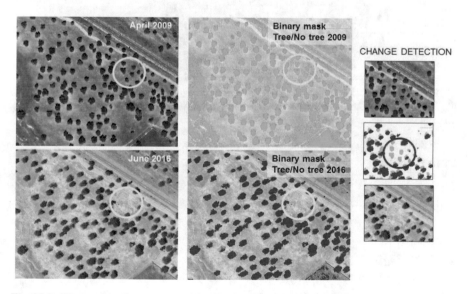

Fig. 25.8 Change detection process to identify dead/surviving *Quercus ilex* trees

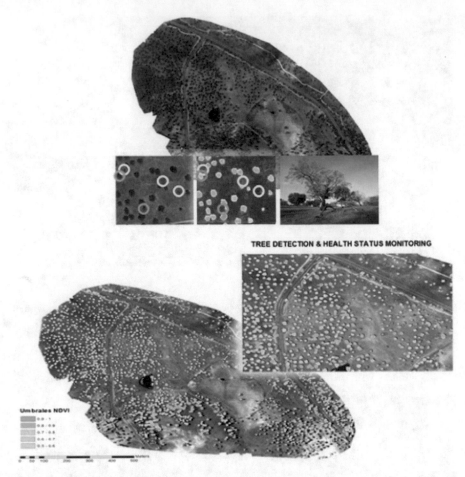

Fig. 25.9 Mosaic of the study area derived from RPAS data and location of trees (upper figure); map of tree status marked with different colors (lower figure)

Data visualization and presentation:

The results of this pilot are typically raster images as shown in Fig. 25.9. Those images show indexes as NDVI, for example, using a choropleth schema.

25.3.2 Data Used in the Pilot

As explained, we used massive and frequently updated data, like Earth observation data and RPAS data from different sources:

- SENTINEL-2: Earth observation data owned by the European Space Agency (ESA).
- Aerial Photograph or Orthophotos: Earth observation data in image format obtained from the National Geographic Institute of Spain.
- RPAS: The RPAS data collected were obtained using a eBee+ platform with SODA camera and multispectral Sequoia camera (Green, Red, RedEdge ,and NIR bands) over the study site.

25.3.3 Reflection on Technology Use

Regarding RPAS and field data, the following results and conclusions have been reached:

- Spatial resolution: it is necessary to use very high resolution (VHR) images ($\leq$50 cm), which allow the identification of individual trees.
- Spectral resolution: it is necessary to use information from the Infrared wavelengths, which allows to assess the status of vegetation.
- Temporal resolution: the evolution of the disease does not require a very high temporal resolution. The update frequency of the Spanish National Plan PNOA (2–3 years), with data available from 2005, is enough.
- A radiometric normalization process is mandatory to be able to work with RGB-NIR images with different acquisition dates in the historic database.

It can be concluded that the use of RPAS is interesting for monitoring *Phytophthora* outbreaks at a local scale. For big areas (the area potentially affected is the half South of the Iberian Peninsula), the use of PNOA aerial photography is proposed. These are very interesting results from the management point of view, as PNOA is a free periodic product provided by the Spanish Public Administration.

With big data tools already available, the methodology developed could be extended to a lot bigger "dehesa" areas, so that it would be possible to estimate the number of *Quercus ilex* trees lost in a period of time in a certain province and region. The methodology is very interesting for the periodical monitoring of the vigor status of "dehesas" (analysis of progression/regression of *Quercus* forests, detection of new outbreaks).

Regarding *Eucalyptus* damages, using RPAS & field data give results that so far allow us to state that it is possible to assess defoliation and assign treatment priorities at tree level, the low density of Eucalyptus crowns and the variety of the understory makes it impossible to automatically detect trees, so this task cannot be automated. The goal of extending the model to plantation level and obtaining risk maps has consequently not been reached.

When using Sentinel and field data, no correlation was found between Sentinel data and defoliation produced by *Gonipterus*.

25.4   Business Value and Impact

25.4.1   Business Impact of the Pilot

The pilot reached its defined business goals, and this was validated by a set of KPIs supporting the exploitation potential of the technology pipeline (Table 25.1).

25.4.2   Business Impact of the Technology on General Level

See Table 25.2.

25.4.3   How-to-Guideline for Practice When and How to Use the Technology

The methodologies developed by this DataBio pilot are very useful for monitoring dense forest stands; however, the utility shown when applied to scattered stands (Holm oak) or sparse trees (*Eucalyptus*) has been less. Therefore, this pilot can be considered a good demonstrator of the limits of current technology.

It should be noted that drone flights can be relatively expensive applied to large areas; therefore, the developed methodologies can be applied in two different ways:

* With drone data for reduced areas (At plot level)
* With satellite data on dense masses (at regional or national level).

25.5   Summary and Conclusions

The pilot explained in this chapter shows how it is possible to use field data combined with drone images to obtain relationship equations between the different pixel data and the state of health of forest stands. Once these local models are obtained, it is possible to extend them to larger areas at the regional or national level.

Also, if there is a big gap in resolution between satellite and drone data, we have seen how it is possible to design debugging and improvement methods for orthophotos.

In conclusion, the technical results have been very interesting, but the choice of species, despite its economic interest, has led us to work on the edge of Earth observation technologies.

All this information is utterly developed in DataBio public Deliverable D2.3 Forestry Pilots Final Report [2].

Table 25.1 KPIs of the pilot application

KPI short name	KPI description	Goal description	Base value	Target value	Measured value	Unit of value	Comment
Surface	Surface analyzed and processed	Plots analyzed	0	Municipalities	Several plots	Plots	Number of hectares of forest land monitored using different EO data sources. Satellite data has not enough definition. In the last period, some problems have arisen related to RGB and NIR images
New protocols	Generation of new protocols for diseases control	Number of protocols	0	2	2	Protocol	
New products	Generation of new products	Number of products	0	2	2	Product	
Users	Estimation of potential users	4 PPAA	0	2	2	Customer	

Table 25.2 Pilot results

Results	Pilot exploitation
The use of RPAS is interesting for monitoring *Phytophthora* outbreaks at a local scale	**Product**: maps of vigor/decay status from field & MS data from RPAS flights Reliable solution for forest managers of small surfaces
For big surfaces (the area potentially affected is the half South of the Iberian Peninsula), the use of aerial images is proposed for monitoring *Phytophthora* at a regional scale	**Product**: map of surviving /dead trees Reliable solution for forest managers from the Public Administration These works have been developed in collaboration with the Spanish Ministry of Agriculture and Environment MAPA in the framework of the "Phytophthora Working Group" coordinated by MAPA. All Spanish Autonomous Communities affected by Phythopthora participate in this Working Group, as well as research centers and Universities A trial has been presented to the Working Group, and it has been considered of great interest. It is being assessed in order to apply the methodology to different study areas in Spain, and could be extrapolated to the whole area affected by *Phytophthora* in Spain and Portugal
Limitations encountered for the operational application: need of space and IT resources for processing the PNOA aerial photograph (RGB-NIR), radiometric normalization, image segmentation (eCognition). There is also the need of developing more automatic processes	Technology solutions available and ready to be implemented within the framework of DataBio project and consortium

Acknowledgements The authors would like to thank everybody who contributed to the work presented in this chapter. Particularly, we would like to acknowledge the inputs of Ence-Energía & Celulosa and the participation and support of TRAGSA Group Managers: Clara Alonso Fernández-Coppel, Rosario Escudero Barbero, Mariano Navarro de la Cruz, Manuel López Hernández and Luis Ocaña Bueno.

We also want to acknowledge the following organizations for their provision of open source EO and field datasets: the European Space Agency, Instituto Geográfico Nacional de España and Ministerio de Agricultura, Pesca y Alimentación (Subdirección de Política Forestal y Lucha contra la Desertificación)

References

1. DataBio Website, https://www.databio.eu/en.
2. DataBio public deliverable D2.3 Forestry Pilot Final Report4.1 Platform and Interfaces, https://www.databio.eu/wp-content/uploads/2017/05/DataBio_D2.3-Forestry-Pilot-Final-Report_v1.1_2020-03-04_VTT.pdf.

Chapter 26
Monitoring of Bark Beetle Forest Damages

Petr Lukeš

Abstract In this chapter, we present a multi-source remote sensing approach for country-wise monitoring of bark beetle calamity to support government decision making processes. In the first part, we describe the forest health monitoring system, which is based on the analysis of satellite big data–Sentinel-2 observations collected every five days. We propose an automated processing chain for high-quality cloud-free image synthesis for user-defined acquisition periods. Such a processing chain is applied to yield yearly cloud-free images of the entire Czech Republic from 2015 onwards. Based on this data, we assess forest health trends using Sentinel-2 derived vegetation indices and in situ data of forest status. Finally, we demonstrate the benefits of multi-source remote sensing for timely and objective mapping of bark beetle spread by combining several data sources, including planet high-resolution satellite data, Sentinel-2 forest health maps and other maps of forest conditions. Detected bark beetle sanitary logging and dead standing wood polygons are used by the Ministry of Agriculture of Czech Republic in their decision processes regarding the management of affected forest areas.

26.1 Introduction, Motivation and Goals

In recent years, there is significant forest health decrease in the Czech Republic (Fig. 26.1), with similar trends of rapid increase in forest harvested area observed also for other European countries [1]. Forest loss and forest health decay can be attributed to various factors, both biotic and abiotic. These are independent of the forest owner and his/her management practices, resulting in loss of forest value compared to the unaffected forests (Fig. 26.2). One of the serious obstacles for finding a solution to the situation is the lack of timely and objective information about the forest conditions, especially for forest plots of small sizes. Under the ongoing bark beetle calamity, such information should be ideally updated multiple times a year,

P. Lukeš (✉)
Forest Management Institute of the Czech Republic, Nádražní 2811, Frýdek-Místek, Czech Republic
e-mail: lukes.petr@uhul.cz

© The Author(s) 2021
C. Södergård et al. (eds.), *Big Data in Bioeconomy*,
https://doi.org/10.1007/978-3-030-71069-9_26

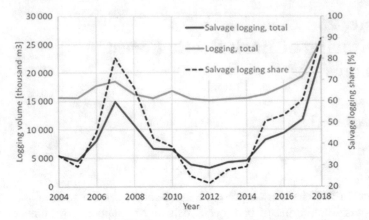

Fig. 26.1 Percentage share of salvage logging (dashed line—secondary y-axis on right) to total logging (green line—primary y-axis on left) and salvage logging (red line—primary y-axis on left)

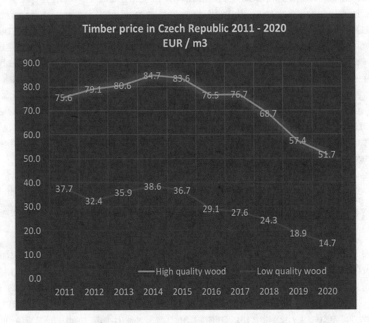

Fig. 26.2 Decrease of timer price in Czech Republic between 2011 and 2020 (green—high quality wood, blue—low quality wood) as a result of oversupply of raw wood due to the bark beetle calamity

ideally as a convenient Web service for a broad range of users and the government decision making. As an example, the government in Czech Republic compensates the forest damage and potential loss to forest owners by the means of direct subsidies and indirect tax reliefs. In order to correctly identify the affected forest owners and their eligibility for subsidies/tax relief, the Ministry of Agriculture of Czech Republic must precisely spatially locate the affected areas. For this purpose, the field surveys were traditionally used. These are, however, local, costly, and subjective.

The main goal of this pilot is the development of Web-based mapping services for government decision making in the field of forestry which would help in the ongoing unprecedented outbreak of bark beetles. The services should objectively describe the current health status of the forests and allow for timely pro-active management in the forests with regards to the allocation of both the harvesting resources and finances into the most affected regions.

26.2 Pilot Set-Up

We developed a processing chain for satellite data interpretation for forest health assessment and started its routine deployment at FMI's infrastructure. Country-wise forest health trends are obtained in two relatively independent steps:

Step 1—Sentinel-2 satellite data preprocessing and cloud-free mask synthesis.

Step 2—Retrieval of absolute values of forest leaf area index (LAI) and its trends.

The key to assessing the health status of forests from remote sensing data is the availability of a high quality (i.e., cloud-free) image mosaic that is generated from all-available Sentinel–2 satellite observations. This is a basic prerequisite for any remote sensing data interpretation. The methodology for forest health assessment proposed here presents a novel processing chain for automated cloud-free image synthesis based on the analysis of all available Sentinel–2 satellite data for a selected sensing period (e.g., the vegetation season from June to August) via three successive processes:

(1) batch downloading of all-available Sentinel-2 observations,
(2) atmospheric corrections of raw images (so-called L2 process), and
(3) automated synthetic mosaic generation (so-called L3 process, or space-temporal image synthesis (see Fig. 26.3)).

Due to its high computational and data storage requirements, the processing chain is implemented on IT4Innovations supercomputer facility (© 2018 VŠB-TU Ostrava), which enables for distributed computing on many computational nodes. In the first step, Sentinel-2 scenes are automatically downloaded from Copernicus Open Access Hub (global Copernicus data access point) and CESNET (collaborative ground segment of Copernicus implemented in Czech Republic). Next, the atmospheric and topographic corrections are performed for each Sentinel-2 image using sen2cor tool (ESA). Then, each pixel in image mosaic is evaluated independently in the time series of images. Selection of the highest quality pixel, having lowest cloud cover and being in vegetation growing season, is based on a decision tree using the values of vegetation index sensitive to biomass (e.g., the normalized difference vegetation index, or NDVI). In addition to highest NDVI value, several other rules are applied in the form of a decision tree: these include cloud masking

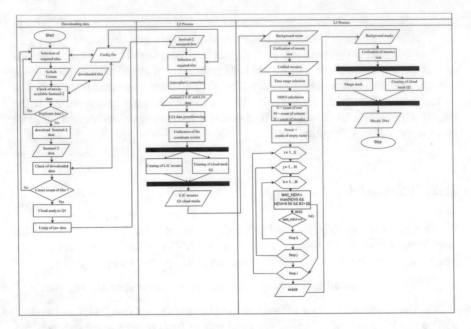

Fig. 26.3 Flowchart of the processing chain from automated production of cloud-free satellite images from source Sentinel-2 observations

and a priori assumptions on reflectance range in visible and near-infrared regions. An example of the synthetic cloud-free mosaic and the individual dates is shown in Fig. 26.4.

In the presented methodology, health status is not assessed as absolute amount of leaf biomass (having LAI as proxy for leaf biomass), but as its change over time. The basic premise is that the health status can be objectively determined only by observing the relative change in LAI over time. In the first step, we calculate selected vegetation indexes (e.g., normalized difference vegetation index—NDVI, red edge inflection point—REIP, and normalized difference infrared index—NDII), and image transformations (e.g., components of tasseled cap transformation) and compare their sensitivity against in situ data from sampled plots (e.g., LAI and ICP Forests plots). For each dataset, linear regression models between in situ data and Sentinel-2 indices were calculated and evaluated. For indices yielding best linear fit, the neural network was trained and applied per pixel to retrieve prediction LAI maps.

In summary, we propose a forest health classification system, which will evaluate forest health on pixel level as a change in LAI values over time and classify each pixel in the following five categories:

I.   Significant increase: increase in LAI by 1.5 and higher,
II.   Moderate increase: increase in LAI from 0.5 to 1.5,
III. Stable conditions: change of LAI between −0.5 and 0.5,
IV.   Moderate decrease: decrease in LAI from −1.5 to −0.5,

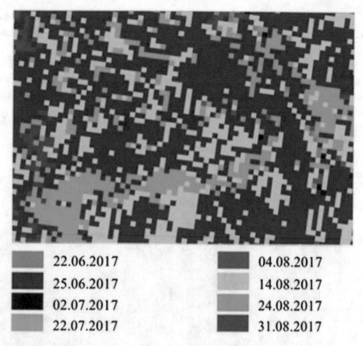

	22.06.2017		04.08.2017
	25.06.2017		14.08.2017
	02.07.2017		24.08.2017
	22.07.2017		31.08.2017

Fig. 26.4 Automated per-pixel selection of best quality observation from the time series of Sentinel-2 observations

V.   Significant decrease: decrease in LAI higher than −1.5.

The countrywide assessment of forest health is carried out on cadastral level, where the area of forest stands of classes IV and V are evaluated for the total forest area of cadastre for stands of age between 0 and 80 years. This condition is put due to the fact that it is not possible to distinguish between sanitary logging and planned logging for old-grown forests—both will be reflected by a sharp decrease in LAI values. Each cadastre is assigned to one of the following categories:

I.   Category 1: 0–5% of class IV and V forests—healthy stands,
II.   Category 2: 5–10% of class IV and V forests—predominantly healthy stands,
III.   Category 3: 10–15% of class IV and V forests—moderate conditions of stands,
IV.   Category 4: More than 15% of class IV, and
V.   forests—damaged stands.

Maps of retrieved LAI from 2015 to 2018 and the between year changes are being routinely published on FMI's mapserver (https://geoportal.uhul.cz/mapy/Map yDpz.html). This allows easy access of the maps for end users—stakeholders in the forestry sector in Czech Republic (Ministry of Agriculture, Forests of the Czech Republic, Military forests, etc.). This Web-based mapping solution is capable of combining different map sources for background map layer (topographic maps, orthophotographs, base maps, cadastral maps) on both desktop and mobile Web browsers (user geolocation is available for mobile platforms). Example of a country-wide LAI map for 2018 is shown in Fig. 26.5, LAI change between 2018 and 2017 in Fig. 26.6.

The forest health maps were also published as Web-mapping service (WMS) on FMI's mapserver (WMS URL: https://geoportal.uhul.cz/wms_dpz/service.svc/get). This allowed dissemination of the results to the broad forestry community (Fig. 26.7). The following map layers were made available:

- Leaf area index maps of 2015, 2016, 2017, 2018, 2019, and 2020.

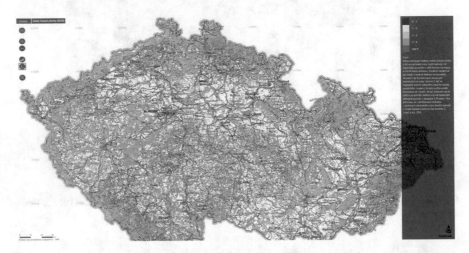

Fig. 26.5 Example of FMI's mapserver with forest health map of 2018

Fig. 26.6 Example of FMI's mapserver with forest health map trends between 2017 and 2018 (change in leaf area index)

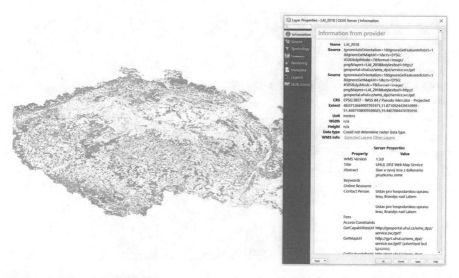

Fig. 26.7 Web-mapping service of forest health maps (Leaf area index map of 2018) running in QGIS 3.6.0 environment

- Leaf area index change maps of 2016–2017, 2016–2018, and 2017–2018.

In addition to routine publication of LAI maps and its trends on a Web-based portal, the data has been used by the Ministry of Agriculture for allocation of available harvesting resources to fight the unprecedented bark beetle outbreak that forests in Czech Republic are currently undergoing. Here, the Sentinel-2 based LAI maps were combined with the timely clear cut and standing dead wood detection from planet commercial satellite data of high spatial (<5 m) and temporal (daily), canopy height model of stereo-orthoimagery, and tree species map to identify the most affected cadastres, where the sanitary logging occurs. Combination of these unique data sources allowed us to detect the recent salvage logging and dead wood in mature spruce forests—areas affected by the unprecedented bark beetle calamity in Czech Republic (see Fig. 26.8 for more details).

The resulting analyses—polygon layers of timely detection of salvage logging and dead wood—are published on (1) FMI's mapserver, (2) WMS service of the mapserver, and (3) specialized Web-based portal "Kurovcovamapa.cz" (see Fig. 26.9).

26.3 Business Value and Impact

The Web-mapping service for government decision making will in the future be extended to allow publication of the results to the broad forestry community of

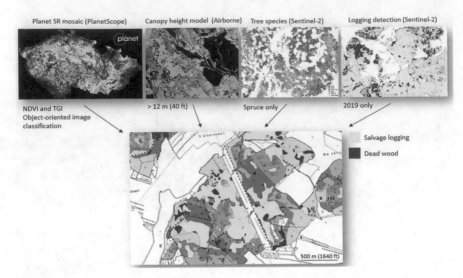

Fig. 26.8 Workflow of multi- source remote sensing approach for bark beetle monitoring (www. kurovcovamapa.cz)

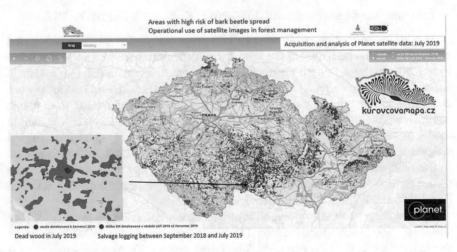

Fig. 26.9 Web-based portal "Kurovcovamapa.cz" for the broad public allowing easy access to timely information in the bark beetle calamity in Czech Republic

the Czech Republic as dedicated WMS services and specialized map portals (e.g., "Kurovcovamapa.cz"). According to those maps, the Ministry of Agriculture of Czech Republic issued a "Public decree"—a legislation instrument to help forest owners by reducing the regulation of their obligations under the Czech forest law, so that they can manage the bark beetle calamity in the most affected regions (Fig. 26.10). The decree is regularly updated several times per year to reflect the

Ministerstvo zemědělství
Odbor hospodářské úpravy a ochrany lesů

18918/2019-MZE-16212

Spisová značka.:   14LH7893/2019-16212
Č.j.:   18918/2019-MZE-16212

000311166084

Vyřizuje: JUDr.Ing. Jiří Staněk, CSc.
Telefon: 221812381
E-mail: Jiri.Stanek@mze.cz
ID DS: yphaax8

Adresa: Těšnov 65/17, Nové Město, 110 00 Praha 1

V Praze dne : 3. 4. 2019

VEŘEJNÁ VYHLÁŠKA

OPATŘENÍ OBECNÉ POVAHY

Ministerstvo zemědělství jako ústřední orgán státní správy lesů, věcně příslušný podle ustanovení § 49 odst. 2 písm. e) zákona č. 289/1995 Sb., o lesích a o změně a doplnění některých zákonů (lesní zákon), ve znění pozdějších předpisů (dále jen „lesní zákon"), v souladu s § 171 a násl. zákona č. 500/2004 Sb., správní řád, ve znění pozdějších předpisů (dále jen „správní řád"), vydává podle § 51a lesního zákona následující

opatření obecné povahy,

kterým Ministerstvo zemědělství rozhodlo o následujících opatřeních odchylných od ustanovení § 31 odst. 6, § 32 odst. 1 a § 33 odst. 1 až 3 lesního zákona:

1. V lesích na území České republiky, s výjimkou lesů na území národních parků a jejich ochranných pásem, se stanoví, že na kůrovcové souše se až do 31. prosince 2022 nevztahuje povinnost vlastníka lesa přednostně zpracovat těžbu nahodilou; povinnost vlastníka lesa aktivně vyhledávat kůrovcové stromy, provádět jejich včasnou těžbu a účinnou asanaci zůstává zachována;

2. V lesích na území, které je tvořeno katastrálními územími, jež jsou uvedena v příloze č. 1 tohoto opatření obecné povahy, která je jeho nedílnou součástí

2. 1. se stanoví, že vlastník lesa není povinen používat jako obranná opatření lapače a klást lapáky; povinnost vlastníka lesa aktivně vyhledávat kůrovcové stromy, provádět jejich včasnou těžbu a účinnou asanaci zůstává zachována;

Fig. 26.10 Legislative instrument (the public decree) with an annex that defines the list of cadastral units selected according to the analyses presented in this chapter

actual situation of the forests (Fig. 26.11). All these measures will help reduce the overall loss for forest owners due to climate change and the ongoing bark beetle calamity in the Czech Republic.

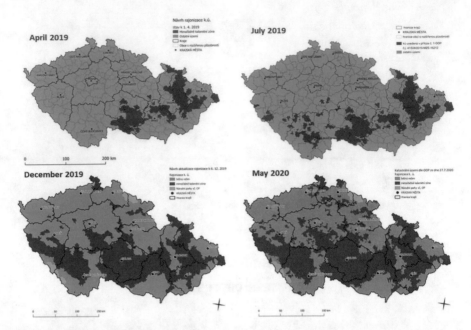

Fig. 26.11 Cadasters of bark beetle calamity identified using multi-source remote sensing approach, green—cadasters unaffected by bark beetle calamity, red—calamity zones of bark beetles. Cadasters are updated typically three times per year in the vegetation season

26.4   Conclusions

The exploitation of described DataBio pilot results was successfully achieved via:

- Map portal of the Forest Management Institute (https://geoportal.uhul.cz/mapy/MapyDpz.html)
- Web-mapping service for online publication of forest health layers in GIS and Web environment (https://geoportal.uhul.cz/wms_dpz/service.svc/get)
- Specialized Web portal "Kurovcovamapa.cz" (https://www.kurovcovamapa.cz/)

Moreover, based on the timely detection of recent salvage logging and dead wood, the Ministry of Agriculture of the Czech Republic issued a public decree to apply different forest management regimes in the areas with ongoing beetle calamity. The areas are updated regularly and are based on the outcomes of the DataBio pilot. Presented work thus demonstrates the potential of the integration of multi-source remote sensing (e.g., the multispectral Sentinel-2 data, high spatial and temporal resolution planet data, and the ancillary forestry data) for the decision making in the forestry sector.

Reference

1. Ceccherini, G., Duveiller, G., Grassi, G. et al. (2020). Abrupt increase in harvested forest area over Europe after 2015. *Nature 583,* 72–77. https://doi.org/10.1038/s41586-020-2438-y.

Chapter 27
Conclusions and Outlook—Summary of Big Data in Forestry

Jukka Miettinen and Renne Tergujeff

Abstract In this chapter, we summarize the findings from the forestry pilots conducted during the DataBio project. Although the pilots demonstrated the functionality of big data in forestry through several practical applications and services, they also highlighted areas where further development is needed. More effort is needed particularly in ensuring smooth connections between the technical components of the processing pipelines, as well as designing the best business solutions within the big data service chain and between the service providers and users. Overall, the challenge for the coming years is to establish operational big data processing pipelines that meet the requirements and expectations of forestry stakeholders.

27.1 Introduction

As discussed in the forestry introduction chapter (Chap. 22), new technologies that have emerged over the past decade enable utilization of novel big data approaches in forest monitoring. At the same time, the requirements for forest monitoring information have widened. Indicators of carbon balance, biodiversity, and forest health, to name just a few, have an increasingly important role in forest management, alongside the traditional forest characteristics (e.g., height, volume, species distribution). The forestry pilots of the DataBio project set out to investigate and demonstrate ways to maximize the benefits of big data in forestry, providing users with timely datasets and analysis results that would meet their specific information requirements.

The preceding chapters (Chaps. 23–26) have presented four DataBio pilots utilizing big data for forest monitoring and management. The use of datasets varied from crowdsourced field data to satellite observations. The selected use cases included forest structural variable estimation (e.g., tree height and basal area), health monitoring as well as bark beetle and storm damage mapping. Geographic coverage of the use cases varied from local forest estate level to national level. Stakeholders involved in the pilots ranged from Earth observation (EO) service companies

J. Miettinen (✉) · R. Tergujeff
VTT Technical Research Centre of Finland, 02044 VTT, Espoo, Finland
e-mail: jukka.miettinen@vtt.fi

© The Author(s) 2021
C. Södergård et al. (eds.), *Big Data in Bioeconomy*,
https://doi.org/10.1007/978-3-030-71069-9_27

and private forestry businesses to government organizations and academic institutions. Technical solutions included local processing, interlinked cloud storage and processing platforms, and online user interfaces.

As much as the DataBio project pilots confirmed the usability of big data in forestry and the functionality of already existing technical solutions, they also revealed some weak points in the value adding chain where more effort is needed to fully utilize the potential of big data in the forestry sector. In the following, the lessons learnt from the DataBio forestry pilots have been synthesized from the perspective of (1) technical solutions and (2) business solutions. In both of these areas, functioning solutions for individual components within big data value adding chain already exist, but more effort is needed in smooth connections between the components, as well as the interface between the big data service providers and users.

27.2 Lessons Learned from DataBio: Technical Solutions

Due to the high data volumes and processing requirements in big data analysis, traditional data processing and analysis approaches (i.e., image-by-image analysis on personal computers) are not sufficient to fully exploit the benefits of the data. Redesign of the processing and delivery pipelines was needed to match with today's data volumes and modern processing infrastructures. In the DataBio project, big data processing chains were divided into four main steps:

1. Acquisition and storage
2. Preparation
3. Analytics
4. Visualization and user interaction.

As presented in Chaps. 23–26, functioning solutions for all of these individual steps in the forestry sector were identified, developed further and piloted. The Finnish Forest Centre demonstrated the functionality of their crowdsourcing field observation application, allowing innovative data acquisition for big data applications concentrating on forest damage monitoring. VTT and Spacebel demonstrated the usability of their online platforms Forestry TEP (https://f-tep.com/) and EO Regions! (https://www.eoregions.com/) for data preparation and analytics. TRAGSA, Technical University of Denmark and the Forest Management Institute (Czech Republic) exhibited the usability of their data analytics algorithms with EO big data. Many of these activities were linked with Wuudis (https://www.wuudis.com/), a commercial service for forest owners, timber buyers and forestry service companies, providing, e.g., visualization tools and supporting the linkage between the users and big data service providers. Likewise, the Forestry TEP and EO Regions! have their own user interfaces, and the Forest Management Institute (Czech Republic) provided their maps through the online service "Kůrovcová mapa" (www.kurovcovamapa.cz).

From a technical perspective, the weakest link in big data utilization in the forestry pilots was considered to be the connection between different datasets, platforms, and applications. As the legacy of traditional, localized processing approaches, many processing and analysis applications are optimized to work with locally stored datasets. Although in some cases, typically in large institutions with sufficient storage and processing power, big data-based operational systems may be set up locally, the only way to fully and effectively unleash the benefits of big data for the wider forestry stakeholder community is through interplatform connections. The aspects of inter-platform operability will need to be developed further in the future to ensure that technical difficulties do not start to hinder further uptake of big data solutions by forestry stakeholders.

In the DataBio forestry pilots, the interplatform connections based on established infrastructural configurations, like the connection between Forestry TEP processing and analysis platform and the CREODIAS (https://creodias.eu/) data and storage platform, worked well. However, various technical problems were encountered in connections between independently operated processing, application or visualiza-tion systems. Further development of smooth interoperability of different platforms should be a key goal for technical development in the near future. The future big data solutions, covering storage, processing, analysis and visualization capabilities, would optimally lean on interconnected online platforms. Large storage and processing facilities on the cloud, like the Copernicus Data and Information Access Services (DIAS), will provide the core EO and supporting datasets. Other national or interna-tional databases may store, e.g., field data or other auxiliary datasets usable specif-ically in the forestry sector. Forestry application platforms, like the Forestry TEP and EO Regions!, will provide processing tools, algorithm development interfaces and ready-made products with user interfaces optimized for forestry stakeholders. Further still, these platforms can provide analysis results (e.g., on structural forest characteristics, damages or forest health) to various forestry services (like Metsään.fi or Wuudis), which utilize up-to-date data for forest management and operations planning and user interaction. This entire service chain would benefit from smooth interplatform operability.

27.3 Lessons Learned from DataBio: Business Solutions

In the DataBio forestry pilots, big data solutions were piloted with the users of an online forest management support platform Wuudis, forestry sector partners, asso-ciated partners and other stakeholders, to evaluate the business potential and end user interest in the products. The pilots demonstrated a high demand for frequently updated forestry information on forest structural characteristics (e.g., tree height, basal area), forest health, storm damages, and other. However, it also became evident that significant progress is needed in the business practices and market development. Need for improvement was identified in two major areas, before the full potential of big data in forestry can be efficiently unleashed:

1. Business practices within the big data value chains.

2.   Operating practices in the forestry sector.

The big data economy with multiple operators working together in a single value chain is something new to most forestry sector stakeholders. Whether it be a national database that has opened up to the public, a commercial EO satellite operator, a processing and application platform operator or a private forestry company (to name only a few stakeholders), they all need to define the value of their work and information in a new way. This process takes time. A typical example of the new complications is the interplatform operations. They do not create only technical challenges, but they also require new types of business arrangements.

In the DataBio forestry pilots, progress was made in creating business connections between the platforms involved. A business agreement was set up between Wuudis Solutions and Spacebel, regarding the distribution of the Wuudis services to the forest users of the EO Regions! platform and the commercialization of Spacebel's earth observation forest products in the Wuudis platform. This type of business arrangements is needed between collaborating services or other big data providers who operate on a commercial basis, before the technical benefits discussed in the previous section will materialize. Delays in setting up business agreements will slow down the uptake of big data in the forestry sector as surely as technical problems.

The other major area of challenge in the field of business is the slow development of management practices in the forestry sector in many countries. Traditional management practices are largely based on manual field work and static management plans. It may take some time to convince forestry stakeholders of the benefits of big data for their operations. This is best achieved by providing high quality services and products that meet the requirements of the stakeholders. For this, it is essential to (1) know the requirements of the stakeholders to be able to provide the right kinds of products, (2) create smooth user experience when accessing and using the information/products, and (3) actively promote the possibilities of big data in the user community. Local promotional activities and locally designed services are in an essential role in marketing, due to the varying forestry practices in different countries. Over time, forestry stakeholders will realize the benefits of online service provision of frequently updating information based on big data.

27.4   Future Outlook

Overall, the DataBio forestry pilots (1) demonstrated the usability of big data in forestry through several practical applications and services and (2) highlighted areas where further development is needed to increase the benefits of data-driven solutions for forestry stakeholders. Although big data solutions in forestry are far from being fully developed, it is clear that big data is here to stay. The technological development that has already enabled the collection of massive data volumes from both remote sensing and field measurements, and their processing on online platforms, will only accelerate in coming years. The information available to be extracted from the massive volumes of data is too valuable to be ignored by the forestry sector. The challenge for the coming years is to establish operational big data processing pipelines that meet the requirements and expectations of forestry stakeholders.

The future will tell how fast big data solutions will replace traditional practices in forestry. In some countries, this may require even legislative changes, to allow utilization of remote sensing based solutions in official reporting. In any case, the great benefits of big data to the forestry sector are clear. At the same time, the reporting and monitoring requirements are constantly increasing with growing demands, e.g., on forest carbon flux and forest management sustainability monitoring. Big data approaches through online platforms provide the means to answer these demands. Big data also provides possibilities for entirely new and exciting types of forest monitoring approaches based on artificial intelligence, which were not yet within the scope of the DataBio project. It is up to all of us forestry stakeholders to find the best solutions to make big data benefit the entire forestry sector, our common environment and the whole world.

Part VII
Applications in Fishery

Chapter 28
The Potential of Big Data for Improving Pelagic Fisheries Sustainability

Karl-Johan Reite, Jose A. Fernandes, Zigor Uriondo, and Iñaki Quincoces

Abstract The use of big data methods and tools are expected to have a profound effect on the pelagic fisheries sustainability and value creation. The potential impact on fuel consumption, planning and fish stock assessments is demonstrated in six different pilot cases. These cases cover the Spanish tropical tuna fisheries in Indian Ocean and the Norwegian small pelagic fisheries in the North Atlantic Ocean. The areas encompassed by these pilots have an annual capture production above 13 million tonnes.

Fisheries provide jobs and income to coastal communities and are expected to contribute to long-term European food security and economic growth. No other bioeconomy sector appears to be as regulated and monitored as fisheries, with numerous data inputs collected (i.e., catch and effort, stock sampling, ocean environment, fishing vessel activity, sales and transactions) to better understand and control the industry. Still, there is little coordinated use of big data technologies in the sector.

Fuel consumption is a challenge for most fisheries, as it represents 60–70% of the total annual costs of a vessels' activity [1–4]. Ocean-going pelagic fishing vessels employ both energy efficient gear, such as purse seines, and energy intensive gear, such as trawls. The vessels are frequently searching for fish between fishing operations, since schooling pelagic species are migratory. The vessels have been engineered to become very flexible in their production, routing, and consumption of energy onboard [5], and several methods have been proposed for adapting vessels' operations to these variations [6, 7]. Still, the crew often operate the vessel based on habits and preferences for certain configurations of the power system.

K.-J. Reite (✉)
SINTEF Ocean, Brattørkaia 17 C, 7010 Trondheim, Norway
e-mail: karlr@sintef.no

J. A. Fernandes · I. Quincoces
AZTI, Marine Research, Basque Research and Technology Alliance (BRTA), Herrera Kaia—Portualdea z/g., 20110 Pasaia, Gipuzkoa, Spain

Z. Uriondo
University of the Basque Country (UPV/EHU), Alameda Urquijo s/n, 48013 Bilbao, Spain

© The Author(s) 2021
C. Södergård et al. (eds.), *Big Data in Bioeconomy*,
https://doi.org/10.1007/978-3-030-71069-9_28

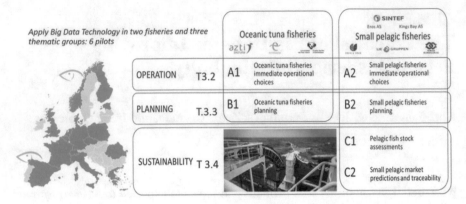

Fig. 28.1 Overview of fisheries pilots

Fishing trip planning and routing are important factors for reducing the fuel consumption within fisheries and achieving better margins. Decisions about when, where, and how to harvest are taken by expert fishers based on their own experience, information gathered from industry contacts and publicly available data. In most cases, such information is limited to meteorological forecasts, catch reports and communication with a small number of collaborating fishermen. The subjectively perceived market development is an important factor for fisheries planning, but there are no tools to assist fishermen in this respect.

Fish stock assessment is traditionally carried out based on measurements from yearly campaigns. These campaigns follow a preestablished pattern and apply both test fishing and hydroacoustic observation to sample the spatial distribution of fish in the ocean. The data from these campaigns are used in statistical models for stock estimation and resource management advice. The International Council for the Exploration of the Sea (ICES) determines quota recommendations for the national authorities, which have jurisdiction over these fish stocks. Great effort is expended in the collection of this critical data, but its spatial and temporal coverage is limited by the associated costs. In addition, consideration of market variations is important for fisheries planning to optimize the value created by fisheries. These market variations are caused by the relationship between supply and demand, which is influenced by multiple factors such as fisheries effort, fish distribution, quotas, weather conditions, competing products, and economic factors.

Part VII focuses on two separate types of pelagic fisheries: The Spanish tropical tuna fisheries in Indian Ocean and the Norwegian small pelagic fisheries in the North Atlantic Ocean. The areas encompassed by these pilots have an annual capture production above 13 million tonnes. Six separate pilot cases have been defined, addressing three separate viewpoints: (i) immediate operational choices, (ii) fishing vessel trip and fisheries planning, and (iii) fisheries sustainability and value, as shown in Fig. 28.1.

The two "Operation" pilots aim at providing crew with information to help them perform current operations in the most energy efficient way possible, while avoiding unscheduled maintenance. To achieve this, the vessels' energy systems are monitored, and various methods are employed to supply crew direct advice.

The goal of the two "Planning" pilots is to provide information that benefits fisheries planning. The information will be provided based on extensive historical datasets of fisheries activity (e.g., VMS, GPS tracking), catch statistics (e.g., logbooks and public records), oceanographic conditions (e.g., SST, salinity, chlorophyll), meteorological conditions, and FAD data (e.g., GPS data, echosounder data, SST). The hypothesis is that large amounts of historical data combined with ocean environment near real-time information can be used to accurately forecast species distribution, by using machine learning or other techniques. This will reduce fuel consumption through targeted effort and more efficient engine operation.

The two "Sustainability" pilots investigate how long-term fish market forecasts can benefit long term fisheries planning, in particular for best timing of different fisheries to maximise economic revenue. These pilots also investigate how oceanographic simulations using fishing fleet data, as an inexpensive biomass and physical property source of measurements, can benefit pelagic fish stock assessments.

These pilots require a large amount of data to reach their goals. In addition to its **volume**, data collected on a large scale from a diverse set of sensors, published records and regional observation systems, also exhibits other unique characteristics as compared with data collected for a single purpose and from a single source. This data is commonly **unstructured** and requires more **real-time** analysis [8]. Many of these aspects are present in the fisheries pilots. The pilots are likely to end up producing over 5 TB of data per year and coming from many different sources. Such sources include earth observations, sensors onboard fishing vessels (i.e., acoustics, machinery, operations), simulations (i.e., meteorological, oceanographic, and marine biology) and human annotations. The update frequency, regularity, and volumes of these sources are on very different scales, affected by simulation times, vessel communications, and satellite orbits. The lack of data acquisition standardization on board vessels and data structuring poses another challenge for these pilots (Table 28.1, Fig. 28.2).

Table 28.1 Data production by DataBio fisheries pilots

Dataset type/variety	Dataset	Volume (GB)	Velocity (GB/year)	Start date
Market data	Catch reports, economic figures	<1 GB	<0.07	20120101
Vessel data (including buoys with sonar data)	ESAS	67	18.8	20140901
	Eros	64	20.5	20140903
	KingsBay	78	19.6	20140826
	LiegFi	1.5	17.1	20180731
	Echebastar immediate pilot data	55	10.0	20140301
	Echebastar UE fleet data	903.8	602.6	20170101

(continued)

Table 28.1 (continued)

Dataset type/variety	Dataset	Volume (GB)	Velocity (GB/year)	Start date
EO, CMEMS, Met	Ocean physics (current/temp)	310	115.5	20160101
	Wave data (open ocean)	283	165.7	20161209
	Biogeochemistry	99.8	37.1	20160101
	Wind	325	121.2	20160101
	Coastal waves (Met)	1203	802	20160918
	Oceanic Tuna EO + research data	938.8	625.9	20170101
SINMOD	Oceanographic modeling (4 km)	385	1752	20180614
	NOAA atmospheric + SINMOD input	N/A	3500	N/A
Hydroacoustics	SIMRAD EK80 series-echosounder	3.3	121.2	N/A
	(SIMRAD SX90 Sonar	N/A	5402	N/A
	Simrad SN90 Sonar + echosounder	317	Per cruise	20180612
WP3 Total	All fishery pilot data assets	5004.2	5815.4	

Fig. 28.2 Echebastar company tuna fishing vessel within the DataBio project

Acknowledgements The authors would like to thank the Norwegian Fishermen's Sales Organization for Pelagic Fish, for providing access to important data and for participating in discussions. The fishing vessel owners Liegruppen Fiskeri, Echebastar, Eros, Ervik & Sævik and Kings Bay, as well as part of their crew, have participated in collecting operational data and discussions about the fishing operations and the fishing industry. Jefferson Murua (AZTI) has provided significant feedback for the improvement and quality of this chapter. The work leading to this chapter has been possible mainly through the DataBio project funded from the European Union's Horizon 2020 research and innovation programme under grant agreement No. 732064.

References

1. Suuronen, P., Chopin, F., Glass, C., Løkkeborg, S., Matsushita, Y., Queirolo, D., & Rihan, D. (2012). Low impact and fuel efficient fishing—looking beyond the horizon. *Fisheries Research, 119*, 135–146.
2. Rojon, I., & Smith, T. (2014). On the attitudes and opportunities of fuel consumption 512 monitoring and measurement within the shipping industry and the identification and 513 validation of energy efficiency and performance interventions, p. 18.
3. Parker, R. W., & Tyedmers, P. H. (2014). Fuel consumption of global fishing fleets: Current understanding and knowledge gaps. *Fish and Fisheries, 16*(4), 684–696.
4. Fernandes, J. A., Santos, L., Vance, T., Fileman, T., Smith, D., Bishop, J. D., et al. (2016). Costs and benefits to European shipping of ballastwater and hull-fouling treatment: Impacts of native and non-indigenous species. *Marine Policy, 64*, 148–155.
5. Aursand, I. G., Digre, H., Ladstein, J., Kyllingstad, L. T., Erikson, U. G., Tveit, G. M., Backi, C. J., & Reite, K. J. (2015). Development and assessment of novel technologies improving the fishing operation and on board processing with respect to environmental impact and fish quality (DANTEQ).
6. Reite, K.-J., Ladstein, J., & Haugen, J. (2017). Data-driven real-time decision support and its application to hybrid propulsion systems. In *Proceedings of the International Conference on Offshore Mechanics and Arctic Engineering—OMAE* (Vol. 7B-2017, p. V07BT06A024--V07BT06A024). https://doi.org/10.1115/OMAE201761031
7. Skjong, S., Kyllingstad, L. T., Reite, K. J., Haugen, J., Ladstein, J., & Aarsæther, K. G. (2019). Generic on-board decision support system framework for marine operations. In *Proceedings of the international conference on offshore mechanics and arctic engineering—OMAE*.https://doi.org/10.1115/OMAE2019-95146

8. Hu, H., Wen, Y., Chua, T. S., & Li, X. (2014). Toward scalable systems for Big Data analytics: A technology tutorial. *IEEE Access, 2,* 652–687.

Chapter 29
Tuna Fisheries Fuel Consumption Reduction and Safer Operations

Jose A. Fernandes, Zigor Uriondo, Igor Granado, and Iñaki Quincoces

Abstract This chapter demonstrates the potential of tuna fishing fleets to reduce their fuel oil consumption. In the "Oceanic tuna fisheries, immediate operational choices" pilot, the data monitoring system on vessels periodically upload data to the server for shore analysis. The data analytics employs fuel oil consumption equations and propulsion engine fault detection models. The fuel consumption equations are being used to develop immediate operational decision models. The fault detection models are used to plan maintenance operations and to prevent unexpected engine malfunctions. The data-driven planning software allows probabilistic forecasting of tuna biomass distribution and analysing changes in fishing strategies leading to fuel consumption reduction. These changes in fishing strategies can be summarized as a transition from hunting to harvesting. Vessels do not search for fish, but instead take less risks and fish, where it is more likely that the fish can be found and is easier to capture. Buoy data are increasingly used to improve stock assessments and have the potential to allow better monitoring and planning of fish-quotas fulfilment.

29.1 Introduction

As the catches of tropical tunas have almost reached their limit, this fishery needs to reduce its costs and carbon footprint to achieve objectives such as improved margins and less environmental impact. **Fuel consumption** may represent up to 50% of a tuna vessel's total **operational costs**, thus, representing one of the main concerns for fishing companies [1, 2]. Moreover, world fishing industry emissions per landed fish tonnes have increased by 21% recently [3]. Large pelagic fish, such as tuna species, is highly migratory. Because of this, vessels targeting tuna species **tend to have higher**

J. A. Fernandes (✉) · I. Granado · I. Quincoces
AZTI, Marine Research, Basque Research and Technology Alliance (BRTA), Herrera Kaia—Portualdea z/g., 20110 Pasaia, Gipuzkoa, Spain
e-mail: jfernandes@azti.es

Z. Uriondo
University of the Basque Country (UPV/EHU), Alameda Urquijo s/n, 48013 Bilbao, Spain

© The Author(s) 2021
C. Södergård et al. (eds.), *Big Data in Bioeconomy*,
https://doi.org/10.1007/978-3-030-71069-9_29

and more variable fuel consumption costs than others fishing for coastal species [4]. However, it is also worthwhile highlighting that this sector provides 25,000 direct jobs and 54,000 indirect jobs in the European Union while contributing to food security.

The tropical tuna fishing industry uses Earth observation (EO) data, to characterize the environmental conditions of the surrounding areas to locate fishing grounds with less effort (i.e. time, fuel and consequent costs). High digitalization of tuna vessels means that their capacity to record and to use existing EO data has increased [5]. However, due to the large volume, diversity of sources and quality of recorded data, they are rarely used for further analysis and remain intact and unstructured. *Big data* methodologies seem to be the solution to deal with such large volume of heterogeneous data and turn it into useful information. Solving these problems demands new system architectures for data acquisition, transmission, storage and large-scale data processing mechanisms [6]. *Big data* processing techniques, enhanced by machine learning methods, can increase the value of such data and their applicability for industry and management challenges. *Machine learning* has already proved its potential in marine sciences applied to fisheries forecasting [7–9]. However, big data use by the fishing industry is behind the state-of-the-art and day-to-day applications when compared with the other shipping industries [10–12].

The aim of oceanic tuna fisheries pilots is to improve economic sustainability of oceanic tuna fisheries while reducing their emission footprint. This double objective can be achieved through reducing fuel use and therefore economic costs. Visualization of historical environmental and vessel behaviour will help tuna companies detect improvement strategies. The system aims also to provide advice on potential strategies companies can follow. Purse seine is the fishing gear that contributes the most to yellowfin and skipjack tuna catches globally.

29.2   Oceanic Tuna Fisheries Immediate Operational Choices

The pilot's main targets are on onboard energy efficiency to reduce fuel consumption and on condition-based maintenance of the propulsion system to reduce ship downtime and increase safety on board. This is done via optimization of the propulsion system operation to minimize fuel consumption. To reach these goals, ships are recording energy performance data with onboard systems and uploading the data periodically to cloud services. The data are available for analysis by onshore services, like company machinery superintendents. Data analytics have been used to analyse the recorded data and obtain ships' energy consumption equations that are used for operational decision-making. The propulsion engine performance data have been analysed with machine learning techniques to develop models that inform of engine condition deviation from a healthy state. This deviation information is used to proactively participate in engine maintenance and inform in advance the ship's technical staff about forthcoming problems or undetected problems. In this way, minor faults can be detected in advance and be solved without compromising vessel safety and operation before potentially becoming big failures.

Different solutions have been developed in this pilot for the technical staff on shore and the crew on board. IBM has implemented their event-based prediction (PROTON, PROactive Technology Online) component onboard two ships on a dedicated computer. VTT has employed their OpenVA component to develop the user interface (UI) for IBM PROTON onboard ships and for onshore analysis of data collected onboard. VTT has developed and implemented a server-based visualization and analysis tool to be used by fishing company technical staff on shore. EXUS has used their analytics framework to develop an engine fault detection tool based on historical engine performance data. EXUS has also developed the UI of the software and applied some of the solutions developed by VTT for the data collection and processing from Google Drive. The solutions have been tested by Echebastar Fleet on their vessels, while EHU (University of Basque Country) has coordinated the partner work in the pilot and also developed the fuel oil consumption equations based on the historical vessel performance data (fuel consumption model). The equations developed have been implemented in the pilot "Oceanic tuna fisheries planning" for energy saving decision-making.

The energy efficiency target has been pursued through a ship fuel consumption model that is used together with weather models to provide an efficient route from point A to point B. The energy efficiency model also assists the crew in deciding which propulsion mode (constant speed or variable speed) and which ship speed are most suitable from an energy efficiency perspective to get from point A to point B. The developed models use common parameters but have specific coefficients for each ship and offer great accuracy in fuel consumption prediction depending on ship speed. Skippers use this information for decision-making when deciding where to go next during fishing operations. Offline software for monitoring ship performance has been developed and implemented. The offline monitoring software is used by the ship owner's technical staff from shore to collaborate with the crew on board for a more efficient fishing operation.

VTT has worked with their OpenVA platform to develop useful and user-friendly visualization tools for the recorded data (ship owner technical staff). The ships are uploading a daily file with operational data to Google Drive. The visualization tool opens these files, makes necessary calculations and obtains several performance indicators (Fig. 29.1).

The propulsion system has been modelled by EXUS to define a machinery healthy state condition (i.e. the model provides engine parameters based on some inputs defining engine condition). This baseline condition is used as a reference to monitor changes of the engine's condition and predict faults in advance, prior to their occurrence. In this way, machinery fault chances will be reduced along with main engine

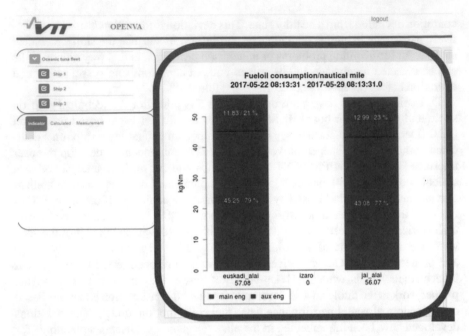

Fig. 29.1 VTT OpenVA visualization for vessel key performance indicator (KPI) comparisons

downtime. Note that the main engine is the ship's only propulsion engine. Hence, with reduced main engine downtime and machinery failures, the overall safety on board will increase. Energy efficiency will also improve with a good condition of the engine. IBM has used their PROTON system to develop an event-driven application for main engine predictive monitoring. The system has been installed on two ships on dedicated computers. IBM PROTON receives engine operation and performance data from the ship's data logger. When the engine is operating in a steady condition, performance data (i.e. pressures and temperatures) are processed in order to detect possible deviations from normal operation conditions. If an event is detected (i.e. deviation from normal condition in engine) a warning will be issued to inform crew. If the event remains and the condition gets worse, an alarm is issued to the crew to be aware and check the evolution of the faulty variable. VTT has implemented a user-friendly visualization interface for the crew on board. The interface enables the vessel crew to assess machinery performance and improve maintenance planning (Fig. 29.2).

29.3 Oceanic Tuna Fisheries Planning

An important element of this pilot is improving the computational capacity to accommodate the data acquisition, processing, analysis and visualization components. Our solution is divided into three parts (Fig. 29.3). The first part is dedicated to the

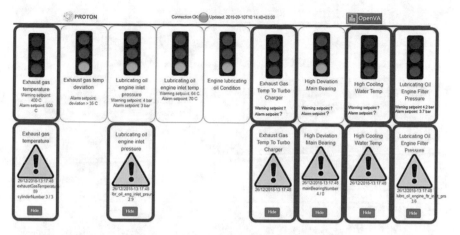

Fig. 29.2 IBM PROTON Dashboard visualization implemented by VTT

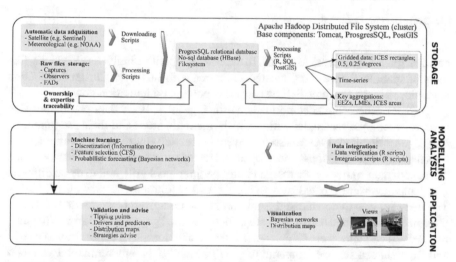

Fig. 29.3 Conceptual diagram of oceanic tuna fisheries planning

storage of data and components (software) making up this platform solution. This first part provides not only the storage, but also the computational capacity to run these components. Modelling and analysis are the second part of the solution, with existing and new algorithms for advice on species distribution and vessel behaviour. Finally, the third part deals with results from previous parts to visualize them and provide fisheries operators and managers with user-friendly advice.

The big data cluster is designed specifically for storing and analysing huge structured and unstructured data volumes in a distributed environment. All the servers are virtual machines, which allow to change the number of processors, RAM memory

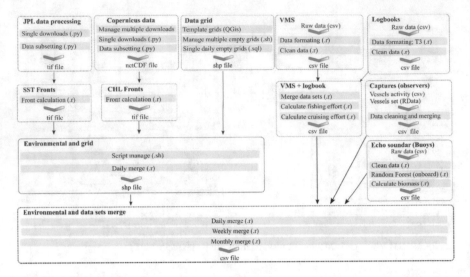

Fig. 29.4 Data processing flow scheme

and disk space for the work assigned. In the case of repetitive high workload jobs, like merging spatially data, an HPC cluster configured with the Rocks Cluster distribution is employed if the Postgres SQL server lacks enough computation power.

The data processing scheme is shown in Fig. 29.4. Each data processing step is represented by a rectangle, within which the main steps and their programming language are detailed. Two data sources were processed: environmental variables and tuna fisheries data. Previously developed and tested scripts were used when possible. Then, both data sources were merged into a geographical grid, and to do so different grid templates were created by varying the cell resolution (i.e. 0.5°, 0.25°, 0.1°). The first step was to download the environmental data from Copernicus and JPL at the time frames needed and in daily steps for the studied geographical area. After that, two derived variables were calculated (i.e. fronts of chlorophyll concentration (CHL) and fronts in sea surface temperature (SST)). Finally, the environmental variables were merged with the grid template in a daily time step.

Tuna fisheries data processing also started with the raw data collection. Due to the different sources of fisheries data, different formats and errors were present, making it necessary to clean and reformat the different raw data sets. VMS and logbook data were combined to calculate the fishing and cruising effort by vessel. Observer data came in two parts: vessel activity and set information. The former has trip information such as trip start and end date, speed, and latitude and longitude, among others. The latter has catch information, in our case species and kg fished. The last source of data comes from the echo-sounder buoys attached to fishing aggregating devices (FADs), these data sets provide accurate information on buoy geo-location and rough estimates of fish biomass aggregated underneath.

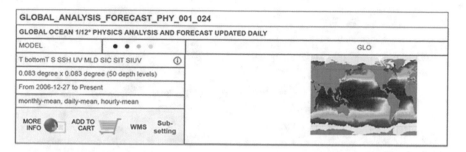

Fig. 29.5 Example of Copernicus data catalogue for a model projection covering all the world's oceans

Finally, environmental variables and tuna fisheries data were merged with the possibility of using different time scales such as daily, weekly or monthly intervals.

For the historical retrieval of environmental data the APIs provided by the data providers (motu-Python for Copernicus and Python script using OPeNDAP) invoked from bash scripts were used. These products provided environmental variables (i.e. temperature, salinity, currents, nutrients and chlorophyll) at resolutions that ranged from 4 km^2 to 50 km^2 (Fig. 29.5).

Each product was chosen based on its temporal and spatial resolution, choosing those covering the tropical areas where these fleets operate, and the level of observational data processing focusing on the processed data, i.e. Level 3 and 4 if available, and environmental parameters provided in each catalogue. Each catalogue provides several variables or potential predictors for identified areas with high probability of tuna occurrence. Chlorophyll (Chl-a), sea surface height (SSH) and sea surface temperature (SST) have been identified as good potential tuna distribution predictors as they enable detection of oceanic fronts and productivity changes (i.e. available food for fish). While these indicators focus on tuna distribution based on their feeding behaviour, oxygen, thermocline depth and gradient or subsurface temperature are good predictors of fish physiology vertical and horizontal constraints (i.e. oxygen availability for efficient energy use). On the other hand, weather forecasts limit fleet distribution by avoiding areas with strong winds and swell where fishing operations are unviable (Fig. 29.6).

Ocean fronts are the interfaces between water masses. These hydrographic features have been recognized to enhance primary and secondary production and promote the aggregation of commercial pelagic fish species. The ocean fronts for SST and chlorophyll were calculated with the Belkin and O'Reilly algorithm (Belkin and O'Reilly, 2009) implemented in the grec R library (R Core Team, 2018) (Fig. 29.7).

In order to store, access and process non-raster data sets the PostgreSQL with PostGIS spatial extension was used. This database was fed with all the fishery and vessel data for use in the machine learning pipeline. Machine learning approaches that are characterized by having an explicit underlying probability model (i.e. provides a probability of the outcome, rather than simply a forecast without uncertainty) are being evaluated for application in Indian Ocean tuna fisheries. Bayesian networks

Fig. 29.6 Example of Copernicus data downloaded using SPACEBEL component from a catalogue. Chlorophyll in the left and Oxygen in the right

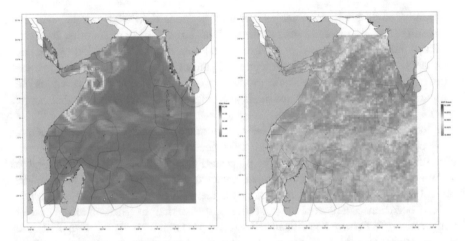

Fig. 29.7 Example of sea fronts. Chlorophyll in the left and sea surface temperature in the right

(BNs) are a paradigm suitable to deal with uncertainty, providing an intuitive interface to data. These intuitive properties of Bayesian networks and their explicit consideration of uncertainties enhance domain experts' confidence in their forecasts [7–9, 13]. This machine learning approach was used here to forecast the likelihood of finding high tuna biomass.

A pipeline of supervised classification methods which include selection and discretization of features, and the learning of a Naïve Bayes classifier (i.e. a type of Bayesian network) was applied [7]. The application of this methodology selected the following features or predictors: Chl-*a*, net primary production, temperature, salinity, oxygen, nutrients and current velocity. It was first applied only to past captures from public sources yielding poor results. However, the results improved significantly when additional data from a tuna company were used. This highlights the importance of working in close partnership with industry. The final model could correctly forecast the areas lacking tuna 80% of times (i.e. helps to identify areas to be avoided that would waste fuel). The model was also able to forecast areas of high biomass

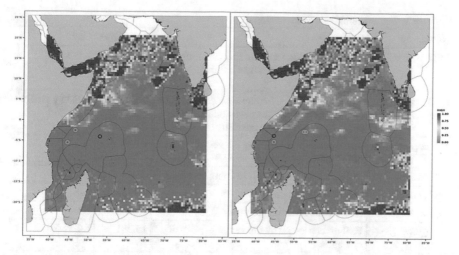

Fig. 29.8 Map showing areas of higher probability of finding high tuna biomass. Green circles show successful fishing attempts and red circles failed fishing attemps. Thin black lines show exclusive economic zones (EEZs) territorial waters where only country fleets and other authorized fleets can fish

with only a 25% of false positives, so it was right 75% of the times. The model was validated using tenfold stratified cross-validation (Fig. 29.8).

The historical vessel performance data have been collected and analysed to calculate key performance indicator (KPI) values and obtain the vessel sailing energy consumption model. The vessels analysed started operation in years 2014 and 2015, and historical data since the start of operations have been used for KPI calculation. The data used for KPI calculation correspond to the period 2015–2018 (Table 29.1).

All three ships used as a reference to obtain the KPIs have undergone repairs during the period analysed (2017–2018), which had an impact on KPIs estimations. When vessels go for repair work, they are usually stopped for a 30- to 60-day period. During the repair period the vessel is not sailing. Hence, variables like fuel consumption and sailed nautical miles suffer a reduction in years when repair works take place. When analysing the KPIs it is necessary to consider impacts from regulatory changes in the Indian Ocean during the period analysed. For example, during 2017 new regulations by the Indian Ocean Tuna Commission (IOTC) entered into force regarding tuna fishing. Quotas for yellowfin tuna were established and ships had to stop temporarily fishing during the year. This means that years with quotas are very different from an energy expenditure viewpoint compared with the rest of unrestricted fishing years. Due to this, a clear decline in total fishing days and total sailed nautical miles was observed after 2016.

Although a marked decline in sailed miles and sailed days is clearly observed in Fig. 29.9, there is no parallel tendency in catches. Instead, catches have increased even considering that sailing days and consumed fuel has been reduced. Thus, fuel oil consumed per kilogram of catch has been noticeable reduced. All five ships of

Table 29.1 Fishery pilots assessment criteria

Name	Description	Base value	Unit
SFO_NM	Propulsion engine specific fuel oil volumetric consumption per sailed nautical mile while fishing	71.25	L FO/Nm
LFO_kgCatch	Ship specific fuel oil volumetric consumption per kilogram of fish caught (total fuel oil consumption including auxiliary engines)	0.57	L FO/kg Catches
FO_consumption	Total fuel oil consumed by the vessel per year of operation	4,826,262	L
SOGave	Average ship velocity in steaming condition	8.95	knot
kgCatches	Total fish caught per year	8,373,460	Kg
Sailed_NM	Sailed nautical miles per year	66,153	Nm
LFO_day	Fuel oil consumed by the vessel per day of operation	15,281	L/day
Day_trip	Average value of days spent per fishing trip (from departure to return to harbour)	25.09	day/trip
NM_trip	Average value of sailed nautical miles per fishing trip (from departure to return to harbour)	5410.9	Nm/trip

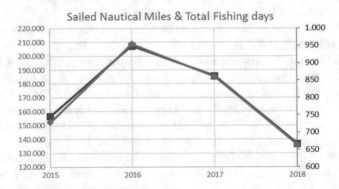

Fig. 29.9 Total sailed nautical miles and fishing days (three vessels)

this fleet reduced their fuel consumption in 2017 by an average 19% (range 4–30% reduction). However, it is not possible to distinguish how much of this improvement in the ratio of catch to fuel consumption is due to DataBio technologies or other continuous and ongoing initiatives to improve their operations and sustainability such as the MSC certification, bioFADs or new Indian Ocean fisheries management regulations (Fig. 29.10).

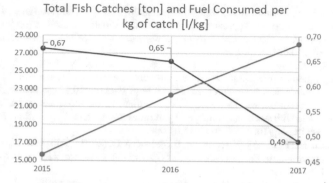

Fig. 29.10 Total consumed fuel oil and fuel oil consumed per kg of catch (three vessels)

Acknowledgements The authors would like to thank the tuna fisheries company Echebastar S.L, for providing access to important data and for participating in discussions. VTT, IBM, EXUS have contributed to the development of the fuel consumption and engine failure software. These contributions have been valuable for the presented work and for this chapter. Jefferson Murua (AZTI) has provided significant feedback for the improvement and quality of this chapter. The work leading to this chapter has been possible mainly through the DataBio project funded from the European Union's Horizon 2020 research and innovation programme under grant agreement No 732064.

References

1. Basurko, O. C., Gabiña, G., & Uriondo, Z. (2013). Energy performance of fishing vessels and potential savings. *Journal of Cleaner Production, 54*, 30–40. https://doi.org/10.1016/j.jclepro. 2013.05.024.

2. Suuronen, P., Chopin, F., Glass, C., Løkkeborg, S., Matsushita, Y., Queirolo, D., & Rihan, D. (2012). Low impact and fuel efficient fishing—looking beyond the horizon. *Fisheries Research, 119*, 135–146. https://doi.org/10.1016/j.fishres.2011.12.009.

3. Parker, R. W., Blanchard, J. L., Gardner, C., Green, B. S., Hartmann, K., Tyedmers, P. H., & Watson, R. A. (2018). Fuel use and greenhouse gas emissions of world fisheries. *Nature Climate Change, 8*(4), 333–337. https://doi.org/10.1038/s41558-018-0117-x.

4. Parker, R. W., & Tyedmers, P. H. (2014). Fuel consumption of global fishing fleets: Current understanding and knowledge gaps. *Fish and Fisheries, 16*(4), 684–696. https://doi.org/10.1111/faf.12087.

5. McCauley, D. J., Woods, P., Sullivan, B., Bergman, B., Jablonicky, C., Roan, A., Hirshfield, M., Boerder, K., & Worm, B. (2016). Ending hide and seek at sea. *Science, 351*(6278), 1148–1150. https://doi.org/10.1126/science.aad5686

6. Hu, H., Wen, Y., Chua, T. S., & Li, X. (2014). Toward scalable systems for Big Data analytics: A technology tutorial. *IEEE Access, 2*, 652–687. https://doi.org/10.1109/ACCESS.2014.233 2453.

7. Fernandes, J. A., Irigoien, X., Goikoetxea, N., Lozano, J. A., Inza, I., Pérez, A., & Bode, A. (2010). Fish recruitment prediction, using robust supervised classification methods. *Ecological Modelling, 221*(2), 338–352. https://doi.org/10.1016/j.ecolmodel.2009.09.020.

8. Fernandes, J. A., Lozano, J. A., Inza, I., Irigoien, X., Pérez, A., & Rodríguez, J. D. (2013). Supervised pre-processing approaches in multiple class variables classification for fish recruitment forecasting. *Environmental Modelling & Software, 40,* 245–254. https://doi.org/10.1016/j.envsoft.2012.10.001.

9. Fernandes, J. A., Irigoien, X., Lozano, J. A., Inza, I., Goikoetxea, N., & Pérez, A. (2015). Evaluating machine-learning techniques for recruitment forecasting of seven North East Atlantic fish species. *Ecological Informatics, 25,* 35–42. https://doi.org/10.1016/j.ecoinf.2014.11.004.

10. Agra, A., Christiansen, M., Delgado, A., & Hvattum, L. M. (2015). A maritime inventory routing problem with stochastic sailing and port times. *Computers & Operations Research, 61,* 18–30. https://doi.org/10.1016/j.cor.2015.01.008.

11. Christiansen, M., Fagerholt, K., Nygreen, B., & Ronen, D. (2013). Ship routing and scheduling in the new millennium. *European Journal of Operational Research, 228*(3), 467–483. https://doi.org/10.1016/j.ejor.2012.12.002.

12. Fagerholt, K. (2004). Designing optimal routes in a liner shipping problem. *Maritime Policy & Management, 31*(4), 259–268. https://doi.org/10.1080/0308883042000259819.

13. Granado, I., Basurko, O. C., Rubio, A., Ferrer, L., Hernández-González, J., Epelde, I., & Fernandes, J. A. (2019). Beach litter forecasting on the south-eastern coast of the Bay of Biscay: A Bayesian networks approach. *Continental Shelf Research, 180,* 14–23. https://doi.org/10.1016/j.csr.2019.04.016.

Chapter 30
Sustainable and Added Value Small Pelagics Fisheries Pilots

Karl-Johan Reite, J. Haugen, F. A. Michelsen, and K. G. Aarsæther

Abstract This chapter describes four pilot cases covering the Norwegian pelagic fisheries for small fish species in the North Atlantic Ocean, such as mackerel, herring and blue whiting. The pilot cases aim to improve sustainability and value creation. Big data methods and tools have been used to demonstrate the potential impact on fuel consumption, fisheries planning and fish stock assessments. Specifically, the pilots have targeted immediate operational choices, short-term fisheries planning, fish stock assessments and longer-term market predictions.

30.1 Introduction

The main challenges for the small pelagic fisheries are related to both the fisheries management and the fisheries itself. Within the fisheries management, one seeks to maximize the production by optimizing the fishing quotas and regulations. At the same time, the resources available for this task are limited. For the fisheries itself, the shipowners want to maximize the value of their fish quotas while minimizing the costs associated with owning and operating their vessels.

The governing bodies (EC and national EU and EEC member states) require fishermen and landing sites by law to report catch data for monitoring purposes. The Norwegian small pelagic fisheries fleet follows the Norwegian law of wild caught fish ('Råfiskloven'), which monopolizes the sale of fish from vessels through sales associations with geographic and species-based areas of monopoly. These sales organizations collect detailed information about species, volume, time of capture, time of lading and price for the entire regional market. This data source is the foundation for the small pelagic fisheries planning and market prediction pilots.

The variations of demands for propulsion and electric energy onboard these ships [1] have led to the development of ships with very advanced energy and propulsion systems. A downside of this development is that the operation of these vessels has become more complex, making it difficult sometimes to take advantage of the

K.-J. Reite (✉) · J. Haugen · F. A. Michelsen · K. G. Aarsæther
SINTEF Ocean, Brattørkaia 17 C, 7010 Trondheim, Norway
e-mail: karlr@sintef.no

© The Author(s) 2021
C. Södergård et al. (eds.), *Big Data in Bioeconomy*,
https://doi.org/10.1007/978-3-030-71069-9_30

possibilities within the systems. The crew is also often engaged in fishing operations, where management of a power plant is not a priority, making decision support systems important [2]. Collecting extensive energy performance of ships and delivering advice based on big data technology is therefore a focus for one of the small pelagic fisheries pilots.

Short-term planning of the fisheries is mainly based on the fishermen's expectations about where they can most efficiently do their fishing. These decisions are mainly based on own experiences, meteorological forecasts and current fisheries activity as it is perceived through catch reports, available AIS data and communication with friendly fishermen on other vessels. Developments in the market situation are considered based on expectations for the amount of catches from other vessels and fish quality. These factors are subjectively considered by the individual fishermen.

Long-term planning involves such decisions as, for instance, catching more herring in the spring to have more time for mackerel fisheries in the autumn, due to expectations of being able to achieve higher mackerel prices in the autumn if one has time to make smaller catches. These decisions are very complex, based on a range of uncertain factors and currently with few tools available for decision support.

The small pelagic fisheries pilots focus on small pelagic species harvesting in the North Atlantic Ocean, with the Norwegian pelagic fishing fleet as the main stakeholder. The stakeholders are represented by the pelagic sales association (Norges Sildesalgslag) and companies which own fishing vessels with fishing rights in the North Atlantic. SINTEF Ocean has established the SINTEF Marine Data Centre in order to test, develop and deploy big data tools such as Apache Mesos, CouchDB and GlusterFS for storage and analysis of the available data.

The small pelagic fisheries pilots are highly dependent on big data, for both modelling the ocean environment and the fish stocks. The datasets, stakeholders and analytic needs are illustrated in Fig. 30.1. The data needed include satellite data

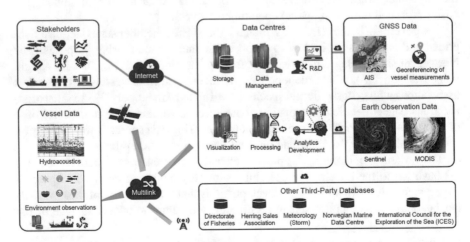

Fig. 30.1 Overview of datasets, stakeholders and components in pelagic fishery

(meteorological and oceanographic), model data (predictions and hindcasts), local measurements (shipborne instruments) and reports on fish catches, for example:

1. Information about all pelagic catches landed in Norway since 2012 is provided by the sales association. This includes information such as price, quantity, catch location, species and size distribution.
2. The ship-owning companies provide onboard measurements (e.g. echo sounders, navigation, machinery and propulsion).
3. Oceanographic hindcasts and daily forecasts are provided by the oceanographic model SINMOD.
4. Satellite-based oceanographic measurements are provided, for example, by CMEMS and NOAA.
5. Meteorological forecasts and hindcasts are provided by the Norwegian Meteorological Institute.

An architectural approach has been chosen, with a focus on the use of case pilots ranging from immediate energy optimization to trip planning and market predictions. The number of potential big data technologies usable for fisheries is vast. The available components and technologies were organized in the framework developed by the Big Data Value Association (BDVA). The potential components were identified during the pilot specification phase and also in the BDVA framework. This selection was refined as the pilot implementation was planned in more detail, ending up with a common architecture design for the pelagic pilots with focuses on the components needed for a minimal viable system, illustrated by the components in the red boxes in Fig. 30.2. The dataset representations are standardized to use JSON for thin data

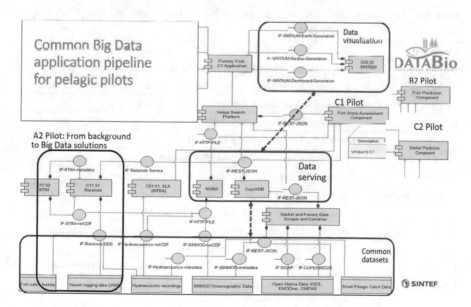

Fig. 30.2 Common architecture for small pelagic pilots

(i.e. catch reports, market and position data) and metadata, while NetCDF is used for large volume data like EO, hydroacoustic and oceanographic data. A combination of search (VESPA) and database technology (CouchDB) is suggested for use in data collation and discovery, both using JSON data representation.

As the small pelagic fisheries pilots had overlapping needs for data centre resources, the provisioning of such resources was a priority. The SINTEF Marine Data Centre was formed for such tasks and therefore chosen as a basis to develop the necessary shared resources for the pilots. The infrastructure includes storage servers, hosting of services and building nodes for software development. A central part of the SINTEF Marine Data Centre is the use of DC/OS for service provisioning and task distribution. This is based on a collection of masters, agents, load balancers and a single bootstrap node. This installation acts like a resource for deploying services in a scalable and repeatable way. It also has functionality for making services available from the Internet without exposing internal systems. The most important services provided by SINTEF Marine Data Centre for the small pelagic pilots are shown in Table 30.1.

The file storage uses the GlusterFS system for creating posix-compliant, replicated network storage. Periodic and dependent jobs are run using the Chronos service running on DC/OS. Vessel data are written by the vessels to an external server ("Incoming"). The data are then fetched to the file storage behind corporate firewalls for further curation, monitoring and analysis. Access is governed by public key cryptography. The high-performance computing cluster unity is used for simulating and predicting oceanographic processes and properties, such as salinity, temperature,

Table 30.1 Services and containers used in SINTEF Marine Data Centre for storage and analysis

Id	Description
Datafetcher	Responsible for writing vessel data to Gluster
Serverconfig	Responsible for keeping the in-house server configurations updated
Vesselconfig	Responsible for keeping the external server configurations updated
Incoming	External server which the vessels can send their data to
GeoServer	Serves GIS data to map clients
Postgis	Serves GIS data primarily to GeoServer
Glusteraccess	Provides access to the Gluster file storage
Chronos	Responsible for running periodic and dependent jobs
Vesseldatamonitor	Functionality for performing health checks on incoming vessel data
Artifactory	Provides storage of and access to built software
Haproxymain	HAProxy instance providing controlled access to some external services
Aptly	Distribution of Debian installation packages to vessels
CouchDB	Database for local caching of incoming data from external sources
Stimanalysis	A basic Docker container containing various analysis capabilities for analysing vessel operational data

nutrients, plankton and fish stock migrations. This system uses earth observation data, as well as catch reports from the sales association [3, 4].

30.2 Small Pelagic Fisheries Immediate Operational Choices

This pilot aims to improve the operation of relatively complex machinery arrangements onboard small pelagic fishery vessels based on measurements of current state and historic performance. The energy needs of the vessel for propulsion power, deck machinery, fish processing and general consumption are met by the same power generation system, which on newer vessel can be configured to produce and distribute power in a variety of ways. The vessel machinery systems may meet crew requirements in a variety of ways but lack feedback on efficiency or suggested actions to reconfigure power production and distribution. Even if the increasing number of sensors can provide valuable information for crew, fishermen's main focus will always be fish harvesting and not the fine-tuning of complex machinery systems. This can lead to higher fuel consumption than necessary.

The four participating vessels have been equipped with instrumentation for continuous collection of navigation data, power production, fuel consumption and high-frequency motion data, as well as fuel and loading condition data where available. The collected data have been analysed and the vessels integrated into the SINTEF Marine Data Centre infrastructure. The signals recorded onboard the vessels are augmented with synthetic signals for decision support in order to cope with the inherent heterogeneous nature of data collected from different fishing vessels. Datasets are heterogenous due to different engine system layouts, different choices of suppliers for propellers, prime movers and auxiliary engines. The new synthetic signals enable the four vessels to slot into a data collection and processing pipeline in the SINTEF Marine Data Centre. This integration of heterogeneous vessel data, or sensor platforms, into a common system has highlighted the need for feedback of both analysis techniques and synthetic signal generation, but also of updated decision support databases to the vessels from SINTEF Marine Data Centre. The introduction of new signals, real or computed, may necessitate an update from the data centre to the vessels of both signal definitions, analyses and the database on which the decision support is based. The already collected data should not be forgone when making such updates, and a new decision support database should be populated from the data centre to the vessel with new signals and new analyses and decision support possibilities.

The first technological hurdle for the pilot is the implementation of harvesting and retrieval of data from the vessels. The retrieved data are of high value for the future and must be kept securely stored, as if it is lost there is no way of recovering it. The pilot has therefore integrated the measurement system onboard the vessels with the SINTEF Marine Data Centre to store all collected data securely for future use

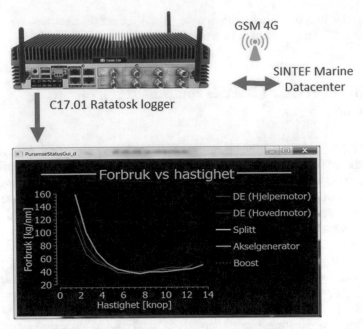

Bridge display: fuel consumption vs. speed

Fig. 30.3 Schematic view of the integration of the vessel's logging computer with SINTEF data centre and screenshot from bride—decision support system for DataBio vessels

and to establish the ability to curate data and update the database of the vessels with synthetic signals derived from the original data as seen in Fig. 30.3. The installed system onboard the vessels accumulates the data and makes a statistical database of the vessel's experienced operations. This database is continuously monitored with the current operation mode in order to give crew a quick feedback when it is practical to operate the vessel in a more efficient manner [5]. This relies on the assumption that the optimum, practical, attainable, operational configuration of the power plant onboard the vessel can often be deduced from its historical data.

30.3 Small Pelagic Fisheries Planning

The main objective of this pilot is to evaluate the effect of utilizing big data technologies in pelagic fisheries planning. The pilot's work focuses on developing services that can help improve vessel operation planning with better fishing ground targeting and improved timing of the fishing execution. The working hypothesis of the pilot is the causality between oceanographic parameters, such as temperature and low-trophic organisms (e.g. *Calanus spp.* copepods), with the location and migration patterns of pelagic species. Therefore, a useful service would be to visualize oceano-graphic and biology parameters together with historical catch data of various species. The primary pilot's goal was to create a Web portal enabling end users to browse through this information on a map. This includes the ability to select a time period of reported catch data for specific pelagic species, which then are displayed on a map that includes oceanographic attributes. A playback feature lets the user see the time evolution of the selected attributes.

The fishing operation region for which the pilot provides decision support includes large portions of the Norwegian Sea and the North Sea, totalling approximately 1.5 million square kilometres. Pelagic fisheries usually only operate in small subregions of this area, depending on targeted species.

The consortium involved in this pilot consists of:

- SINTEF Ocean is a contract research organization committed to technical research within marine applications. SINTEF Ocean leads the pilot and is also the main contributing research organization.
- Norges Sildesalgslag (Norwegian Fishermen's Sales Organization for Pelagic Fish) is a sales organization, owned and operated by fishermen (a coopera-tive), selling fish on a first-hand basis from fishermen to buyers—for further sales/export. They contribute with knowledge and accumulated data on fish catches.
- The fishing vessel owners Liegruppen Fiskeri, Eros, Ervik & Sævik and Kings Bay operate in fisheries targeting pelagic fish species in the North Atlantic. Their role in this pilot is to contribute with their knowledge about fisheries planning and to serve as an end user for the pilot's Web portal.

Important activities in the pilot have been to identify *Data Sources*, select appro-priate components/assets and configure necessary *Data Management* and *Data Processing Architecture*. This work facilitated the primary goal of the project, namely provisioning of the Web portal and its *Data Visualization*. Definitions of key perfor-mance indicators that directly quantify the fishery operation performance were quickly dismissed, because any evaluation of such indicators depends on unmeasur-able and non-deterministic factors. Any potentially improved measurement of fishery efficiency can only be speculatively attributed to the introduction of the pilot service. As a consequence, "key performance indicators" were instead defined as measurable progress/completeness of the technological components used in the pilot.

The following technologies have been found relevant for this pilot:

- **SaltStack** provides configuration management of data centre servers, facilitating version control and remote access.
- **Docker** provides containerization and facilitates version control of onshore systems.
- **SINMOD** provides biomarine simulations and simulation of fish migrations.
- **DC/OS** provides container orchestration and communication.
- **CouchDB** provides storage of and access to catch data.

- **GlusterFS** provides replicated and distributed storage of and access to collected data and the results of biomarine simulations.
- **KRAKK** provides data scraping functionality, especially for data from Sildelaget.
- **GeoServer** provides an open source server for sharing geospatial data.
- **Python** scripts that make use of RESTful API and GDAL for ingesting SINMOD oceanographic and biology data rasters into GeoServer.
- **Python Flask** is used as a Web Server Gateway Interface (WSGI) Web application framework to develop the Web portal.
- **uWSGI** is used for serving the Web portal.
- **Crossfilter, D3.js, dc.js and Leaflet** are important JavaScript libraries for presenting data in the Web portal.

The implementation of this pilot is based on a number of data sources:

- **Catch data** are made available by Sildelaget through an API developed by Sildelaget for DataBio. This API makes available all pelagic catches landed in Norway since 2012, and it is continuously updated as new catches are landed. This provides locations, amounts and price for each catch. The catch data from Sildelaget is proprietary datasets that will not be available after the project. On the other hand, the Norwegian Directorate of Fisheries recently open sourced catch data historic records.
- **SINMOD** oceanographic and biological hindcast and forecast data for the Norwegian Basin, including temperature, salinity, ice thickness and concentration, NO3, *Calanus finmarchicus*, *C. glacialis and* chlorophyll. These parameters were provided both historically, since 2012, and regularly with short-term forecasts two days into the future with a spatial resolution of 4 km in polar stereographic projection. The SINMOD data source relies on several satellite and buoy-based inputs, and see the next pilot for details (Fig. 30.4).

The SINMOD operationalization produces NetCDF4 files that largely follow the Climate and Forecast Convention 1.5. Nonetheless, there have been several issues related to standardized naming conventions of the variables, consistent spatial resolution, as well as correct projection parameters between the historic and predictive datasets. The process of making SINMOD data available to the map service involves extraction of selected depths and timepoints so that only relevant data are being served by GeoServer. Instead of using the NetCDF plugin of GeoServer, we rather used GDAL to manually reproject NetCDF files into the destination projection as GeoTIFF files. File handling logic was developed to facilitate ingesting large datasets. GeoServer's built-in colorbar legend currently lacks the necessary flexibility to show customized styling in a satisfactory manner, which again warranted manual customization. GeoWebCache, the built-in tile caching integration, does not play well with periodic regeneration of new rasters. This is at least true when using GeoWebCache REST API and CQL filters to selectively "reseed" updated datasets. We experienced intermittent issues with newly ingested rasters, where it cached transparent tiles, probably because tiles were cached before their ingestion into the

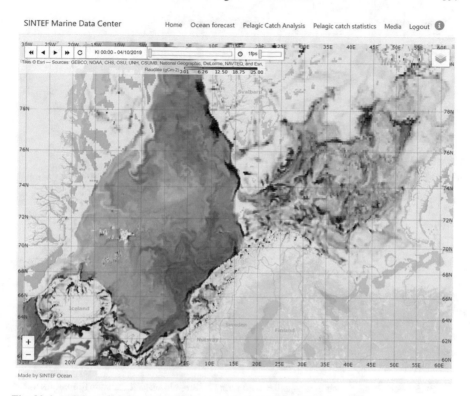

Fig. 30.4 Web portal: Calanus finmarchicus concentration distribution

PostGIS database was done. This issue was not easily reproducible, nor did it produce any error messages, causing undetected issues with the Web map service (Fig. 30.5).

We chose tiled WMS to serve the raster data. The styling of the layers was done on the server side, so no styling configurations were needed in the Web application Leaflet. Designing styles that work globally for a single attribute all year round is challenging, because of the span of interesting values changing throughout the year. WMS playback was achieved using a Leaflet plugin, but the flexibility in zoom levels with different tiles made it challenging for the plugin to buffer many timepoints in a manner that enabled good user experience. Some browser caching occurred, as well as server-side caching, but a different choice of technology or data format may have improved the UX smoothness.

We estimate the impact of the new service provided by this pilot to be minimal; that is, the pilot's end users do not yet actively use of the Web portal for their fishery planning. The reason for this is multifaceted. First, the time period for which the service has been available, with fair service reliability, is very brief still. The user experience in these initial versions of the Web application can be frustrating, due to sluggishness and lack of responsiveness. There is a lack of fundamental features that could be of interest for the user to check for specifically interesting phenomena. For

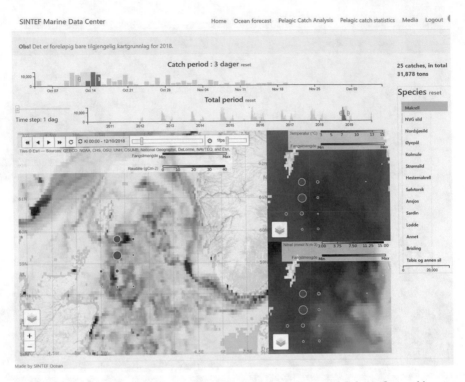

Fig. 30.5 Web portal: Catch data together with temperature, nitrate and Calanus finmarchicus

example, a simple extension would be the ability to select a region and provide key information/analysis on demand. The portal was specifically designed for desktop application use, but in hindsight it should have been readily available on all platforms, including smartphones and tablets. The UX design could also have been more targeted to specific use cases. For instance, by providing several subpages, each designed to provide a very limited set of information. One such tailored design could lower the threshold for use.

The pilot was designed on top of systems and infrastructure designed for use in production. DC/OS are made ready for production use cases, which includes scalability, load-balancing, resource management, etc. What the pilot technology design does not cover is situations in which users employ low bandwidth networks, which is often the case for ocean-going fishing vessels. Therefore, the Web portal is more practical and applicable in an onshore, by-the-computer setting, with high-quality bandwidth. We believe that despite these initial challenges, the concept of collating information and providing insight into multi-origin data in a clear manner still has great potential for improving fishery planning. Establishment of a minimally viable product that the end user is interested in could spawn the foundation for future applications that have a large impact on how fishermen make use of big data and technology in planning their operations.

30.4 Small Pelagic Fish Stock Assessment

Pelagic fish stock assessments are traditionally based on a combination of research cruises with dedicated research vessels, catch statistics and non-spatial stock models. These methods are criticized for low cost efficiency, being based on too few measurements and unable to adapt to rapid climate change effects. The objective of this pilot has been to demonstrate that the combination of information from a great variety of assets can be used to produce better population dynamics estimates for pelagic species. Specifically, crowd-sourced data collection effort from fishing vessels combined with public/private data assets, biomarine modelling and data analytics are assumed to be able to increase both the accuracy and precision of fish migration and stock assessments.

The pilot has concentrated on three research questions:

1. How can hydroacoustic data be cost-efficiently collected from a fleet of fishing vessels?
2. How can a fleet of fishing vessels be part of a crowd-sourced data collection system?
3. How can biomarine modelling and spatio-temporal modelling of pelagic species be used for stock assessments?

To cost-efficiently collect hydroacoustic data from fishing vessels, the integration against existing hydroacoustic sensors was important. Due to the large variations in equipment and interfaces, as well as lack of interface possibilities for much of this equipment, this proved to be a serious challenge. The pilot created a preliminary interface against one type of equipment, but cost-efficient integration against the hydroacoustic equipment of a substantial part of the fishing fleet is not solved.

To make a fleet of fishing vessels part of a crowd-sourced data collection system, cost-efficient installation and maintenance in the vessel are needed. The most important challenges are the variation in vessel systems, sensors and their set-up, as well as how these change over time. This pilot addressed these challenges by using configuration management systems using version-controlled configuration descriptions. This gave a way to perform remote maintenance, updating and reconfiguration, as well as simplify initial installations.

To model the fish stocks and their behaviour, both adequate biomarine models and correction of these, based on measurements, are needed. This pilot developed a preliminary migration model of one pelagic species. Also, a preliminary method for correcting this model using data assimilation was developed, and this correction was performed based on historical data. The results showed that more data for correction are needed, and this has become the focus of new research initiatives.

The consortium involved in this pilot consists of:

- SINTEF Ocean is a contract research organization committed to technical research within marine applications. SINTEF Ocean leads the pilot and is also the main contributing research organization.

- INTRASOFT International offers IT solutions to a wide range of international and national public and private organizations. INTRASOFT has performed comparisons of different methods for classification of hydroacoustic measurements.
- Norges Sildesalgslag (Norwegian Fishermen's Sales Organization for Pelagic Fish) is a sales organization, owned and operated by fishermen (a cooperative), selling fish at a first-hand basis from fishermen to buyers—for further sales/export. They contribute with knowledge and accumulated data on fish catches.
- The fishing vessel owners Liegruppen Fiskeri, Eros, Ervik & Sævik and Kings Bay operate in fisheries targeting pelagic fish species in the North Atlantic. Their role in this pilot is to contribute with their knowledge about fish migration patterns and how this is observed from the fishing vessels, as well as the technical installations available onboard the fishing vessels.

This DataBio pilot has been aimed at assessing if and how stock assessments of pelagic fish species could benefit from low-cost data collection during fishing vessels' day-to-day normal operations, combined with biomarine simulations and migration pattern simulations of pelagic fish species. To this end, this pilot aimed at developing a demonstration version of an infrastructure consisting of both vessels and shore systems.

Relating to the above specified research questions, the following technologies have been found to be relevant for this pilot and its implementation:

- **SaltStack** provides configuration management of both shore servers and vessel equipment, facilitating version control and remote access.
- **Ratatosk** provides onboard data acquisition, data exchange and monitoring of these functions.
- **STIM** provides efficient analysis of collected data (except for hydroacoustic data).
- **Docker** provides containerization and facilitates version control of onshore systems.
- **SINMOD** provides biomarine simulations and simulation of fish migrations.
- **Ratacoustics** provides integration between hydroacoustic equipment and Ratatosk.
- **DC/OS** provides container orchestration and communication.
- **CouchDB** provides storage of and access to catch data.
- **GlusterFS** provides replicated and distributed storage of and access to collected data and the results of biomarine simulations.
- **KRAKK** provides data scraping functionality, especially for data from Sildelaget.

The implementation of this pilot is based on a number of data sources:

- **Catch data** are made available by Sildelaget through an API developed by Sildelaget for DataBio. This API makes available all pelagic catches landed in Norway since 2012, and it is continuously updated as new catches are landed.
- **Hydroacoustic data** are found to be important for correcting the biomarine models and the fish migration model. Some data have been collected using ad hoc methods, but creating general tools for large-scale deployment has proved to be challenging.

- **Vessel operational data** are important for determining what the hydroacoustic data represent in both time and space. Also, for example, ship motions can be important for interpreting the data. The vessels Eros, Kings Bay, Ligrunn and Christina E are contributing with such data.
- **Global ocean tidal components** M2, S2, N2, K2, K1, O1, P1, Q1, Mf, Mm and SSa at the open boundaries of the SINMOD model are imported from [6], which is based on [7].
- **Boundary conditions** for the large-scale 20 km model are acquired from the Mercator Global Ocean model system.
- **Atmospheric input for the large-scale models** is acquired from NOAA Global Forecast System.
- **Atmospheric input for the local scale models** is provided by the Norwegian Meteorological Institute from the 2.5 km MetCoOp EPS system.
- **Sea surface temperatures** are downloaded from the product METOFFICE-GLO-SST-L4-NRT-OBS-SKIN-DIU-FV01.1 [8].

The selected technologies seem to be adequate for the tasks, and there are no obvious benefits associated with making technology changes. But as there are possible alternatives for most of them, the final choice is as much dependent on preferences and existing tools as on the task itself. Without loss of benefits, one may, for example, replace SaltStack with Ansible, Puppet or Chef; Docker could be replaced by Mesos Containerizer; DC/OS could be replaced by Mesos or Kubernetes; CouchDB could be replaced by another database or file storage; GlusterFS could be replaced by Ceph. But for now, no clear benefits are seen from making such changes in the choice of technologies.

One possible exception is with the hydroacoustic data collection, where a Simrad echo sounder was used in the project. This echo sounder facilitates two main approaches for collecting hydroacoustic data in a systematic manner. One is to use the record functionality in the graphical user interface, and the other one is through a subscription-based application programming interface. The first approach is simplistic in that a vessel crew member basically pushes a record button and the system will record data. The downside is that it requires human intervention from the crew, and real-time processing is cumbersome. At the beginning of the project, it was deemed as a risky approach. Therefore, it was decided that API-based data acquisition was a more robust and long-term investment and better suited as an extension of the existing data acquisition system (Ratatosk), as visualized in Fig. 30.6. The subscription API is a comprehensive implementation that enables access to processed and unprocessed data streams and parameters using Ethernet User Datagram Protocol (UDP). Our approach is to implement this subscription API and make the data streams available to the Ratatosk logging component, enabling both real-time processing and storage to file. Most of the functionality towards the subscription API is in place, but the adaptations to connect to the Ratatosk component are currently lacking.

The currently available hydroacoustic echo sounder dataset, see snapshot in Fig. 30.7, has been used as a preliminary comparison of classification methods. The dataset consists of five hydroacoustic frequencies (18, 38, 70, 120 and 200 kHz),

Fig. 30.6 Extension of the
vessel logging system to
facilitate logging of
hydroacoustic data

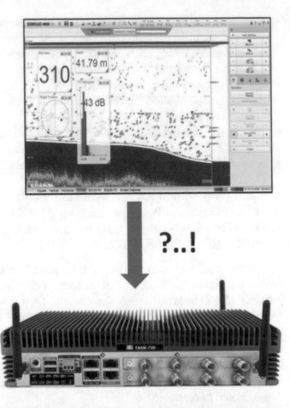

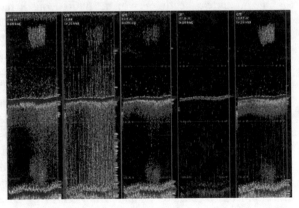

Fig. 30.7 Snapshot excerpt
of echo sounder dataset

which are computed into mean volume backscatter strengths. Four different algorithms have been tested on the dataset: Naïve Bayes, k-nearest neighbours, support
vector machine and principal component analysis. The goal is divided into two tasks:

i.   Identify and remove seabed echoes and determine fish shoal presence.

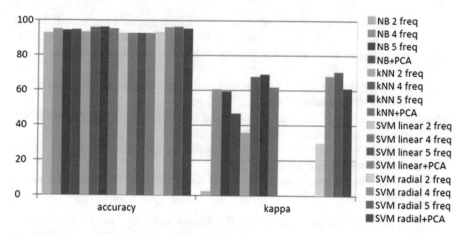

Fig. 30.8 Comparison of classification methods

ii. Discriminate plankton from fish, identify fish species, and perform a biomass evaluation.

Figure 30.8 shows that accuracy is high for all tested methods, but this is due to the few positives of the dataset. Kappa is a more sophisticated metric that shows how much the algorithm improves the average expected accuracy. Kappa shows more varying results when comparing the different methods.

For simulation of the marine ecosystem and the migratory behaviour of selected species, the tool SINMOD was used. This tool perfectly suits the task, as it is able to integrate the simulation of oceanography, low-trophic biology and how this affects higher-order processes. For demonstration purposes, a preliminary fish migration model for herring (*Clupea harengus*) was developed, based on simple behavioural rules and corrected by reported catches. Even if very simplified, the model was able to recreate migration patterns. The model will need to be developed further before it can provide actual value for fish stock assessments, but the results are promising.

The aim of this pilot was to demonstrate that the combination of data collection, existing datasets and biomarine simulations can benefit pelagic fish stock assessments. The business value of this pilot will only materialize once the developed methodologies and technologies become integrated into the fish stock assessment process. At that time, the business impact of reducing the inherent uncertainty associated with stock assessments and thereby improving management and production of the oceans can be very large. If, for example, the production (and thus the catch) of pelagic fish species could be increased by say 10% as a result of this work, this would amount to approximately a € 60 million increase in first-hand value of pelagic fish species in Norway alone.

As stated above, alternatives exist for many of the technologies used in this pilot. Still, the combination of provided functionalities is a good fit for the pilot's objectives. Most notably, the abilities of such a system are to:

- Adapt to the great variations of sensors and configurations onboard fishing vessels, as well as introduced changes over time. This includes both hydroacoustic equipment and operational sensors, such as motion reference systems and GPS.
- Handle a large fleet of vessels in a structured way, with respect to installation, configuration, maintenance and data collection.
- Simulate oceanography, marine biology and fish migrations, while assimilating available data for model and output corrections.
- Extract useful information from hydroacoustic equipment with respect to, for example, fish species and amount of fish.
- Provide systems for data flow, analysis and storage which are suitable for large-scale deployment.

Most of the systems and infrastructure developed in the pilot are ready for use in production, and many of these are easily available. But for such a system to really have an impact on fish stock assessment, improvements are needed in the interpretation of hydroacoustic data and the fish migration modelling.

30.5 Small Pelagic Market Predictions and Traceability

Norwegian fishermen in the pelagic sector work in fisheries for different pelagic species. The timing for these fisheries is to some extent determined by the availability of fish species and their migrations. In addition, to some extent, the shipowners make strategic decisions about when and where to do their fishing based on expectations of both market development and fishing possibilities. These are important choices, but there is a lack of tools helping the fishermen select the right one.

Preliminary exploratory analyses for mackerel showed expected seasonal variations, as well as other variations so far unexplained. Figure 30.9 shows daily average mackerel price variations and daily catch from 2012 to 2019 for Norwegian mackerel landings. Only the second half of each year is plotted, as this is the main season for this fishery. The size of each point marker reflects the amount of daily/weekly catch. The seasonal variations are obvious, while the variations with other variables in this dataset other than time are not.

The goal of this pilot is to enable fishermen to make the right strategic decisions, which can make a substantial difference in both profitability and landed quality.

The consortium involved in this pilot consists of:

- SINTEF Ocean is a contract research organization committed to technical research within marine applications. SINTEF Ocean leads the pilot and is also the main contributing research organization.
- Norges Sildesalgslag (Norwegian Fishermen's Sales Organization for Pelagic Fish) is a sales organization, owned and operated by fishermen (a cooperative), selling fish at a first-hand basis from fishermen to buyers—for further sales/export.

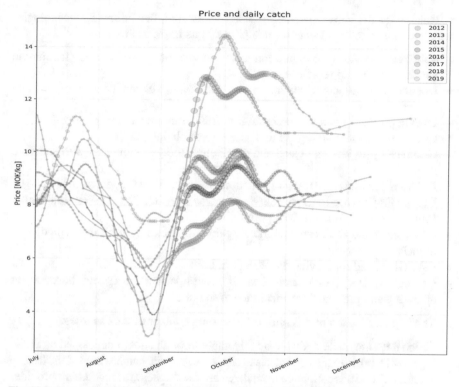

Fig. 30.9 Seasonal variations of Norwegian mackerel prices from 2012 to 2019

They contribute with knowledge and historic and present data on mackerel catches and price.

• The fishing vessel owners Liegruppen Fiskeri, Eros, Ervik & Sævik and Kings Bay operate in fisheries targeting pelagic fish species in the North Atlantic. Their role in this pilot is to contribute with their knowledge on mackerel fisheries and the pelagic market.

This pilot has developed a Web portal to provide fishermen with the tools to analyse historical data. In addition, machine learning has been employed to predict the development of pelagic market segments, so that the fisheries may be targeted based on the species that will allow the highest yield given a predicted economic outlook. The Norwegian mackerel market has been used as a case benchmark, as this is an important pelagic species with large price fluctuations. The basis for the market predictions has been to combine different data sources relevant for price development, such as time, season, predicted catch volume and financial data. Machine learning and predictive analytics have been used to model the relationship between market development and other factors. These models can then be used to provide predictions for how the market will develop in future.

Relating to the above specified research questions, the following technologies have been found to be relevant for this pilot and its implementation:

- **SaltStack** provides configuration management of shore servers, facilitating version control and remote access.
- **Docker** provides containerization and facilitates version control of onshore systems.
- **DC/OS** provides container orchestration and communication.
- **CouchDB** provides storage of and access to catch data.
- **GlusterFS** provides replicated and distributed storage of and access to collected data.
- **KRAKK** provides data scraping functionality, especially for data from Sildelaget.
- **Python Flask** is used as a Web Server Gateway Interface (WSGI) Web application framework to develop the Web portal.
- **scikit-learn** and **Keras** are important Python libraries used for training prediction models.
- **uWSGI** is used for serving the Web portal.
- **Crossfilter, D3.js, dc.js and Leaflet** are important JavaScript libraries for analysing and presenting results in the Web portal.

The implementation of this pilot is based on a number of data sources:

- **Catch data** are made available by Sildelaget through an API developed by Sildelaget for DataBio. This API makes available all pelagic catches landed in Norway since 2012, and it is continuously updated as new catches are landed. This provides locations, amounts and price for each catch. Each catch is typically defined in terms of approximately 70 variables, such as catch size, where it is caught, sale price, storage method and sales method.
- **Catch areas and other definitions** are provided by the Norwegian Fisheries Directorate, such as definitions of various codes representing fish species, catch areas, conservation methods, storage methods, seller, vessel and so on. These data are necessary to interpret the data from Sildelaget.
- **Historical value exchange rates** are made available by the Norwegian bank DNB. These data are potentially valuable for interpreting and forecasting market variations [9].
- **World Bank, EMODnet, Comtrade, Eumofa, Eurostat, ICES and Statistics Norway** offer various data which can be of interest when developing price forecasts for pelagic species. Data scrapers have been developed for these data sources to use in price prediction pipelines.

The selected technologies seem to be adequate for the tasks, and there are no obvious benefits from making additional technology changes. But as there are possible alternatives for most of them, the final choice is as much dependent on preferences and existing tools at the time, as on the task itself.

In a case study, the possibilities for direct predictions of the mackerel prices were investigated. The focus was on long-term predictions, aiming to enable fishermen to adopt long-term successful strategic decisions. As the market is greatly influenced by unpredictable psychological factors, the results were not expected to be good. This can be compared to predicting the stock market, which understandably is a close-to-impossible task.

A Web portal was developed to allow fishermen to investigate how prices have developed with factors such as species, landed quanta, year, time of year, moon phase and catch location. This Web portal is based on providing the possibility to filter historical catch data along the relevant factors. For example, by selecting only last year's catches of mackerel using a short time window, and then slide this window to see how the prices varied with time. Also, similar procedures can be employed to consider variation with moon phase. Or one can use the opposite approach and select only the catches giving the highest prices to investigate under which circumstances high prices were achieved (Fig. 30.10).

The service developed in this pilot is, as far as we know, the first of its kind. It is notably difficult to estimate the potential business impact. Even if one can investigate how fisheries have historically performed, any changes in fishery timing would influence the market, and we do not know how efficient the fishery could be predicted for alternative timings. As an example, in 2015, the price distribution for herring in the spring (66,000 tons) and in the autumn (119,000 tons) is shown in Fig. 30.11. If one assumes that the market would not be affected by shifting the fisheries to autumn, and that the fisheries could be performed in autumn without affecting other fisheries, a 10% shift of this fishery to autumn would approximately generate an extra 700,000 €.

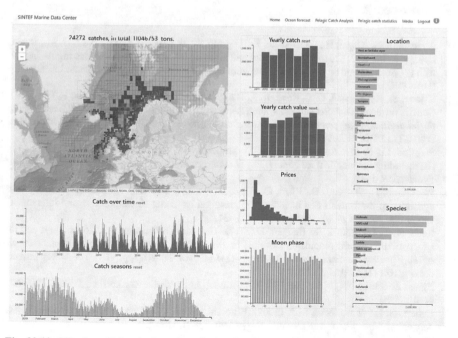

Fig. 30.10 Filtering of historical catch and price data facilitated in the Web portal

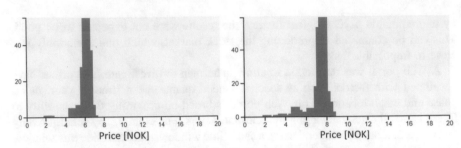

Fig. 30.11 Changes in Norwegian mackerel prices between spring and autumn 2015

Acknowledgements The authors would like to thank the Norwegian Fishermen's Sales Organization for Pelagic Fish, for providing access to important data and for participating in discussions. The fishing vessel owners Liegruppen Fiskeri, Eros, Ervik & Sævik and Kings Bay, as well as part of their crew, have participated in collecting operational data and discussions about the fishing operations and the fishing industry. INTRASOFT has investigated the use of machine learning algorithms for fish school classification. These contributions have been valuable for the presented work and for this chapter. Jefferson Murua (AZTI) has provided significant feedback for the improvement and quality of this chapter. The work leading to this chapter has been possible mainly through the DataBio project (Horizon 2020 grant agreement No. 732064) and in collaboration with the SMARTFISH H2020 project (Horizon 2020 grant agreement No. 773521).

References

1. Aursand, I. G., Digre, H., Ladstein, J., Kyllingstad, L. T., Erikson, U. G., Tveit, G. M., Backi, C. J., & Reite, K. J. (2015). Development and assessment of novel technologies improving the fishing operation and on board processing with respect to environmental impact and fish quality (DANTEQ).
2. Skjong, S., Kyllingstad, L. T., Reite, K. J., Haugen, J., Ladstein, J., & Aarsæther, K. G. (2019). Generic on-board decision support system framework for marine operations. In *Proceedings of the international conference on offshore mechanics and arctic engineering—OMAE*. https://doi.org/10.1115/OMAE2019-95146
3. Slagstad, D., & McClimans, T. A. (2005). Modeling the ecosystem dynamics of the Barents sea including the marginal ice zone: I. Physical and chemical oceanography. *Journal of Marine Systems, 58,* 1–18.
4. Wassmann, P., Slagstad, D., Riser, C. W., & Reigstad, M. (2006). Modelling the ecosystem dynamics of the Barents Sea including the marginal ice zone: II. Carbon flux and interannual variability. *Journal of Marine Systems, 59,* 1–24.
5. Reite, K.-J., Ladstein, J., & Haugen, J. (2017). Data-driven real-time decision support and its application to hybrid propulsion systems. In *Proceedings of the international conference on offshore mechanics and arctic engineering—OMAE* (Vol. 7B-2017, p. V07BT06A024). https://doi.org/10.1115/OMAE201761031
6. Egbert & Erofeeva. (2019). The TPXO 7.2-model. https://volkov.oce.orst.edu/tides/TPXO7.2.html
7. Egbert, G. D., & Erofeeva, S. Y. (2002). Efficient inverse modeling of barotropic ocean tides. *Journal of Atmospheric and Oceanic Technology, 19*(2), 183–204.

8. Copernicus. (2019). Copernicus marine environment monitoring service. https://nrt.cmems-du.eu/motu-web/Motu
9. DnB Markets. (2019). Historical valuta exchange rates. https://www.dnb.no/bedrift/markets/valuta-renter/valutakurser-og-renter/HistoriskeValutakurser/Hovedvalutaerdaglig/Historikk

Chapter 31
Conclusions and Future Vision on Big Data in Pelagic Fisheries Sustainability

Jose A. Fernandes and Karl-Johan Reite

Abstract The digitalization of the fisheries sector has been limited. However, in this book, the potential for making the sector more competitive and resilient through higher digitalization has been demonstrated using pelagic fisheries as an example. COVID-19 has recently shown the resilience advantages of having a more digitalized industry that makes larger use of big data and artificial intelligence. Moreover, these technologies can help us to mitigate climate change due to lower emissions and to adapt to climate change-induced changes of species distribution. One of the challenges is the accessibilty of enough cost-effective information. This can be achieved if fishing vessels becme also scientific data gathering platforms in a circular data economy. Then the fishing vessels are both users and providers of environmental data.

31.1 Conclusions

The fishery pilots have demonstrated the potential of big data to boost performance in the fishery sector. They have worked in both types of pelagic fisheries: the oceanic tuna fisheries (Spanish) and the small pelagic fisheries (Norwegian). Six separate pilot cases have been defined, addressing key concerns such as the cost of fuel, vessel maintenance, fish finding, fish markets and fish stock management. Therefore, the pilots cover three separate viewpoints: immediate operational choices (i.e. in each vessel during their operations), short-term planning (i.e. fishing vessel trip and fisheries planning) and long-term planning (i.e. fisheries sustainability and value creation).

End users have been actively participating and giving feedback during the whole project period, with participation from the very start in the project's kick-off meeting.

J. A. Fernandes (✉)
AZTI, Marine Research, Basque Research and Technology Alliance (BRTA), Herrera Kaia—Portualdea z/g., 20110 Pasaia, Gipuzkoa, Spain
e-mail: jfernandes@azti.es

K.-J. Reite
SINTEF Ocean, Brattørkaia 17 C, 7010 Trondheim, Norway

Six fishing companies have been involved in the project to test the framework and give feedback to ensure the most useful implementation. The fault detection and energy efficiency tasks have been developed for on-board and shore applications with fishing company machinery surveyors. End users from the fishing industry have advised researchers about which data integration and visualization are most useful for operations planning. Several fishing vessel owners have provided feedback about the project objectives and how they can benefit them. Other end users are national and international organizations interested in fisheries sustainability. There have been several opportunities to show the project progress and receive feedback from these organizations, which include regional fisheries associations, International Council for the Exploration of the Sea (ICES) fisheries experts, Food and Agriculture Organization of the United Nations (FAO) stakeholder meetings and the European Fisheries and Aquaculture Research Organizations (EFARO).

Tuna oceanic pilots have achieved the installation of vessel fuel consumption monitoring systems collaboratively with end users and the big data cluster for data storage, integration, processing and visualization in AZTI facilities.

Oceanic tuna fisheries immediate operational choices have data monitoring in place in all vessels, with periodical uploads of data to servers for shore analysis. Data analytics have produced fuel consumption equations and propulsion engine fault prediction models. The fault prediction models provide helpful advice for maintenance operation planning and for preventing unexpected engine malfunctions, thus increasing safety.

Oceanic tuna fisheries planning has all the components deployed and partly operational with data feeding the system in terms of environmental data, vessel data and fish catch data. Data integration is in progress using PostgreSQL database and R scripts. Data analytics have been performed using a machine learning pipeline and forecasting maps contrasted with historical data (for statistical validation) and experts' judgement.

Small pelagic pilots have established services and pipelines to facilitate the pilots' objectives, as well as developed demonstration versions of the associated end user tools. The implementations are running partly on-board fishing vessels, partly on the SINTEF Marine Data Centre infrastructure.

Small pelagic fisheries operational choices have instrumentation on-board four pilot vessels. This instrumentation collects data from a range of sensors and systems. The collected data are aggregated on-board each vessel, where it forms the basis for the decision support system. All vessel instrumentation is also connected to the SINTEF Marine Data Centre and data are collected automatically from the vessels when they are in range of shore-based cellular data networks. This enables updating and correcting the on-board database in case of failures, or if in retrospect one is able to apply corrections to individual sensor signals.

Small pelagic fisheries planning have operationalized the SINMOD in-house ocean model system that simulates physical and biological processes. It now provides daily forecasts of the physical and biological parameters important for the distribution and availability of small pelagic fish species. This information is made available through an online service, available for the shipowners participating in the project.

Small pelagic fish stock assessments has improved on board vessels data acquisition for hydroacoustics. SINMOD has been extended with fish population simulations. Models for automatic classification of acoustic signals have been developed with an accuracy over 90% to differentiate fish aggregations from other acoustic signals.

Small pelagic market predictions and traceability has developed components performing data scraping and caching in local databases. This is operational for download and data collation for Sildes and the Norwegian Fisheries Directorate. A web service has been developed, which allows the users to search, filter and analyse historical catch and market data. This enables shipowners to better perform long-term planning, such as deciding how to split the year's fishing between different fish species.

31.2 Future Vision

The fisheries pilots have shown the great potential benefits for the fishing industry from big data and its associated methods and tools. At the same time, it seems evident that even though we piloted with some of the most modern and technologically advanced fleets in the world (Spain and Norway), they are in many respects quite immature in terms of digitalization. To achieve many of the possible benefits, several developments must first take place. One of the main hurdles to overcome is posed by the lack of standardization of data exchange, in particular for on-board fishing vessels. This implies that for each vessel, specific tailoring is required to interface vessel sensors, which is costly for large scale data collection in large fleets with many vessels. Another issue recently identified is the closed nature of many of the sensors and devices on board. This complicates data capture and storage. Also, future work should focus on forecasting biomass by species, instead of aggregated predictions, including non-targeted species estimates to avoid incidental fishing. This would help with quota management and compliance of target and bycatch species. Integration with commercial systems can help develop multi-vessel approaches and incorporation of biomass estimates from echosounder buoys. Furthermore, fuel consumption models should be incorporated into a decision support system to forecast potential benefits and costs of alternative fishing routes. Moreover, with enough digitalization, this system could coordinate multiple vessels from a variety of gears in different fisheries. This further digitalization could also make fishing vessels become oceanographic data capturing platforms that improve the capacity to observe the marine ecosystems. This added capacity could be used to improve biomass distribution forecasting in a kind of circular data economy, where the users of processed data are also providers of raw data.

Acknowledgements The developments in DataBio have formed the base for several ongoing projects, such as SusTunTech (Horizon 2020 grant agreement No. 869353), SMARTFISH H2020 (Horizon 2020 grant agreement No. 773521) and FishGuider (Norwegian Grant agreement No.

296321). In return, these projects have helped in the framing and publishing of the future vision in this chapter.

Part VIII
Summary and Outlook

Chapter 32
Summary of Potential and Exploitation of Big Data and AI in Bioeconomy

Caj Södergård

Abstract In this final chapter, we summarize the DataBio learnings about how to exploit big data and AI in bioeconomy. The development platform for the software used in the 27 pilots was a central tool. The Enterprise Architecture model Archimate laid a solid basis for the complex software in the pilots. Handling data from sensors and earth observation were shown in numerous pilots. Genomic data from crop species allows us to significantly speed up plant breeding by predicting plant properties in-silico. Data integration is crucial and we show how linked data enables searches over multiple datasets. Real-time processing of events provides insights for fast decision-making, for example about ship engine conditions. We show how sensitive bioeconomy data can be analysed in a privacy-preserving way. The agriculture pilots show with clear numbers the impact of big data and AI on precision agriculture, insurance and subsidies control. In forestry, DataBio developed several big data tools for forest monitoring. In fishery, we demonstrate how to reduce maintenance cost and time as well as fuel consumption in the operation of fishing vessels as well as how to accurately predict fish catches. The chapter ends with perspectives on earth observation, machine learning, data sharing and crowdsourcing.

32.1 Technologies for Boosting Sustainable Bioeconomy

Big data and AI have the potential to boost—in a sustainable way—biomass production within agriculture, forestry and fishery. Biomass means raw material for food, biomaterials and energy. For this, data is gathered in several ways: through satellites, airplanes and drones; from sensors in fields, air and ocean as well as from sensors in agriculture machinery, forest harvesters and fishing vessels. In addition, there is other data to be utilized, like weather forecasts and market prices. When all these data sources are integrated, analysed through various models and visualized, huge opportunities are created. These solutions are able to support the end users—farmers, forest owners, fishermen and other stakeholders—in their decisions and thus increase

C. Södergård (✉)
VTT Technical Research Centre of Finland, 02044 VTT, Finland
e-mail: Caj.Sodergard@vtt.fi

417

biomass production as well as decrease costs and the burden on the environment, as demonstrated in the numerous pilots in this book.

As the DataBio pilots in the three sectors utilize similar big data solutions, we created a *development platform* for the software to be used in the 27 pilots as described in Chap. 1. The platform and its assets are on the cloud and can be used by developers of bioeconomy services after the end of the project to accelerate their developments. The platform assets are gathered together in the DataBio hub (https://www.databi ohub.eu/) and consist of 101 software *components*, of which 62 components from 28 partners were used in the 2 trial rounds conducted in 2018 and 2019 for the 27 pilots. The assets also include 65 *data sets*, of which 45 were created in DataBio and partly openly published. In addition, we collected components into 45 software *pipelines* grouped into 7 generic ones. The pipelines consist of components from the project partners and open-source components. They show how the components are interconnected. The descriptions of the *pilot systems* and the trial results are published as publicly available reports on the website (https://www.databio.eu). The reports are cross-linked to the hub providing a more detailed and multi-view description of the single assets, e.g. which components and datasets have been used in which pilot.

The DataBio project significantly matured already existing components during the project by adding, e.g. new user interfaces and new APIs. As a result, the technology readiness level (TLR) of the components grew with 2.7 units during the project being on average 7 on a scale from 1 to 9. When the project finished, many components were well on their way towards TLR 8 that means "system complete and verified". One factor behind this achievement is that we applied in the planning stage a solid *enterprise architecture* model. This modelling was needed as a basis for the extensive and complex software to be constructed for 27 pilots. We adopted Archimate, which is based upon the Unified Modelling Language (UML), to create 580 diagrams, which described interfaces, subordinates and deployment environments of the components as well as the integration of components into pipelines. In addition to serving the system design, the visual models helped to communicate the pilot designs across the project team. As shown in Chap. 9, we developed a measurement system to evaluate how efficient and comprehensive the software models are.

Digital bioeconomy benefits from the rapid development of *sensors* and more widely from the emerging Internet of Things, which is expected to grow annually with two digit numbers and exceed $1 trillion in 2022. Highly accurate sensors measuring environmental conditions at farms have enabled precision agriculture. As pointed out in Chap. 3, our DataBio pilots were able to utilize autonomous, sun-powered and wireless sensing stations from our partners measuring plenty of properties from the air, crops and soil. We also show how smart tractors equipped with telemetry tools can support current farm work as well as enable new business models.

In addition to sensor data, *earth observation* data forms the second underpinning of digital bioeconomy as shown in Chaps. 2 and 4. Almost all DataBio pilots have used freely accessible Sentinel 2 satellite data that is offered by the European Space Agency ESA. A third data category, *genomic data* from crop species of agricultural interest, opens unprecedented opportunities to predict *in-silico* plant performance and traits like yield as well as abiotic and biotic resistance. This has, as discussed in Chap. 6, impressive applications in plant breeding, where *genomic selection* is a new paradigm allowing to bypass costly and time-consuming field phenotyping by selecting superior lines based on DNA information.

With this variety of data sources in bioeconomy, methods for integrating them are crucial. *Linked data* is a one such technology for integrating heterogeneous data. In Chap. 8, we show how we with linked data can query, for example, how fields with a certain crop intersect with buffer zones of water or the amount of pesticides used in selected plots. The semantic RDF database—triplestore—enabling these functions in DataBio has over 1 billion triples making it one of the largest semantic repositories related to agriculture. Such *knowledge graphs* are important in environmental, economic and administrative applications, but constructing links manually is time and effort intensive. Links between concepts should therefore be *discovered* automatically. In DataBio, we developed a system for discovery of RDF spatial links based on topological relations. The system outperforms state-of-the-art tools in terms of mapping time, accuracy and flexibility.

Bioeconomy applications often require *real-time processing* of sensor data as a key pillar. We demonstrate in Chap. 11 how detected situations and events provide useful real-time insights for operational management, such as preventing pest infestations in crops or machinery failures on fishing boats. In addition to being real time, data is frequently sensitive. Data might then not be made available, because of concerns that the data becomes accessible to competitors or to others that could misuse the data. In Chap. 12, we show that it is possible to handle *confidential data* as part of data analytics, combining open data and confidential data in a way that both provides business value and preserves data confidentiality. As an example, we were able to analyse high-precision data on the location and time for fishing catches without the fishery shipping companies revealing to each other where and when they got the catches.

The pilot chapters in this book show how the technologies described above and in Part I – IV of the book were deployed to meet the performance and user experience needs of each pilot.

32.2 Agriculture

As stated in previous chapters, there are high expectations on smart and precision agriculture—the forecasted market value worldwide in 2023 is over 23 billion US dollar. Smart agriculture utilizes big data technologies, Internet of Things and analytics in the various stages of the agriculture supply chain. The examples in this book illuminate the importance of smart agriculture for productivity, environmental impact, biodiversity, food security and sustainability.

In the *precision farming* pilots in Chap. 15, we achieved a significant reduction in costs of up to 15% for pesticides, 30% for irrigation and up to 60% for fertilization. These economic savings are at the same time environmental benefits. Furthermore, in yet another precision farming pilot (Chap. 18), the experiences showed the benefits with optimal variable application of nitrogen fertilizers based on satellite monitoring of the farm fields. It is expected that the precision farming results achieved will be further improved as more data is collected to further train the models. In Chap. 17 on sorghum and potato phenology, big data allowed a more accurate prediction of yield and other plant characteristics in comparison with approaches currently in use. This improved yield prediction will help the farmers, but also the processing industry, to enhance their sales planning. In Chap. 16, we report a four times reduction in breeding time and a five times reduction in breeding costs for sorghum by applying *next-generation sequencing technologies,* and genomic prediction and selection modelling, allowing to select superior cultivars based on genetic merit derived from whole-genome DNA information. This technology can easily be scaled up to other crop species and animal husbandry.

In the *insurance* pilot in Chap. 19, we introduce new computational tools for getting more insight about the risk and the impact of heavy rain events for crops. For example, potato crops are very sensitive to heavy rain, which may cause flooding of the field due to lack of run-off and saturation of the soil. This may cause the loss of the potato yield in just a few days. A more accurate insurance assessment will encourage bigger agricultural investments. The pilot results point on possibilities to strongly reduce manual ground surveys, thus decreasing insurance costs for the farmers. To support the authorities in common agricultural policies (CAP) *subsidies control*, we achieved excellent results as reported in Chap. 20. As an example, we detected fully automatically 32 crops with 97% accuracy on areas of 9 million ha encompassing 6 million parcels in Romania. Overall, the results showed that authorities can benefit from the use of continuous satellite monitoring instead of random and limited controls. While conventionally only about 5% of the applications are cross-checked either by field sampling or by remote sensing, the methodology developed in this pilot allows checking the compliance of the farmer declarations for all agricultural parcels above 0.3 ha.

32.3   Forestry

Big data technologies have potential to replace traditional practices in forestry, even if this may require legislative changes in many countries. The reporting and monitoring of forest carbon fluxes and sustainability are increasingly in demand, and big data online platforms provide optimal tools for this. Big data and AI allow development of entirely new types of forest monitoring. DataBio developed several tools for forest owners and other stakeholders. In the work of Chap. 23, an open version of Finland's national Metsään.fi resource database was developed and got around 11 million visits in a year. The mobile crowdsourcing service *Laatumetsä,* which is connected to Metsään.fi, makes it possible for the forest owner and citizen to easily report forest damages and control quality of implemented forest operations. In 2019, the Big Data Value Association (BDVA) selected this solution as the second best success story of big data projects funded by the European Commission.

As discussed in Chap. 24, DataBio developed *a forest inventory system* that estimates forest variables and their changes based on remote sensing data and field surveys. Overall, the pilot demonstrated the benefits of big data use in forest monitoring through a range of forest inventory applications. In addition, the pilot highlighted (1) the technical transferability of online platform-based forest inventory services and (2) importance of local involvement in fine-tuning services to meet local needs. The pilot presented in Chap. 25 shows that it is possible to use field data combined with drone images to assess the *health of forest stands*. Once we obtain these local models, it is possible to extend them to larger areas at the regional or national level. The chosen tree species, despite their economic interest, required the systems to operate at the limits of the capacity of current earth observation technologies.

In Chap. 25, we report our results on forest observation from satellites for government decision-making. Because of our work, the Czech Republic changed its national legislation with updated calamity zones. The maps produced by the DataBio method help the forest owners to *optimize* timber harvesting, process resources and *fight bark beetle calamity*.

32.4 Fishery

As for the other two sectors described above, the fishery pilots demonstrated that the fishing industry can benefit from big data and AI for a more cost effective and sustainable activity. As discussed in Chap. 29, we were able to demonstrate the potential to reduce maintenance cost and time as well as fuel consumption in the *operation of fishing vessels* with better utilization of sensor information and intelligent data analysis. Both the *energy consumption model* and the *species distribution models* help optimize the route and fuel saving decisions as well as the time at sea. The DataBio engine fault predicting tool was installed on one oceanic Tuna fishing vessel and tested in real operations.

The pilot in Chap. 30 demonstrated the potential of using physical and biological parameters like catch area, season, moon phase and fish species to forecast *catch volumes*. This helps to reduce fuel consumption, stock management and to a certain extent to estimate patterns in fish prices. The decision support system has been installed on several pelagic vessels.

End users have been actively participating and giving feedback during the whole project period. Seven fishing companies have been involved in the project to test the framework and give feedback to ensure the most useful implementation including installation on the vessels.

On the other hand, the fishing industry is still in the beginning of the digital transformation and needs to overcome several obstacles before a wider scale adoption of digital technologies can take place.

32.5 Perspectives

Earth observation data is central in the applications described in this book. The freely available Sentinel satellite images offered by the European Space Agency ESA through the Copernicus Programme are used by most pilots in DataBio with good success. However, it was noted that cloudy conditions in satellite images can disturb the image analysis used for decision support, like determining the harvesting time for a crop. Therefore, it is important to have secondary sources of information as well as strong models and filtering algorithms to compensate for the disturbances.

Machine learning and data-driven artificial intelligence models are largely used for prediction and image recognition, as described earlier in this book. Advances in algorithms, like artificial neural networks and deep learning, have radically raised the accuracy of these methods. However, these data-driven methods require that extensive volumes of *labelled training data* are available. For example, data from several years might be needed in reliable crop detection. Some labelled data, like farmer´s declarations and manual field observations, are costly and time consuming to obtain. As more labelled data gathers—for example, from data sharing practices, modelling and simulations—the methods used in precision agriculture and prediction of yield and fishing catches become increasingly accurate enabling better economy and sustainability. Furthermore, current artificial neural networks need in some applications to be complemented with more transparent understandable methods to create trust in the machine created recommendations. Long-range forecasts like prediction of grain and fish market prices remain challenging. However, the forecasts are continuously improving and might be useful to stakeholders even if they contain uncertainties.

One of the main hurdles in data-driven bioeconomy is the lack of *standardized data exchange and sharing*. For instance, sensors on-board fishing vessels typically demand proprietary interfaces to be built to get access to its readings. Therefore, currently, a lot of resources are needed to collect data from a large fleet of vessels. The European initiatives to create common data spaces and data infrastructures for vertical sectors, like agrifood, are highly needed. It is important to develop them also for other bioeconomy sectors like forestry and fishery.

Crowdsourcing, involving land and forest owners, as well as citizens in general, provides valuable complementing information about natural resources. However, we found that it requires a great deal of motivating actions to get, e.g. forest owners and others visiting and moving around in the forests, to participate.

Big data and artificial intelligence have to be applied to a much larger extent than currently for a more sustainable bioeconomy. The DataBio results can here offer a stepping stone for future developments, where the DataBio pipelines and solutions are scaled up to serve diverse business models and societal needs.

Correction to: Big Data in Bioeconomy

Caj Södergård, Tomas Mildorf, Ephrem Habyarimana, Arne J. Berre, Jose A. Fernandes, and Christian Zinke-Wehlmann ⓘ

Correction to:
C. Södergård et al. (eds.),
Big Data in Bioeconomy,
https://doi.org/10.1007/978-3-030-71069-9

The original version of the book was inadvertently published with wrong affiliation of the editor "Tomas Mildorf" in frontmatter. The affiliation has been changed from "Plan4All Horní Bříza, Czech Republic" to "University of West Bohemia, Univerzitni 8, 301 00 Plzen, Czech Republic".

The erratum book has been updated with the change.

The updated version of the book can be found at
https://doi.org/10.1007/978-3-030-71069-9